MODERN CHLOR-ALKALI TECHNOLOGY

Volume 4

MODERN CHLOR-ALKALI TECHNOLOGY

Volume 4

Edited by

N. M. PROUT
J. S. MOORHOUSE

Staveley Chemicals Limited,
Chesterfield,
Derbyshire

SCI

FOR THE APPLICATION
OF CHEMISTRY AND
RELATED SCIENCES

Published for the
SOCIETY OF CHEMICAL INDUSTRY
by
ELSEVIER APPLIED SCIENCE
LONDON and NEW YORK

ELSEVIER SCIENCE PUBLISHERS LTD
Crown House, Linton Road, Barking, Essex IG11 8JU, England

Sole Distributor in the USA and Canada
ELSEVIER SCIENCE PUBLISHING CO., INC.
655 Avenue of the Americas, New York, NY 10010, USA

WITH 56 TABLES AND 268 ILLUSTRATIONS

© 1990 SCI
© 1990 Howard Schoenberger—Chapter 1
Softcover reprint of the hardcover 1st edition 1990

British Library Cataloguing in Publication Data

Modern chlor-alkali technology.
  Vol. 4.
  1. Chlorine. Production
  I. Prout, N. M. II. Moorhouse, J. S. III. Society
  of Chemical Industry
  665.8′3

Library of Congress Cataloging-in-Publication Data

LC card number 84-643214

ISBN-13: 978-94-010-7004-1   e-ISBN-13: 978-94-009-1137-6
DOI: 10.1007/978-94-009-1137-6
ISSN 0747-7406

CONTENTS

R. Ellis Knowlton
Chemetics International Company,
1818 Cornwall Avenue,
Vancouver, BC,
Canada, V6J 1C7

R. M. J. Withers
Loughborough University of Technology,
Ashby Road,
Loughborough,
Leicestershire LE11 3TU,
UK

J. P. H. Shaw
7 Arley End,
High Leigh,
Knutsford,
Cheshire WA16 6NA,
UK

K. J. Behling
Du Pont de Nemours (Deutschland) GmbH,
Du Pont Str. 1,
D-6380, Bad Homburg VDH,
Düsseldorf,
FRG

D. L. Peet
EI Du Pont de Nemours & Company,
Wilmington,
Delaware 19898,
USA

A. M. Couper, W. N. Brooks, D. A. Denton
ICI Chemicals and Polymers Ltd,
PO Box 7,
Winnington,
Northwich,
Cheshire CW8 4DJ,
UK

J. A. Harrison (Formerly The University of Newcastle upon Tyne)
Eurodisc Technologies Ltd,
First Avenue,
Deeside Industrial Park,
Deeside,
Clwyd CH5 2NU,
UK

D. S. Cameron, R. L. Phillips, P. M. Willis
Johnson Matthey Technology Centre,
Blount's Court,
Sonning Common,
Reading RG4 9NH,
UK

T. Yamashita, Y. Sajima, H. Ukihashi
Asahi Glass Co. Ltd,
1–2 Marunouchi, 2-chome,
Chiyoda-ku,
Tokyo 100,
Japan

J. T. Keating
EI Du Pont de Nemours & Company,
Wilmington,
Delaware 19898,
USA

K. J. Behling
Du Pont de Nemours (Deutschland) GmbH,
Du Pont Str. 1,
D-6380, Bad Homburg VDH,
Düsseldorf,
FRG

Y. Tominaga, T. Kanke, K. Takagi, T. Miyazaki
Asahi Chemical Industry Co. Ltd,
3–2 Yako 1-chome,
Kawasaki-ku,
Kawasaki 210,
Japan

Contents

T. C. Jeffery
PPG Industries Inc.,
PO Box 1000,
Lake Charles, LA 70602,
USA

INTRODUCTION

The papers in this book were submitted for the 1988 London International Chlorine Symposium. This was the fifth symposium organised by the Electrochemical Technology Group of the Society of Chemical Industry and proved as popular as ever, attracting a record number of 294 delegates from 31 countries. Twenty-seven papers were presented during the two and a half-day event covering the latest developments in chlor-alkali technology.

The field of membranes and membrane cells was well represented by some 15 papers, reflecting the importance of membrane technology to the future of the industry. This is particularly relevant in view of increasing environmental pressures and rising costs. However, papers relating to the more traditional mercury and diaphragm cell technologies were also presented, together with a paper concerned with sodium chlorate manufacture. In addition, there were presentations covering the commercial and safety aspects of the chlor-alkali industry.

The Electrochemical Technology Group of the Society of Chemical Industry offer thanks to the many people and organisations whose help ensured the success of this symposium. In particular, we would like to thank:

1. The contributors of the papers.
2. The session chairmen:
    Dr R. G. Smerko (The Chlorine Institute Inc.);
    Mr B. Lott (The Associated Octel Company Limited);
    Mr T. F. O'Brien (United Engineers and Constructors);
    Dr B. S. Gilliatt (ICI Chemicals and Polymers Limited);
    Mr D. Bell (Hays Chemicals Limited).
3. The Chlorine Institute for assistance with printing costs and for active participation.
4. Staveley Chemicals Limited for assistance with printing costs.
5. The organising committee:
    Mr D. Bell (Hays Chemicals Limited);
    Dr D. S. Flett (Warren Spring Laboratory);
    Mr C. Jackson (ICI Chemicals and Polymers Limited);
    Mr B. Lott (The Associated Octel Company Limited);

Dr J. P. Millington (The Electricity Council Research Centre);
Mr J. S. Moorhouse (Staveley Chemicals Limited);
Mr N. M. Prout (Staveley Chemicals Limited);
Dr K. Wall (The Associated Octel Company Limited).

N. M. Prout
J. S. Moorhouse

1

THE WORLD CHLOR-ALKALI INDUSTRY
PAST—PRESENT—FUTURE

Howard Schoenberger

Kline & Company, Inc., USA

INTRODUCTION

This chapter explores past and present trends in the world chlor-alkali industry. It reviews historical production, capacity and economic factors leading up to current industry status. Current and future production issues are discussed along with the possibilities of conversions from caustic soda to soda ash. Factors influencing future chlorine demand are explored, and growth forecasts are presented. Lastly, such key technologies as membrane cells and superconductors are discussed with respect to the chlor-alkali industry of the future.

Many analyze world chlor-alkali markets on a country by country or even a region by region basis. While this may be useful in evaluating macro trends, it tends to overlook some significant factors, namely: (1) world trade is a function of chlorine derivative demand, and the associated surpluses or shortages of caustic soda; (2) corporations, rather than countries, are the entities which produce, buy, sell and trade most chlor-alkalis; and (3) multinational companies may operate under a different set of geopolitical goals than the countries in which they operate.

It would be desirable for a country to have steady chlor-alkali production rates, supplemented by imports and exports as needed to balance demand. However, this is not realistically possible in a free economy. Companies are driven by profit motives. They typically attempt to maximize profits, not social wellbeing. To further complicate trade patterns, these profit motives are more often than not based on the maximization of short-term returns.

Thus, as regional economies fluctuate and exchange rates float, world trade patterns will change. These factors partially explain the tremendous volatility in chlor-alkali prices over time, as seen in Fig. 1. This has been quite apparent in the caustic soda markets of the last 2 years. In the summer of 1986 deep sea caustic soda was selling for as little as $35 a tonne fob the US Gulf Coast. By the fall of 1987, the same material, if and when available, was selling for as much as $300 on the same basis. Today, while almost impossible to obtain, spot caustic soda is priced as high as $500 a tonne.

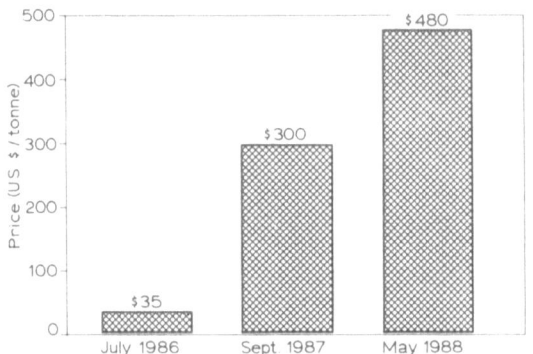

Fig. 1.   Caustic soda export price ($/t fob US Gulf Coast).

Further complicating world trade patterns are artificial trade barriers which preclude certain products from particular countries. Examples which come to mind in the chlor-alkali industry are Mexican tariffs on caustic soda, and Japanese and European tariffs on natural soda ash. Although the merits of these and other tariffs are well beyond the scope of this chapter, they are pointed out simply to illustrate some of the factors affecting international trade.

While the product price swings are extreme cases, they illustrate the point that regional factors can, and do, have a significant impact on world trade. In all probability world trade in chlor-alkalis, and many other commodity chemicals, will become considerably more competitive in the next few years. Several factors which will continue to affect international trade in chlor-alkalis and derivatives include:

—fluctuations in energy costs and currency translations;
—rationalization of aging capacity;
—capacity creep and expansion;
—capacity additions in developing regions;
—continued growth of multinational corporations;
—impact of China and the East Bloc;
—changing technology.

These and other factors will force chlor-alkali producers to be much more cognizant of world markets than ever before. In order to remain competitive companies will have to effectively manage resources. Undoubtedly, this will entail maximization of profits on a multinational basis. While this will be the only way for companies to survive, it brings with it a new set of concerns. How will the growth of multinationals impact regional markets?

CAPACITY SHIFTS

There has been a continuing trend toward the establishment of chlor-alkali capacity in less-developed countries. Much of this development has taken place for nationalistic reasons, although some has occurred to take advantage of lower cost structures. In 1975 total worldwide chlorine capacity was roughly 32 747 000 t, of which 82% was based in the industrialized regions consisting of the USA, Canada, Japan and Western Europe. By 1985 capacity stood at 40 645 000 t, but the industrialized group could only claim 71% of this capacity. What is even more dramatic is that while total capacity increased over the 10-year period at an average annual rate of 2·2%, capacity in the industrialized nations grew at an effective rate of only 0·75% a year.

However, these statistics are somewhat misleading, since the rates of capacity addition and retirement were anything but steady during the time period. While capacities in the developing nations grew at a moderately steady rate, capacities in the industrialized nations actually peaked around 1982 and then dropped. Net industrialized capacity dropped by almost 1·4 million t during the 5-year period from 1980 to 1985. Over the same time frame capacity in the developing nations increased by just over 1·0 million t.

In the USA alone, at least 15 plants with chlorine capacity of roughly 1·8 million t have closed since 1980. Many others have reduced their effective maximum operating rates. In the same period, with the exclusion of Niachlor, only seven capacity increases have been logged, totaling under 350 000 t. Additionally, the number of companies which produce chlorine in the USA has declined from 34 in 1975 to only 23 in 1988. A breakdown of the number of US plants and companies by year is shown below:

Year	Number of	
	Plants	Companies
1975	71	34
1980	65	31
1988	49	23

The chlor-alkali industry is currently operating at almost 100% of capacity, and product pricing is the strongest that it has been in many years. Many companies are debottlenecking plants and incrementally adding capacity. However, it is worthwhile to take a moment and reflect upon the recent past.

The US chlor-alkali industry averaged well over 90% capacity utilization during the period from 1960 to 1974. Chlorine demand was growing at an annual rate of 7% a year during the period. As we all remember, the oil shock of 1974 plunged the world economy into a recession. Chlor-alkali operating rates plummeted from over 95% to well under 75% in less than a year. The situation was further aggravated by forecasts of continued growth. When the bottom fell out of the market, construction of many new world-scale plants was already underway. This led to a prolonged period of over capacity, as seen in Fig. 2. Although chlorine demand has been anything but constant since 1974, effective annualized demand growth from 1974 through 1988 has been under 0·4% a year. Only now, after much capacity rationalization and hardship, are capacity and demand coming back into balance.

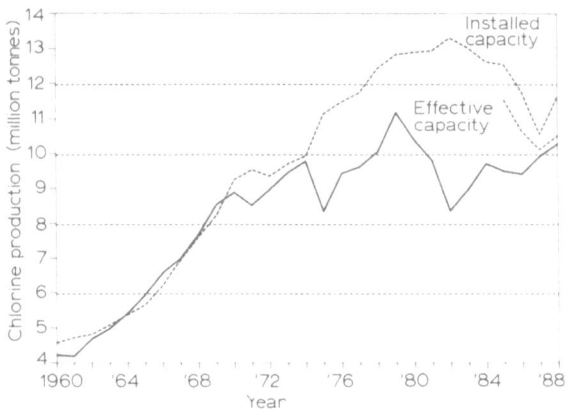

Fig. 2.   US chlorine production and capacity by year (million tonnes).

GLOBAL TRENDS

Trends in chlorine derivatives have further complicated the picture. Much more often than in prior years, such end products as PVC are being manufactured in different regions from the EDC or VCM required for its production. This tends to alter regional ECU balances, which can cause marked shifts in the deep-sea caustic soda markets as producers attempt to balance inventories with demand. Further complicating the picture is the fact that many chlor-alkali producers which had been vertically integrated in PVC divested themselves of those businesses during the early 1980s.

Consolidation and expansion by multinational companies has also complicated the global market. In the last few years there have been several sales or mergers which have seen the USA become more tied to world markets. In the soda ash area General Chemical sold a 49% interest in its natural soda ash business to an Australia-based company. Similarly, other US soda ash companies have evaluated foreign partners. Occidental Petroleum purchased Diamond Shamrock's chlor-alkali business along with that of FMC-Canada. Stauffer's chemicals business was sold, with parts of its chlor-alkali business going to Akzo. The future ownership of other parts of Stauffer's business remains uncertain. DuPont closed a major production facility, which was later sold to, and is being operated by, Occidental. Additionally, other US companies have undertaken ventures which will result in the building of new plants in many parts of the world. Some that come to mind are: Olin in South America, Occidental and others in China, and DuPont in Ireland.

CURRENT STATUS

What does this all mean today? For one thing, as the world becomes smaller, actions in one country can have dramatic implications on a country many thousands of miles away. With greatly improved transportation and communications capabilities, multinational companies operate world-scale chlor-alkali operations. These operations are driven by a prime motive—profit. Computers are used to optimize production strategies to minimize costs and maximize returns. While this sounds good in theory, it will continue to have serious implications on local economies and, more importantly, on individual companies.

World factors had a dramatic impact on the US chlor-alkali industry in the early 1980s. A period of capacity growth was just coming to completion as an economic slowdown occurred. Domestic demand for chlorine, which had been fueled by the construction industry's demand for PVC, came to a halt. This caused caustic soda to become scarce. Most US-based suppliers put their customers on allocation. Some even converted chlorine to EDC, which they stored in salt domes, just to provide for more caustic production than chlorine demand would allow. This point has been reiterated because the severe impact of the past will continue to affect the industry for many years to come. Producers will be much more cautious toward expansion having experienced the worst from past aggressive capacity additions.

This has been seen again during the current period, although for somewhat different reasons. Due to the over expansion of the late 1970s almost no new capacity has been built in the USA. Further, many plants have been 'mothballed', 'rationalized' or 'idled' in recent years. Pick the word you like, basically capacity has been eliminated due to economic forces. However, temporarily closed chlorine plants are rarely restarted, unless extreme care is taken when they are idled. Thus, there is again another apparent shortage of caustic soda in the USA, this time due to reduced capacity and favorable prices abroad.

What does this mean to the chlor-alkali industry? For one thing it means that problems caused by ECU-based production will be exacerbated. More than ever before, balancing will have to be conducted on a worldwide basis. This will require that significantly more market research be conducted before making changes in production capacity. Companies will have to take a much longer-term view of their goals and objectives. Such factors as currency fluctuations and regional economic disparities will make short-term profit maximization a poor objective. Figure 3 presents a graphical analysis of percentage fluctuations of six major currencies with respect to their value in 1975. A more detailed analysis of exchange fluctuations may be seen in the Appendix.

Chlor-alkali producers will need to look not just at their own goals but at their customers' goals. This may mean foregoing short-term profits in order to build long-term customer relationships. During the last 10 years we have seen a dramatic shift in the way many companies purchase chemicals. Some have taken the view that low price is the most important factor in selecting a supplier. This may be a good purchasing strategy in times when there is excess product on the market, as was the case with caustic soda during the early 1980s. World exchange rates made it profitable for European producers to ship caustic soda to the USA. Many short-sighted consumers tossed aside long-standing relationships with domestic suppliers in order to obtain lower cost product.

Now that exchange rates are reversing, many of these companies are finding that they cannot get caustic soda at any price. The foreign material is in short supply, and

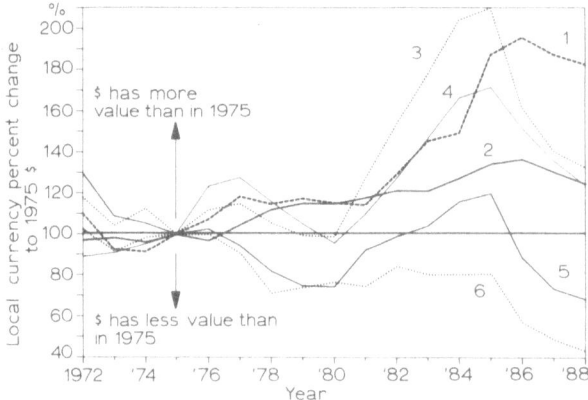

Fig. 3.   Percentage fluctuations in exchange rates relative to US dollar in 1975. (1) Australian dollar; (2) Canadian dollar; (3) French franc; (4) British pound; (5) German mark; (6) Japanese yen.

if available its price is extremely high in dollar terms. To exacerbate the situation, many of the domestic suppliers were forced to make significant reductions in capacity because of competitive world factors. Today there is less available US domestic capacity and many US suppliers do not have enough product available to meet demands of former customers. These producers have been forced to enact order controls, restricting purchases to ensure that they can supply contract requirements.

Evidence of the scope of this reversal is that the US chlor-alkali industry has gone from an average operating rate that was under 70% of capacity in the early 1980s to an effective capacity utilization of currently well over 95%. Historically, producers have added to capacity when utilization exceeded 90% of capacity. However, despite today's high operating rates, producers have been very hesitant to expand capacity due to the increased global nature of the business. Only now are producers starting to incrementally debottleneck facilities, yet excluding Niachlor no significant new grass roots plants have been announced.

OUTLOOK

The future of the chlor-alkali industry will be linked to chlorine demand growth. While most chlorine derivatives are currently experiencing peak demand, the future is questionable. Some markets which may experience significant downturns in demand include:

—pulp and paper
—CFCs
—perchloroethylene
—water treatment
—methyl and methylene chloride

Current high demand by most of these products will likely give way to environmental pressures. The pulp and paper industry is attempting to better understand the mechanisms which cause the formation of dioxins. Public concern may force that industry to turn to alternatives to chlorine bleaching. The same is true for water treatment, where chlorine has been linked to cancer-causing chloromethanes.

The future for chlorine in the production of CFCs is highly questionable as a result of the Montreal Protocol which calls for a world freeze followed by a gradual cutback in the use of several major chlorofluorocarbons, including CFC-11 and CFC-12. While this will reduce chlorine demand, many of the potential substitutes for CFCs also consume chlorine. So the exact effects will not be known for a few years.

Environmental pressures are influencing demand for many chlorinated solvents, including carbon tetrachloride, chloroform and perchloroethylene. These and other chlorine-based solvents are either proven or suspected carcinogens. The use of most chlorosolvents can be expected to decline due to public health concerns.

The only significant strength for chlorine demand appears to be in the resurgence of the polyvinylchloride industry. The general robustness of the worldwide economy has brought with it greater construction demand. Further, PVC has recently found favor in such residential remodeling applications as siding, windows and doors. The US market for remodeling is significantly more recession resistant than is new construction. Vinyl-based products are continuing to capture markets that were formerly dominated by aluminum. The residential siding market, which was once almost totally aluminum-based, is now almost 90% vinyl-based.

Even with this one bright spot, world growth for chlorine will be minimal. Historical US chlorine production, trend lines and forecast demand are shown in Fig. 4. Over the next 5–7 years, chlorine demand is projected to grow by approximately 1% a year, while caustic soda will experience growth of approximately 1·5% a year. While others have predicted chlorine growth rates as high as 2–3% per annum, we believe that such growth rates are predicated on either

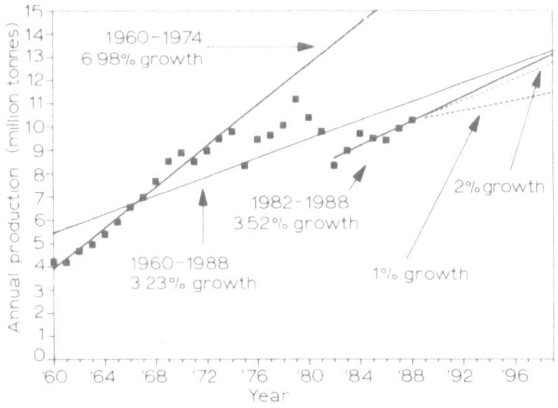

Fig. 4. US chlorine production trend lines 1960–88.

straight line extrapolation of historical data or unrealistically high growth projections for chlorine derivatives. If our more modest growth projections hold true, the current excess demand for caustic soda to co-produced chlorine will continue and worsen for the balance of the century.

SODA ASH

If the industry is unable to develop new markets for chlorine, it seems likely that the current caustic soda shortage will continue indefinitely. One possible solution is the substitution of soda ash for certain noncritical caustic soda applications.

While it is theoretically possible to use soda ash in place of caustic soda, few if any users actively switch from one source of alkali to the other. As an equivalent source of Na_2O a user needs 1·3 t of soda ash to obtain the alkalinity of 1 t of caustic soda. Soda ash is normally a more expensive form of alkalinity than caustic soda due to the following factors:

—higher equivalent alkalinity cost
—larger capital investment
—greater handling cost

Because soda ash is a powder, it costs more to handle than liquid caustic soda. Additionally, dry handling equipment requires a higher capital investment than liquid handling equipment.

Despite these cost penalties, many users are currently evaluating conversions to soda ash. Some pulp and paper producers have already partially converted. Estimates indicate that in the USA alone soda ash has the potential to capture anywhere from 200 000 to 1·0 million t of caustic soda demand. Much of the existing and contemplated conversion has occurred due to lack of caustic soda availability rather than pure economics, but the point of economic conversion for such users as alumina, pulp and paper, and phosphates is rapidly approaching. With contracted caustic soda nearing $300 a short ton and soda ash priced at under $100 a short ton, even the added transportation and handling costs will soon allow for economic conversion.

It may take the chlor-alkali industry a great deal of time to regain customers lost to soda ash. Thus, while high caustic soda prices are beneficial today, they may drive entire segments of demand away from caustic soda for many years to come. While industrial-scale plants to caustify natural soda ash seem highly unlikely at present, that route could prove viable if chlorine demand were to falter significantly. In fact, if new chlorine markets do not develop, it may be the only way to supply the growing world demand for caustic soda.

PRODUCTION ISSUES

Chlorine plants are designed based upon such factors as current densities, capital costs and economies of scale. Mercury-based plants are typically most economic in the 250–450 t/day range, while diaphragm plants with cogeneration and multiple effect evaporators are sized in the neighborhood of 1000 t/day. While some small facilities still exist, most remain in operation for such unique reasons as existing pipeline customers, potassium-based alkali products, regional markets or other location-specific situations. Most industry participants agree than world-scale diaphragm plants currently provide the most cost-effective production route. This is especially true of those with large multiple effect evaporators which are integrated with electric cogeneration.

While Japan mandated the cessation of mercury-based production effective July 1986, mercury plants in most other locations can be expected to continue operating for some time, assuming they do so in an environmentally responsible manner. While technology exists to minimize mercury effluent, due to such factors as energy consumption and environmental considerations, we will probably never see another mercury cell plant built.

Likewise, there are tremendous pressures to reduce the use of asbestos. This could threaten the viability of diaphragm technology as, to the best of my knowledge, no one has perfected a 100% effective substitute for asbestos, although many are actively working on it. Additionally, the chlor-alkali industry will face pressures to minimize waste products. While these rules have become quite strict in many industrialized countries, there is no doubt that many less-developed regions will ultimately implement similar environmental safeguards.

TECHNOLOGY

The current and future availability of new technology will continue to play a significant role in the chlor-alkali industry. Over the next 10 years membrane technology can be expected to play a much greater role in chlor-alkali production than in the past. In the more distant future, there exists a tremendous potential for structural changes due to the development of superconductors.

Membranes

The logical chlor-alkali production process alternative appears to be membrane technology, but adoption has been slow. One obvious reason for this is that very few new production facilities have been built in the last 10 years, but many other reasons exist. Mercury and diaphragm technology are accepted because both are commercially proven processes. However, until recently there was a lot of hesitancy to commit to membrane technology. No large-scale membrane plants had been built, and some of the operating plants suffered from negative publicity. Few managers were willing to risk investing in a technology which was viewed as not fully proven.

We are now at the stage where most industry leaders accept the capabilities of membrane technology, and openly state that any new plant that they would build would be based on that technology. However, this does not guarantee a rapid conversion to the technology.

At current growth rates, free-world demand can at best accommodate the equivalent of two new world-scale plants a year. At that rate it would take many years for membrane technology to become the major production method in terms of tonnage.

However, membranes offer the capability to economically construct smaller specialized plants. That is where they are likely to experience the greatest level of adoption.

Traditionally plants have been built in proximity to cheap power and salt supplies. Plant sizes were more dependent on optimal costs than on the needs of local markets. Membrane technology has the potential to significantly alter the chlor-alkali industry as we know it today. While economies of scale will continue to exist, they are significantly less important for membranes than for the other technologies.

The advantages of membranes are greatest for such near-equal ECU users as pulp and paper producers. These users can run a membrane facility without the need for chlorine liquefaction, and without the need for caustic soda concentrators. For them an entire plant may consist of the membrane cell room and a brine treatment system. Thus, they will save in capital invesment and transportation of finished product. Even if such small users must pay higher electric rates than a large-scale producer, they will most likely be ahead of the game.

The technology will also allow smaller users to produce product on site rather than transport it great distances. This has definite benefits in light of concerns regarding the transportation of hazardous materials.

However, one of membrane's greatest attributes—the ability to produce higher concentration caustic soda than a diaphragm cell—may actually be the greatest deterrent to cell conversion at existing plants. Cogeneration has been a natural fit for the chlor-alkali industry because diaphragm technology provides a ready sink for the steam turbine exhaust. A conversion to membranes would upset the entire facility's steam balance, and would minimize the worth of the cogeneration investment. In some instances, membrane conversion would actually increase total operating costs.

This is not to say that membranes will never see application at world-scale plants. In fact, quite the opposite is true. However, in their present form, they are unlikely to replace existing world-scale diaphragm capacity. Rather they will be attractive for incremental capacity expansions. This is because many existing plants are capacity constrained by limited caustic evaporation capacity. An incremental membrane expansion will greatly conserve scarce evaporator capacity. Additionally, the crystallized salt from the caustic soda evaporators provides an excellent starting salt supply to produce a brine feed for membrane cells.

As a result, membrane technology is expected to dominate all new grass roots capacity, and will gain a sizable portion of incremental cell capacity at large diaphragm plants. However, it seems highly unlikely that a cogeneration-based plant would undertake a wholesale membrane conversion.

Superconductors

One technology that has yet to receive major attention is the application of superconductors for large-scale power generation and storage. Superconductors have the capability of storing electrical energy for later consumption. Energy is stored in DC form, and later converted back to AC for most applications. Chlor-alkali applications have the advantage of using large quantities of electricity in DC form. As a result, the industry is a natural candidate for stored power. Coils could be designed so as to be fully charged at night using off-peak power, and then called upon during peak hours to supply electrical demand for a plant. Such a scenario could enable a chlor-alkali plant to obtain electricity for just slightly above the variable cost of generation.

The potential with such technology would be to allow a small membrane plant to enjoy lower cost power than a large cogeneration facility. An implicit assumption in this scenario is that most utilities have significant surplus capacity during off-peak periods that could be contracted for off-peak use.

Such electrical storage would allow membrane plants to be built at the point of use, even in regions with high electric costs, thus eliminating product transportation costs. The technology has the potential to significantly reduce effective power costs in such high energy cost regions as Japan.

While superconductors are currently only a laboratory-scale curiosity, the potential exists for commercialization. Estimates indicate that a superconductor generator will be able to produce electrical power with only half the loss of conventional generators. Further, the size of a generation facility will be greatly reduced.

While storage systems are far from commercial, indications are that they will have efficiencies of roughly 95%, as compared to 72% for pumped water storage and 65% for storage batteries. It is currently estimated that a 30 MW storage unit will cost $72 million. However, that is based on helium-cooled superconductors. As developments with nitrogen-based superconductors proceed, these costs should drop significantly. Although superconductor systems will not be commercial for some time to come, the chlor-alkali industry would be a logical candidate for adoption.

SUMMARY

The world chlor-alkali industry is entering a period of change. Chlorine demand growth, which once fueled the industry's growth, seems likely to subside. Environmental pressures will increase costs, as will long run increases in energy costs. The industry will be forced to adapt to these changes, along with the problems associated with the possibility of nonbalanced ECU demand. The current shortage of caustic soda will likely remain with us for the foreseeable future. This will most certainly be the case if the industry's demand for chlorine falters due to recession.

The key challenge for the 1990s and beyond will be to develop new chlorine markets. Further, the industry needs to prepare itself for a dichotomy of growth. Large world-scale plants will most likely continue to dominate production on a total capacity basis. However, those plants will expand incrementally using membrane technology. Mid-size plants will likely be forced out of operations as they become noncompetitive. Small specialty plants serving one or two customers will crop up, taking advantage of membrane technology and other developing technologies. This will be especially true for chlorine derivative plants that can captively consume chlorine and sell caustic soda to a market that is forecast to continue to be caustic short for years to come.

Most existing chlor-alkali producers are resistant to building new plants based on the current financial returns. To justify the risks inherent with building a new chlor-alkali facility, most US producers indicate that they must obtain $365 per ECU tonne. (While current long-term contract pricing is only generating roughly $255 to $275 an ECU tonne.)

Perhaps chlor-alkali customers should consider these factors and re-evaluate their purchasing decisions. One means of securing a continued supply and consistent pricing of chlor-alkalis is to captively produce some portion of requirements. Historically many pulp and paper producers have produced their own chlorine and caustic soda, but most exited the business when chlor-alkalis were readily available. Now that demand is bucking capacity, decisions must be made as to who will add capacity and where. Membrane technology is now at the stage where most consider it to be proven technology. This is likely to change the future of the chlor-alkali industry. Many consumers may find it advantageous to install small prefabricated plants rather than purchasing product. There are numerous advantages that membrane technology offers that were never before available.

What this all boils down to is that chlor-alkali producers will have to pay more attention to world markets. That will entail actively monitoring production and trade on a worldwide basis. It will also require monitoring planned capacity additions. This information will all have to be factored into a company's long-range capacity planning, much more so than in the past.

Strategically, producers should work toward developing a large percentage of long-range business ties to balance world spot and short-term contract sales. This is because as such factors as currency fluctuations continue today's low-cost supplier may be tomorrow's over-priced supplier.

APPENDIX: HISTORICAL EXCHANGE RATES PER US DOLLAR

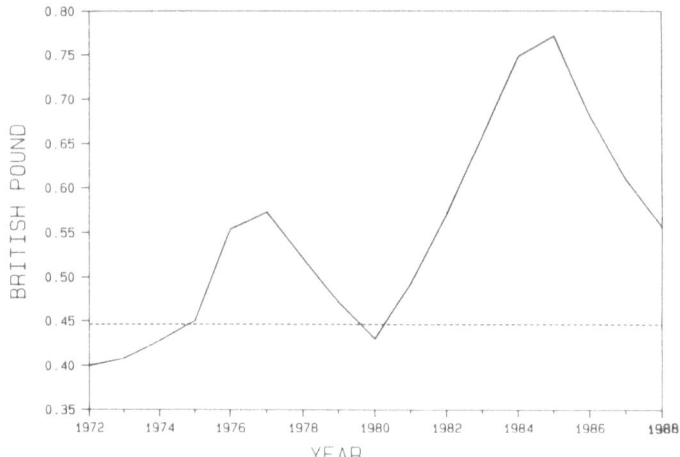

Fig. A1

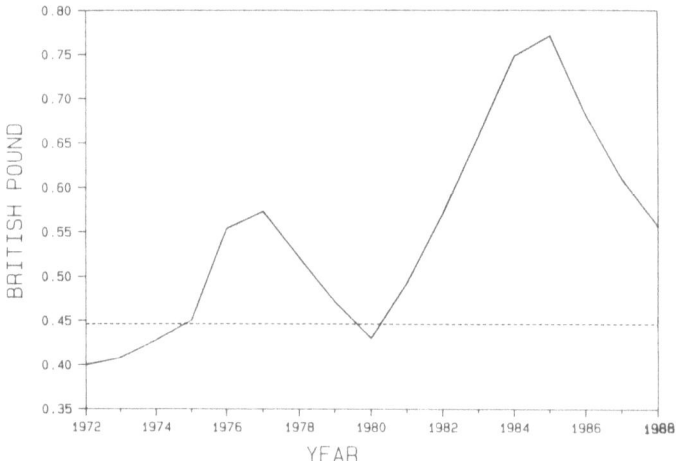

Fig. A2

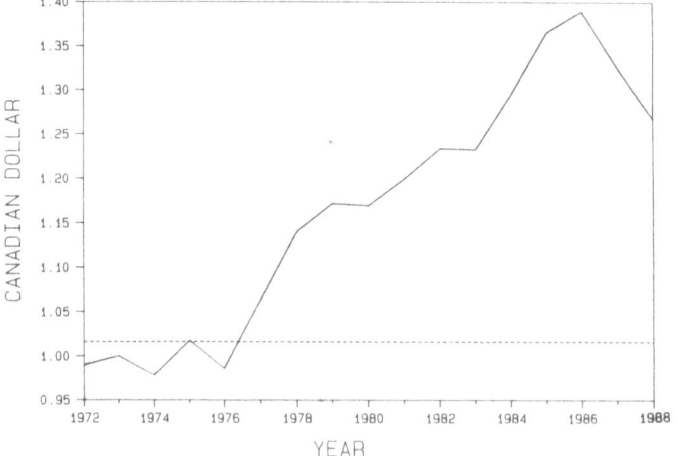

Fig. A3

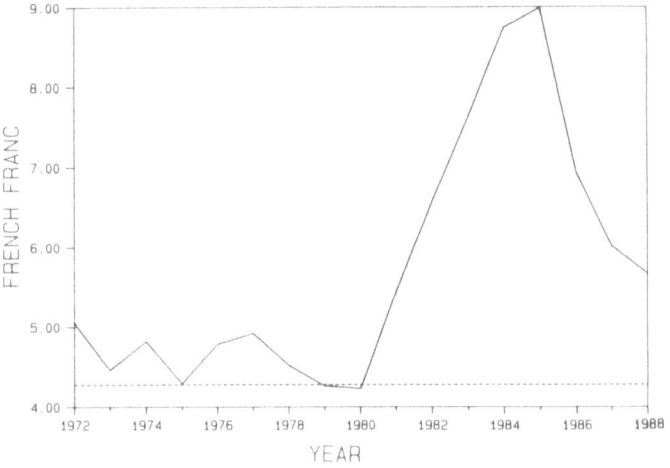

Fig. A4

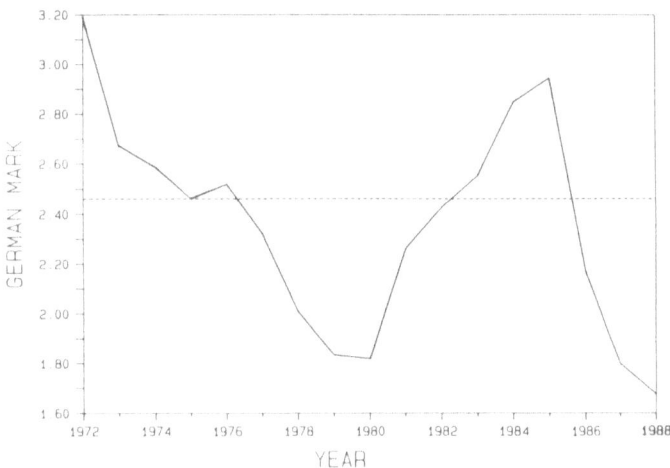

Fig. A5

Fig. A6

2

THE INCREASING REGULATORY CONTROLS AND THEIR EFFECT ON BUSINESS

Clive Stanbrook

Stanbrook and Hooper, Belgium

INTRODUCTION

Responsibility for the initiation of EEC legislation lies with the EEC Commission. Within the EEC Commission, it is Directorate General XI which looks after Environment, Consumer Protection and Nuclear Safety, and the Commissioner answering for the Directorate General is Mr Stanley Clinton Davis.

The spirit in which the Commission is working can best be described by quoting part of the statement which Mr Delors, as President of the Commission, made to the European Parliament on 20 January when he outlined the programme of the Commission for 1988:

> 'The first three months of 1988 will mark the end of the European Year of the Environment, the culmination of a long, successful campaign to increase public awareness of environmental issues. The momentum built up, coinciding with the first full year of application of the Single European Act, will enable the Community's environment policy to make full use of the greater authority conferred by its inclusion in the Treaty. The Commission will capitalize on these assets in 1988 to scale up the work on all the priorities set in the fourth five-year action programme (1987–1992).'

What does this mean for the chlorine industry?

First look at the EEC Directive on major-accident hazards of certain industrial activities in relation to the use and storage of chlorine, and then examine the various controls on discharges of dangerous substances and chlorine derivatives. Finally the Commission's plans for enforcement of environmental legislation will be discussed.

First then:
The 'SEVESO' DIRECTIVE 82/501/EEC, and its effect on the use and storage of chlorine.

You will all know the origin of this Directive, which came into being in 1982. It has once been amended, in March 1987, and as a result of the fire at the Sandoz plant in November 1986 a further amendment has been proposed.

The Directive applies to any person in charge of any operation carried out in an industrial installation involving, or possibly involving, one or more dangerous substances and capable of presenting major-accident hazards. Transport and storage within the establishment are expressly covered.

Industrial installations are defined as installations for the production, processing or treatment of organic or inorganic chemicals using for this purpose amongst others alkylation, carbonylation, hydrolysis, oxidation, and formulation of pesticides and of pharmaceutical products, to mention just a few. Other processing methods are mentioned, including the use of electrical energy and disposal by incineration.

Dangerous substances are defined in the Annexes to the Directive. Annex II deals with storage of chemicals at installations where the distance between them is considered insufficient to avoid, in foreseeable circumstances, any aggravation of major-accident hazards. Chlorine and sodium chlorate appear in the list of nine substances mentioned in Annex II of the original Directive.

There are three distinct obligations in the Directive. First, under Article 3, Member States must ensure that the manufacturer is obliged to take all the measures necessary to prevent major accidents and to limit their consequences.

Next, under Article 4, Member States must ensure that all manufacturers must at any time in the course of inspection or other measures of control be able to prove to the competent authority that they have:

(1) identified existing major-accident hazards;
(2) adopted the appropriate safety measures; and
(3) provided the persons working on the site with information, training and equipment to ensure their safety.

Finally, Article 5 applies to the storage of chemicals at installations where the distance between them might lead to the aggravation of major-accident hazards. Member States must introduce the necessary measures to require manufacturers to provide environmental authorities with detailed information on products stored, installations and plans in the event of major-accident situations.

Articles 3 and 4 apply when 10 or more tonnes of chlorine are stored, and Article 5 now applies to storage of 75 t of chlorine or more.

This last figure was reduced from 200 t by the 1987 amendment and it must be implemented in national law in just 4 months time—in September 1988.

For sodium chlorate Articles 3 and 4 apply when 25 t or more are stored. Notification becomes necessary when 250 t or more are stored where the substance is in a state which gives it properties capable of creating a major-accident hazard.

The proposed amendment, following the Sandoz fire, seeks to extend the list of controlled substances in Annex II from 10 to 28. Compounds of chlorine, hydrogen chloride, and that infamous substance carbonyl chloride would be included. The amendment would seek to cover all storage of listed substances regardless of whether the storage is within the vicinity of the place of manufacture or at a point remote from it. The new proposal would not seek to amend the storage limits of chlorine beyond which the respective obligations apply.

I need hardly remind an audience as expert as this that the administrative burden of the Seveso Directive is quite heavy and that some very clear responsibilities rest not just on producers but importers, distributors and users as well.

Let us now look at a subject which has preoccupied the industry for many years, namely the Water Directives.

THE WATER DIRECTIVES

Although the author was not directly involved at the time when the first proposals were put forward in the mid-1970s, he believes that the nature of the aquatic environment gave rise to lots of questions and arguments—especially concerning the difference between short fast-flowing tidal streams and long, slow-flowing rivers.

At all events in 1976 a framework Directive, 76/464/EEC, was adopted on pollution caused by certain dangerous substances discharged into the aquatic environment of the Community. The Directive classifies dangerous substances in two lists. List I contains certain individual substances selected mainly on the basis of their toxicity, persistence and bioaccumulation, with the exception of those which are biologically harmless or which are rapidly converted into substances which are biologically harmless. Pollution through the discharge of these substances has to be eliminated gradually. Article 6 of the Directive provides for further measures to be enacted by the EEC Council laying down the limit values and quality objectives for List I substances.

List II substances, although they have a detrimental effect on the aquatic environment, can be confined to a given area and depend on the characteristics and location of the water into which they are discharged. The discharge of these substances is controlled by a licensing procedure.

Mercury Discharges

Mercury and its compounds are included in List I. In 1982 the EEC Council of Ministers adopted a Directive, 82/176/EEC, on limit values and quality objectives for mercury discharges by the chlor-alkali electrolysis industry. The Directive applies to inland surface water, territorial waters and internal coastal waters but not to groundwater. Member States had to implement the Directive by 1 July 1983. New limit values came into force on 1 July 1986.

A different criterion may be used where a Member State can prove to the Commission that special or more severe quality objectives are being met and continuously maintained throughout the area which might be affected by the discharges because of the action taken, among others, by that Member State.

Under the Directive it is the Member States that are primarily responsible for exercising environmental control over chlorine production. The main instrument of control is the requirement for prior authorisation by the national authorities for all discharges into water liable to contain a dangerous substance. The authorisation lays down:

(1) the emission standards;
(2) the conditions that must be complied with within a specified period; and
(3) the period of validity.

Member States may grant authorisations for new plants only if such authorisations contain a reference to the standards corresponding to the best technical means available for preventing discharges of mercury.

At this point the author is in some difficulty because he realises that he is talking to people who are in the chlor-alkali industry, and their essential preoccupation is with the basic products of that industry.

In spite of that, however, he feels he cannot deal with questions of control of discharges into the aquatic environment without referring to chlorine-derived or related compounds.

With the exception of mercury and cadmium, List I does not mention individual substances but families or groups of substances, for example organohalogen compounds, carcinogenic substances, etc. The Commission therefore decided to choose individual substances from among the families or groups for further study with a view to submitting proposals on them. After consultation with national experts, four series of substances were chosen. Mercury and its compounds appeared in the first series and led to the adoption of Directive 82/176/EEC discussed above.

Chlordane and Heptachlor

The second series include chlordane, heptachlor, hexachlorocyclohexane, PCBs, and PCTs and hexachlorobenzene. Chlordane and heptachlor were the subject of a Communication to the Council in 1980 in which the Commission stated that no useful purpose would be served by preparing and submitting proposals for Council Directives on these substances as little use was made of them in the Community. In an answer to a Parliamentary Question on the subject, the Council stated that the Commission considered it would be sufficient to keep the use of these substances under review. No further proposal has been submitted on the matter.

Hexachlorohexanes ('HCHs'), Hexachlorobenzenes ('HCBs') and Hexachlorobutadienes ('HCBDs')

In 1984 a Directive, 84/491/EEC, on limit values and quality objectives for discharges of hexachlorocyclohexane was adopted. The EEC Commission has commenced infringement proceedings against Spain for not applying the Directive following complaints that stretches of the River Gallego have been badly polluted by the substance. Waste containing HCH is alleged to have come from a pesticide factory producing Lindane in Sabinanigo, Aragon. The river provides drinking water for some 4000 people living in the area.

From 1 October 1988 new limit values are to be complied with.

A proposal has been made to extend List I to HCBs and HCBDs, the latter appearing in the third series of substances selected for study. The Commission proposes that specific programmes should be drawn up case by case for reducing pollution by HCBs from, inter alia, plants producing chlorine by chlor-alkali electrolysis with graphite electrodes and plants producing trichloroethylene and vinyl chloride.

The proposals on HCBDs essentially concerns industrial plants carrying out operations for the production of perchloroethylene (PER) and carbon tetrachloride by perchlorine action. It is proposed that specific programmes should be drawn up for production plants using HCBD for technical purposes. The Commission 'Impact Statement' on this proposal states that the limit values are to a large extent already being met and that therefore no extra costs should be incurred. The Economic and Social Committee therefore recommend that the limit values should be met before expiry of the 1988/91 deadlines. The only consultation mentioned as having taken place was in the Committee of National Experts.

PCBs and PCTs

Polychlorinated biphenyls and polychlorinated terphercyls appear in the second series. In 1976 a Directive was published prohibiting the use of PCBs and PCTs except in specified circumstances until 30 June 1986. In 1985 a further Directive was adopted amending the 1976 Directive and banning the use after 30 June 1986 of PCBs and PCTs. However, the use of PCBs and PCTs in plant and equipment which were in service on 30 June 1986 is authorised until the disposal of such plant and equipment or until it has reached the end of its tether.

One of the reasons for this prohibition was that substitutes had been developed which were considered less dangerous for man and the environment. From the Report of Jacques Delors to the European Parliament it appears the Commission will propose further measures on the use of PCBs and PCTs in 1988 on the basis of work done in 1987.

Carbon Tetrachloride and Chloroform

These appear in the fourth series. The limit values for carbon tetrachloride came into force on 1 January 1988 in accordance with Directive 86/280/EEC, which had to be implemented by the Member States by this date.

Agreement has been reached in principle on the need for emission limits applying to chloroform. It is anticipated that the emission limits, which are yet to be decided, will hit hardest those firms producing vinyl chloride monomer and whitened paper pulp, and those using chloroform as a solvent.

The UK is reported as agreeing to set emission limits for industrial discharges of chloroform in return for a Commission promise to concentrate in future on those substances which cause more widespread pollution in Community waters. This refers to the list of 129 substances in addition to the List I substances mentioned in the Commission Communication of 1982. Of these 129 substances by far the majority are chlorine compounds, Of the 129 substances 74 are chlorine compounds. The Member States undertook in a Council Resolution in 1983 to provide information on these substances. The view now being taken in some circles is to concentrate on the most dangerous substances rather than attempt to work through the list one by one. It is reported that the Commission will by September 1988 communicate to the Council the most dangerous substances it intends to tackle first.

Groundwater

Groundwater is dealt with by Directive 80/68/EEC. Again substances are divided into two lists.

Member States must prevent the introduction into groundwater of substances in List I, the 'black-list' substances. The introduction into groundwater of List II substances must be limited.

Mercury and its compounds are included in List I. All direct discharge, that is the introduction into groundwater without percolation through the ground or subsoil, of List I substances is prohibited. Any disposal of List I substances which must lead to indirect discharge, i.e. introduction into groundwater after percolation through the ground or subsoil, is subject to prior investigation and licensing. Member States may take any measures they deem appropriate to prevent any indirect discharge of substances in List I by any other activity on or in the ground.

Now to a subject which is very much in the public eye—the question of dumping of waste at sea.

DUMPING OF WASTE AT SEA

The North Sea is now the only marine area in the world where incineration at sea is still carried out and Greenpeace have just published a further study of the environmental effect of these procedures. This question of dumping of waste at sea arouses more real, or imaginary, or emotional or political concern than perhaps any other regulatory issue.

Perhaps this is an area where, as a lawyer, the author can blissfully set aside such wider considerations, but even so we all have to recognise the implications of this question and no apologies are made for referring to the subject, even though at the moment it cannot be said that it is a story of uninterrupted success.

In 1985 the Commission submitted a proposal for a Directive on the dumping of waste at sea. No progress has been made in the Council on adoption of the Directive. Jacques Delors stated on this subject that the conclusions reached at the second conference on the North Sea, held in London in November 1987, imply that the Council should urgently adopt the Commission's proposal on dumping at sea.

The Directive would apply to waste and other materials dumped or incinerated in maritime waters within the jurisdiction of a Member State from ships and aircrafts whether registered in a Member State or not. The 'black list' of substances mentioned for the water Directives is reproduced. An application for or renewal of a permit for the dumping of waste at sea would require details of the chemical composition of the list of substances contained in the waste. Chlorine is one of the substances.

In some earlier comments mention was made apologetically about the need to talk about downstream chlorine-derived products. It is, however, worth remembering two key points.

First, that a principle pervading EEC environmental legislation is that 'the polluter pays'. The water Directives involve administrative costs incurred by applications for and renewals of permits. Costs of production have to allow for treatment plants or the development of more refined methods of production to ensure that discharges from the plants comply with the limit values or quality objectives specified.

Secondly, the vast majority of substances targeted by the Commission for investigation into their effect on the aquatic environment are chlorine compounds. Regulatory action on chlorine compounds will add to their costs. Users will search for cheaper and cleaner alternatives. If these are found, the demand for the chlorine compound will either disappear altogether by virtue of the play of market forces or, as in the case of PCBs and PCTs, the EEC Commission will ban their use on the basis that suitable 'environment-friendly' alternatives exist.

In some senses each of the areas examined becomes not only environmentally more critical but politically more sensitive, and now we must turn to possibly the most sensitive area of all, chlorofluorocarbons.

CHLOROFLUOROCARBONS

In March 1980 the EEC Council adopted a Decision on chlorofluorocarbons (CFCs) in the environment. This Decision implemented Recommendation III of the International Conference on CFCs held in Munich in 1978 from 6 to 8 December. The Decision required Member States to ensure that by 31 December 1981 industries situated in their territories achieved a reduction of at least 30% compared with 1976 levels in the use of CFCs in the filling of aerosol cans.

In 1985 the Community and a number of Member States signed the Vienna Convention for the protection of the ozone layer. Ireland, Spain and Portugal have yet to sign the Convention. A Protocol to the Convention was negotiated in Montreal in September 1987 and signed on behalf of the Community and several Member States. Ireland and Spain have yet to approve the Montreal Protocol. The EEC Commission has proposed that the EEC Council adopt a Decision which would set the timing for Community signature and ratification of the Vienna Convention and of the Montreal Protocol.

The Montreal Protocol to the 1985 Vienna Convention on the ozone layer requires cuts to be made in both production and consumption of CFCs, but leaves signatories the choice of how to achieve them.

The Montreal Protocol is aiming at a 50% drop in the use of CFCs over 10 years, to be achieved in three stages. The initial stage would involve a freeze being imposed on production and consumption at 1986 levels as from 1 July 1989. In 1992 a 20% cut would be imposed. Finally, in 1998 a further 30% reduction would be required.

On the occasion of World Consumer Rights Day on 15 March 1988, the International Organisation of Consumer Unions (IOCU) and the organisation representing consumer associations in the EEC Member States (BEUC) adopted a joint position where they criticised the reductions imposed by the Montreal Protocol as insufficient. According to these organisations only an immediate 85% cut in CFC production and consumption could halt the destruction of the ozone layer between now and the year 2000.

The Commission has tabled a proposal (unpublished as yet) for a Regulation which would apply direct controls on the Community's producers and importers. The Commission proposes to impose import quotas for CFCs from third countries and production controls on producers in the Community.

According to recent reports, the *import quotas* would be the following:

(1) July 1989 to December 1989 791 t of CFCs, weighted according to their ozone depleting potential.
  January 1990 to June 1993 1582 t per annum.
  July 1993 to June 1998 1266 t per annum.
  July 1988 to June 1999 791 t per annum.

(2) After January 1990 imports of CFCs from countries not party to the Montreal Protocol would be prohibited.

(3) From January 1993 products containing CFCs would also be stopped at EEC borders. The product lists has yet to be decided by parties to the Montreal Protocol.

(4) Restrictions would be placed on imports of products manufactured using CFCs. This list, too, is yet to be drawn up.

In practical terms these measures would mean that importers would be obliged to hold an import licence, issued by the Commission with the assistance of a Management Committee of national representatives.

As regards *production controls*, each producer would have to ensure that the ozone depleting potential of the annual CFC production (called the 'calculated level index of production') did not exceed:

(1) from July 1989 to June 1990, his 1986 level;
(2) from July 1993 to June 1994, 80% of the 1986 level; and
(3) from July 1998 to June 1999, 50% of his 1986 level.

Production controls may be tightened up to take into account manufacturers who commence production after 1986.

What are the implications for the wider chlorine industry of the Commission proposals?

The Commission's approach is to control supply rather than demand. This would be achieved by regulating imports from third countries and by imposing production controls on EEC companies. Limiting supply in this way would regulate consumption. The quantities of CFCs available for sale or for internal use by producers in the EEC would be limited.

All producers, importers and exporters would have to inform the Commission regularly and fully of production, sales, trade and stocks held. CEFIC is concerned that data supplied to the Commission will be compiled for each Member State, thus revealing the individual production and/or consumption figures of the small number of companies involved. CEFIC hopes the Commission will only make total EEC figures public. CEFIC is also keen to ensure that industry in other signatory countries provide the same kind of detailed information. If these measures come to fruition the production of CFCs would become more expensive. This cost would probably be passed on to the users, whether they be concerned with the manufacture of propellants in aerosol cans, furniture foam or cleaning materials in the electronic industry. Eventually these users will turn to alternatives. Indeed, to encourage the move to substitutes, the German Minister for the Environment, Mr Klaus Toepfer, who holds the Presidency of the Environment Council until the end of June 1988, is reported as suggesting that products which do not use CFCs should be distinguished by special labels. The demand for chlorine would drop with the consequential effect in terms of profit for the chlorine industry.

As a lawyer and not as an expert on the day-to-day affairs of the chlor-alkali industry, the author is left with the feeling that the Commission is actually falling behind public opinion on this issue and that, perhaps for the very first time, public demand will cause a decline in the use of CFCs and already in the shops we see 'environment-friendly' aerosols on sale.

ENFORCEMENT

In the European Year of the Environment the Commission made a determined effort to use its power to ensure proper implementation of the existing EEC environmental legislation. Jacques Delors stated in his report to the Parliament that:

'The Commission would focus not only on refining rules for individual sectors, but also on effective implementation of the measures adopted in the past. The Commission feels that greater attention will have to be paid to the full application of existing Community Directives and Regulations, i.e. not only to their incorporation into national law but also to their practical enforcement in the field.'

The Infringement Procedure

The EEC Commission is the guardian of the EEC Treaty. Under Article 169 the Commission may take action against a Member State where it considers that a Member State has failed to fulfil its obligations under the Treaty. These are known as infringement proceedings and are usually initiated by complaints from individuals or companies. An infringement action may result in proceedings before the European Court of Justice. If the Court agrees with the Commission that a Member State has not incorporated EEC legislation properly into its national law it will condemn the Member State concerned for failure to comply with its obligations under the Treaty.

The Commission will no longer be satisfied by the fact that a Member State has adopted the relevant regulations. It is increasingly investigating how those Directives are actually applied. An example of this is the action taken over the past year on the water Directives. An insight into the background of these actions was revealed by Mr Clinton Davis in relation to Directive 80/778/EEC on water for human consumption.

Immediately after the 17 July 1985 deadline for full implementation of the Directive, the Commission started to check how effectively the provisions laid down in the Directive were being enforced in the Member States. The inquiries carried out by the Commission revealed that several Member States had granted exemptions. The UK, Denmark and Germany had granted exemptions under a provision in the

Directive for derogations to take account of situations arising from the nature and structure of the ground in the area from which the water supply in question emanated. Italy had granted an exemption under a provision in the Directive allowing a Member State, in an emergency, to exceed the levels indicated in Annex I of the Directive. In addition, the UK submitted a special request to the Commission for a longer period for complying with Annex I in relation to lead and small water distributors.

The Commissioner stated that the Commission was examining the methods chosen by the Member States to enforce all the provisions of the Directive on their territory. With respect to the UK, the Commission instituted infringement proceedings after receiving complaints about a number of limit values being exceeded. Interestingly, in December 1987 it was reported in the press that the UK Water Authorities Association had announced in London that its clients would end up paying most of the £6 billion bill for bringing British drinking water up to the standards required under the 1980 Directive.

Information

In their programme for 1988, the Commission mentions another aspect of enforcement on which work is in progress, namely the collation of information on the environment. A Community Information System on the State of the Environment and on natural resources (CORINE) has been instituted. Its main purpose is to ensure the availability of a sound base of environmental information to industry and policy-makers through the Community as an aid to implementation of legislation and integration of the environmental dimension in other policy areas. The CORINE Programme is already being implemented on a practical level and is being developed.

A further concern of the Commission is to improve access to information held by environmental authorities, whilst ensuring protection to information which can legitimately be regarded as confidential. The Commission is studying the feasibility of a Community 'Freedom of Environmental Information Act' and promises to make appropriate proposals.

As regards chemicals, a notification system was set up by Directive 79/831/EEC enabling the Commission and Member States to monitor the hazards, distribution and uses of chemicals marketed after 18 September 1981. The Commission feel that a similar procedure is needed for the integrated risk assessment of 'existing chemicals'. 'Existing chemicals' are those marketed before 18 September 1981 and listed on the European Inventory of Existing Chemical Substances (EINECS).

A Directive will be proposed to provide a comprehensive structure for risk assessment and regulation of existing chemicals, where such evaluation is needed. This Directive will establish a procedure for treating a priority list of chemicals for immediate attention, as well as setting out the means for gathering information, testing and evaluating the risks to people and the environment. The Directive could also, where necessary, co-ordinate the development of chemical specific control strategies, where this proves to be necessary. A proposal for the systematic assessment of existing chemicals is mentioned in the Commission's Programme for 1988.

Civil Liability

Following the 'Sandoz' accident in Basle in November 1986, mentioned earlier in connection with the Seveso Directive, the EEC Council asked the Commission to look into the need to streamline national civil liability laws relating to the environment. The Commission is expected to report shortly to the Council with a recommendation that a uniform system be applied at EEC level.

CEFIC is reported as insisting that the introduction of civil liability for environmental damage should be confined to damage resulting 'directly and definitely' from the source of pollution.

INTERNATIONAL INITIATIVES

Discussion in this chapter was to be limited to EEC legislation. It would, however, be strange if mention was not made to procedures such as TOSCA, which in the 1970s and the early part of the 1980s caused the world chemical industry a great deal of concern—not only in terms of the technical requirements of the Act but also in terms of the documentation.

TOSCA inspired national legislation in other countries and additional issues such as that of confidentiality of commercial information to come into force.

Beyond the fundamental national development of regulatory procedures we have also seen in the last 10 years or so an increasing amount of surveillance and investigations by various international agencies—some friendly, some not so friendly.

Typically, UN agencies such as UNCTAD, UNIDO and the UN Economic Commission for Europe have studied sectors of the chemical industry and such studies will continue both as impartial investigations into the operations of chemical industries in their own countries and in terms of their impact on developing countries.

These organisations should not be blamed for the work they do but all should ensure that, in your efforts to remain in close touch with your national legislation, you also remain well informed about the world-wide context in which you operate.

The work programme of the EEC Commission has put special emphasis on the importance of international activities in the field of the environment. Legislation and Community action at international level has already been proposed regarding the export and import of dangerous chemicals that are banned or severely restricted in the Community. The question of the export of dangerous industrial processes and plant to non-EEC countries remains a matter of high priority.

The Fourth Action Programme (1987–92) urges the Community to develop legislation on the export of dangerous industrial processes based on the information and experience gained under the Seveso Directive (82/501/EEC). The Fourth Action Programme also recognised that concern over the international movement and production of dangerous chemicals, wastes and plant is growing rapidly. Accidents, though rare, can be devastating (Seveso and Bhopal). The Fourth Action Programme therefore calls for urgent action to develop at international level adequate control measures and notification and authorisation procedures, which whilst providing a high degree of security do not restrict legitimate manufacture and trade in dangerous products.

The Community is urged to act, in collaboration with the OECD and the UN, to promote the rapid development of world-wide codes of practice to supplement the specific legislative measures needed to cover some aspects of these matters. The Commission will take initiatives to this end.

CONCLUSION

Environmental requirements have long been seen as merely imposing regulations and cost on industry. However, without Community-wide regulations unequal conditions of competition would prevail. Industries in different Member States would be able to take advantage of the absence of environmental controls to sell cheaper products or invoke more stringent standards, which might constitute hidden barriers to trade to protect itself from the products of other Member States. To avoid distortions in trade arising from unilateral action by Member States, a measure of technical harmonisation is necessary.

Costs there certainly will be. The Commission recognises that the short-term costs resulting from the introduction of new environmental standards may have adverse effects on the competitiveness of certain enterprises that have to comply with them. The Commission therefore promises to ensure its objectives are made clear to industry and that enterprises are allowed a reasonable time to adjust to new standards.

Adjustment to new environmental standards can in some circumstances be facilitated by financial support by the use of various economic instruments such as taxes, charges, tradeable discharge permits and state aids.

There may be a few raised eye-brows at the mention of state aids. You are asking yourselves: 'How can this be compatible with the competition policy of the EEC?' In recognition of their importance in terms of the common European interest, state aids designed to promote the protection of the environment have been authorised by the Commission, under certain conditions, since 1974. The aim of allowing limited state aids for this purpose is to promote the introduction of, and industrial adaptation to, regulations which ensure effective environmental protection and to lead eventually to the promotion of 'the polluter pays' principle.

Pollution is considered to represent a waste of resources and is often linked to obsolete technologies. The advice to industry now is that strict environmental standards are a necessity, not only in order to achieve an adequate degree of environmental protection and an improved quality of life but also for economic reasons.

As regards Community industry, in particular, the Commission is convinced that, as progress is made towards the completion of the internal market by 1992, the ability of Community industry to compete on the world market will depend to a large extent on its ability to offer 'environment-friendly' goods and services, achieving standards at least as high as its competitors. If such progress is not made, in the view of the Commission, Community producers will lose market share, not only on international markets but also on the internal market.

ACKNOWLEDGEMENT

The author is indebted to his colleagues, Gareth Davies and Nicholas Fernandes, for their background research.

3

THE SAFETY OF LIQUID CHLORINE STORAGE, TRANSPORTATION AND ENVIRONMENTAL PROTECTION IN CHINA'S CHLOR-ALKALI INDUSTRY

Wu Zhengde

Shanghai Chlor-Alkali Complex, People's Republic of China

After the establishment of new China, the development of the chlor-alkali industry was very rapid. The reason for the rapid development initially was the large demand for caustic soda of each economical department. This was followed by the development of chlorine chemical compounds. Before the liberation, the only chlorine chemical compounds produced were hydrochloric acid, bleaching powder and liquid chlorine. After the establishment of new China many kinds of chlorine compounds were produced. At present the market demand of caustic soda and chlorine compounds still cannot be satisfied, resulting in shortage. The actual production of caustic soda and chlorine in 'sixth five plan' in China (not including Taiwan Province) is as in Table 1 [1].

In recent years the market demand for caustic soda and chlorine has increased. The production in China (not including Taiwan Province) is given in Table 2.

Under the 'seventh five plan' the production of caustic soda in China will reach 2 800 000 t by 1990. By 2000 a capacity of 4 500 000 t of caustic soda is planned (including 500 000–600 000 t of high-purity caustic soda). We will actively develop ion-membrane electrolysers and limit the mercury cell process. The mercury cell process will be changed to the ion-membrane process step by step; at the same time we will increase the treatment of mercury pollution. For the balance of chlorine in the manufacture of caustic soda, in the following years we will specially develop chlorine compounds which contain much chlorine, for example PVC, organic chlor-solvents and fine chemicals, etc. In 1987 the production of PVC was 570 000 t (not

TABLE 1
The annual production of caustic soda in China (excluding Taiwan Province)

Year	1981	1982	1983	1984	1985
Production of 100% caustic soda (1 000 t)	1 923	2 073	2 123	2 222	2 349
Production of 100% chlorine (1 000 t)	1 702	1 835	1 879	1 966	2 079

Wu Zhengde

TABLE 2

Production of recent years

Year	1986	1987
Production of 100% caustic soda (1 000 t)	2 511	2 722
Production of 100% chlorine (1 000 t)	2 222	2 409

including Taiwan Province), and we estimate that it will be 1 100 000–1 200 000 t in 1990. The percentage of chlorine consumed by PVC will be increased from 20% to 25%.

In 1983 the percentage of caustic soda manufactured by the diaphragm process was 91·2%, by the mercury process was 6·5%, and by the causticisation of sodium carbonate process was 2·3%. This means that 2% of the total production of chlorine must be excluded. Consumption in 1983 of chlorine for manufacturing chlorine compounds was as given in Table 3 [2].

In order to prevent any accidents involving liquid chlorine during manufacture,

TABLE 3

Consumption of chlorine in 1983

Products	Consumption (%)
I. Total of inorganic chlorine compounds	51·31
This includes:	
Liquid chlorine	25·06
Hydrochloric acid	19·58
Sodium hypochlorite	1·62
Bleaching powder and liquid, etc.	2·42
Iron trichloride	1·05
Calcium hypochlorite	0·66
II. Total of organic chlorine compounds	31·77
This includes:	
PVC	19·85
Chloral	2·11
Chlorobenzene	1·84
Chloroform	1·35
Synthetic lubricating oil	1·10
Carbon tetrachloride	1·10
Chloro-carbons	0·87
Dichloromethane	0·73
III. Total of pesticides	6·86
This includes:	
Dichlorvos	4·01
Diplerex	1·67
Methyl-1605	0·77
Acephetemet	0·18
Chloropirin	0·17
IV. Other compounds	10·06
Total	100·0

transportation and storage, we strictly implement the regulations of design, manufacturing process and inspection.

Liquid chlorine is stored in pressurised containers of a number of differing sizes. Drums contain 500 kg or 1 t, tank cars contain 40 t and storage tanks contain 10 or 20 t, etc. The design of pressurised containers must be according to the design regulations. The maximum allowable operating pressure is 15 kgf/cm^2 for liquid chlorine containers, fixed pressurised storage tanks and buffer pressurised containers. The welded seams of containers must be inspected by non-destructive testing using ultrasonics, X-ray or r-ray, etc. The containers must be annealed at 600–650°C after welding. The appearance of containers must be inspected frequently. If any dents or gouges are seen then they cannot be used. According to the pressure vessel regulations, the thickness of liquid chlorine containers must be determined annually by ultrasonic instruments and the corrosion condition inspected with film in a camera. Every 6 years we inspect the containers with X-ray non-destructive testing methods. The data are entered into a computer to calculate the results. According to the program, the computer tells us the thickness of the container. If the thickness of the container is less than given in the design data, then the computer warns us and we discuss how to treat it. The purpose of testing is to test the container's strength and the tightness of the piping line. We take 19–20 kgf/cm^2 as the liquid testing pressure. Moveable containers must also bear the mechanical loading so we also take 19–20 kgf/cm^2 as the liquid testing pressure.

According to the regulations for liquid chlorine containers, when the customer has used up the liquid a pressure of >0.5 kgf/cm^2 must remain in the container to permit a sample to be taken for analysis. Only liquid chlorine is permitted in containers for liquid chlorine.

It is not allowed that a container of liquid chlorine be directly connected to the reactor. That is to say, the chlorine must pass through a buffer tank before entering the reactor. Otherwise, when the liquid chlorine is consumed, the reactants may suck back into the container. Then when we next put liquid chlorine into the container the reactants may decompose and increase the temperature, resulting in explosion of the container. The remaining chlorine gas in a container can be tested with ammonia water to confirm that it is a liquid chlorine container.

During operation the valves of the cylinders must be strictly inspected. For prevention of accidents a fused material containing 50% Bi, 27% Pb, 13% Sn and 10% Cd is poured into a slot of a fused plug. The composition and purity of the metal must be inspected. The purity must be more than 99.9%. When the temperature exceeds 65°C the metal melts, the chlorine leaks and an explosion is prevented.

The explosion of a liquid chlorine cylinder is mostly caused by overfilling. When the liquid chlorine expands the containers crack. Experimental data for the explosion of a 1-t cylinder are as shown in Table 4 [3].

From Table 4 the maximum pressure of explosion was 153 kgf/cm^2. When the pressure exceeded 153 kgf/cm^2 the cylinder cracked. The quantity of liquid chlorine filled into a container must be carefully controlled. The filling ratio of a liquid chlorine cylinder is 1.25 kg/litre whilst that of a tank car is 1.2 tons/m^3. After filling it is necessary to reweigh each container. The allowable error is ± 1%.

TABLE 4
Data for chlorine cylinder explosion

Number of cylinder	Pressure of yield strength (kgf/cm²)		Explosion pressure (kgf/cm²)	
	Design	Testing data	Design	Testing data
8-2	80	95	112	135
79-941	80	111	112	145
79-827	80	110	112	153
15-18	80	98	112	143
4413	80	102	112	140
850	80	102	112	132

During the transportation of liquid chlorine cylinders or tank cars, workers must be trained, receive safety education and use the protective clothing. During transportation, liquid chlorine cylinders must not be mixed with other dangerous chemicals. In summer time, when the weather is hot, they can only be transported before 10 am and after 4 pm. Also in the storehouse liquid chlorine cylinders or storage tanks must not be mixed with dangerous chemicals.

During the manufacture of chlorine, dangerous, explosive trichloro-nitrogen (NCl_3) is formed. NCl_3 has a molecular weight of 120·37, is a yellow oily liquid or a rhombic crystalline solid, its specific gravity is 1·653 g/cm³, its melting point is $-40°C$ and its boiling point is 71°C. It has a pungent odour, and it is an irritant to eyes and lungs. At 95°C the liquid or gaseous NCl_3 decomposes and explodes. It is also a vigorous oxidising agent; when it is in contact with organic chemicals it will create an explosion. It does not dissolve in cold water but will dissolve in hot water, carbon tetrachloride, benzene, carbon disulfide and other solvents. It is formed when water containing NH_3 is used in the brine purification process. NH_3 reacts with chlorine to form NH_2Cl_2, $NHCl_2$ and NCl_3, so that it is necessary to control as follows:

Brine water: Inorganic ammonium salt $\leq 0·2$ mg/litre
                                      Total ammonium salt $\leq 1·0$ mg/litre
Analyses are performed once per week.

Salt: Inorganic ammonium salt $\leq 0·3$ mg/litre
                                      Total ammonium salt $\leq 1·0$ mg/litre
Purified brine water: Inorganic ammonium salt $\leq 1·0$ mg/litre
                                      Total ammonium salt $\leq 4·0$ mg/litre
Analyses are performed once per day.

In the process of chlorine purification we use water to wash the chlorine. NCl_3 is dissolved, which decreases the NCl_3 contained in the liquid chlorine. According to the Soviet standard TOCT 1288-72, $NCl_3 \leq 0·002\%$ (wt) in liquid chlorine. For safety the residue of evaporating liquid chlorine must be neutralised by alkali or lime water. We have modified the testing method of the Soviet TOCT 6718 and BS 3947 (1976). NCl_3 and hydrochloric acid are reacted in a closed hydrochloric acid separator to form ammonium chloride by the reaction $4HCl + NCl_3 \rightarrow NH_4Cl + 3Cl_2\uparrow$. Then the concentration of the ammonium ion is detected by spectrophotometer. From this we can derive the concentration of NCl_3 even in liquid or gaseous form. This method is very simple, gives correct data, takes less than 1 h and analyses down to ppm grade. The percentage of sample absorption is $>90\%$, the percentage of recovery NCl_3 is $>85\%$ and the repeated frequency of analysis is $>95\%$. This testing method is a rapid and practical process.

In China the regulations governing the working atmosphere state:

1. For a workshop the maximum allowable concentration of chlorine is 1 mg/m^3.
2. For a residential area the maximum allowable concentration of chlorine at any moment in time is 0.1 mg/m^3.
3. The long-term concentration of chlorine which man can tolerate is 3 mg/m^3.

In China, in September 1979, XX chemical plant suffered an accident. Residues of chloro-paraffin remained in a liquid chlorine cylinder. When the works operated the chloro-paraffin decomposed. The temperature increased, resulting in cracking of the liquid chlorine cylinder and storage tank. 10.2 t of liquid chlorine leaked in 30 min. At that time the wind velocity was 3.7 m/s from the south-east. An area of 7.35 km^2 was affected. The author calculated the concentration of chlorine which has travelled different distances, according to Soviet author И. С. Ройзен's [4] pollution formula, and compared the results with actual investigation reports:

$$C_0 = -\frac{100M}{VX^2} e^{-20h/X}$$

C_0 = concentration (mg/litre)
M = toxic gas evolved (t/day)
V = velocity of wind (m/s)
X = distance between point at which the chlorine was tested and the point at which toxic gas evolved (m)
h = height above ground (m)

When the height above ground at which toxic gas evolved is 1 2 m, $e^{-20h/X} \simeq 1$ within 1000 m.

The parameters given in Table 5 were calculated [5].
1. According to the report, when the concentration of chlorine was 2500 mg/m^3 a man breathed it and died suddenly. At a distance of 72.7 m downwind by calculation the man had died suddenly. This agrees with the investigation report as falling within 80 m, to be called the centre of explosion. It is a very dangerous distance.
2. According to the calculation, the concentration of chlorine which is allowable in the workshop is 1 mg/m^3 at a distance of 3638 m as R in sector area $S = \frac{1}{2}R^2Q$ in the downstream wind direction. This also agrees with the investigation reports.
3. According to the calculation, the maximum allowable concentration of chlorine for a residential area is 0.1 mg/m^3 at a distance of 11.5 km.

So safety is very important in the manufacture, transportation and storage of chlorine. It is also related to environmental protection. During the manufacture and use of chlorine we must strictly implement operating regulations. When we

TABLE 5

Calculated Cl$_2$ concentrations

Region of pollution	Extent of region: upwind distance — downwind distance (m)	Distance in downwind direction (m)	Concentration of Cl$_2$ calculated from data (mg/m^3)
1. Centre of explosion	40–80	50	5 292
2. Dangerous area	80–800	400	82·7
3. Seriously polluted area	500–1 100	1 000	13·2
4. Polluted area	1 100–2 200	2 000	3·3
5. Slightly polluted area	2 200–3 000	3 000	1·4

construct a new chlor-alkali plant it must be located at a certain distance for safety and environmental protection. At the same time a proposal for safety and environmental protection must be given. Old plant must also include these proposals.

REFERENCES

1. Ming, Wang, *Year Book of World Chemical Industry*. Scientific and Technical Information Research Institute, Ministry of Chemical Industry, May 1989. (In Chinese.)
2. Guang Qy, Yang & Tao, Tao, *Chemical Industry in China in this Century*. China Society and Science Publishers, 1986, p. 144. (In Chinese.)
3. Sheng, He Yun, Safety technology of liquid chlorine. *Chlor-Alkali Industry*, No. 9, p. 14, 1987 (internal issue).
4. РОЙЗЕН, И. С. ТЕХНИКА БЕЗОПАСНОСТИ И ПРОТИВОПОЖАРАЯ ТЕХНИКА В ХИМИЧЕСКОЙ ПРОМЫШЛЕННОСТИ. ГОСХИМЗДАТ. МОСКВА-ЛЕНТРАД, 1951.
5. Zhengde, Wu, Study on environment pollution of chlorine. *Chlor-Alkali Industry*, No. 1, p. 27, 1981 (internal issue). (In Chinese.)

4

THE PROGRESS OF CHLORINE CELLS IN CHINA

Ma Guiguan

Tianjin Chemical Works, People's Republic of China

1 BRIEF REVIEW

Prior to the founding of new China there were only a few chlor-alkali plants. In 1949 the total capacity for caustic soda on the China mainland was 15 000 t/year. Since then the chlor-alkali industry has gone through vigorous expansion. Up to 1987 the total output of caustic soda amounted to 2·7 million t/year, i.e. 180 times the capacity in 1949.

Formerly Allen-Moore and Billiter type diaphragm cells were used in production. In 1957 new vertical cells incorporating deposited asbestos diaphragms were developed and spread rapidly to many plants.

To meet the demand for high quality caustic soda, in 1952, horizontal mercury cells with double chambers were installed in Liaoning Province and gradually were expanded to a capacity of 150 000 t/year.

In 1973 diaphragm cells with DSA-coated titanium anodes were successfully trial manufactured in Shanghai. Next year the first batch of forty 30-I type cells was erected and put into production. Thereupon a mass growing period of DSA diaphragm cells started. In 1984 caustic production using DSA cells amounted to 800 000 t.

As to mercury cells, after a continuous research period of several years, in 1981 metallic anodes were employed to replace graphite anodes. Cell structures simultaneously underwent a lot of technical innovations. The capacity of a single electrolyser was enlarged 3·5 times. However, out of consideration of environmental pollution from mercury emissions as well as the high consumption of electric power, the mercury process is no longer developed in quantity. The caustic production capacity is maintained at approximately 130 000 t/year.

2 THE DESIGN AND OPERATING CHARACTERISTICS OF CURRENT ELECTROLYTIC CELLS

Several types of representative cells are described here.

TABLE 1

30-III type diaphragm cells: design and operating characteristics

Item	
Number of anodes	48
Active anode area (m^2)	30·7
Current load (kA)	40–52
Current density (kA/m^2)	1·3–1·7
Cell voltage (V)	3·2–3·4
Current efficiency (%)	93–95
Power consumption (d.c.) (kWh/t NaOH)	2 270–2 440
Cell liquor, NaOH (g/litre)	120–130
Caustic production per electrolyser (t/day)	1·35–1·74
Oxygen in anode gas (% vol.)	1–2
Diaphragm life (months)	8–10
Anode life (years)	5–7

2.1 30-III Type DSA Diaphragm Cells (Table 1)

This type of cell is developed from 30-I and 30-II type metallic anode cells and is now commonly used by most producers. The cell is composed of an anode assembly, a cathode assembly and a cell cover. The outer dimensions are 2·2 × 1·1 × 2·1 m. The anodes are of mesh plate made from expanded titanium sheet which is welded onto a titanium-clad copper bar. The surface of the anode is coated with electrically active material containing ruthenium as the main component. The current is fed from the copper plate at the bottom to the anodes by means of Ti–Cu bars bolted onto a copper plate. The steel cell bottom is fastened onto the copper conductor plate. A thin titanium plate is laid on the steel bottom to separate it from the anolyte. Recently, in certain plants rubber lining has been used instead of titanium. The annular clearances between the anode bars and the holes in the cell bottom are sealed with anti-corrosive rubber washers to avoid leakage.

The cathode is made of woven wire net bags. Both ends of the bags are welded onto a screen which is in turn welded onto the wall of the cathode box. The space inside the net bags and that between the side screen and the wall form the cathode chamber. The outer part forms the anode chamber, in which anodes are erected vertically by means of an adjusting mould. The cell cover is made of rubber-lined steel.

The diaphragm material is mechanically pretreated asbestos causticized with cell liquor. During the diaphragm deposition operation the cathode box is immersed in the asbestos slurry tank and evacuated to vacuum, whilst stirring, so that a thin, even and smooth layer of asbestos is formed on the cathode screen. When the deposited asbestos has reached the required thickness (controlled by vacuum and suction time) the cathode box is lifted clear of the slurry tank whilst continuing the application of vacuum for several minutes to make the diaphragm compact. The cathode is then dried at 100–120°C. In some industrial plants a modified diaphragm using a PTFE emulsion is used and operates fairly well.

The voltage distribution of the cell in operation in a certain plant is shown in Table 2.

TABLE 2

Voltage distribution of 30-III diaphragm cell
(operating conditions: feed brine NaCl concentration, 315 g/litre;
current density, 1·5 kA/m²; cell temperature, 85–90°C; catholyte
NaOH concentration, 120 g/litre; NaCl concentration, 190 g/litre;
current efficiency, 96%)

Item	
Reversible anode potential	1·22
Reversible cathode potential	0·85
Anode overvoltage	0·068
Cathode overvoltage	0·22
Diaphragm IR	0·44
Brine IR	0·30
Structure and intercell bus	0·14
Total	3·24

2.2 16 and 8 m² DSA Diaphragm Cells (Table 3)

There are about 200 chlor-alkali plants on the China mainland. Among them are some 150 small plants, each with an annual capacity less than 15 000 t, which are distributed throughout the whole mainland (except Tibet). Most of them employ small graphite anode cells. Since metallic anode cells have already demonstrated vast advantages over the old-fashioned graphite cells, the small chlor-alkali plants tend to improve and retrofit their old installations. For this purpose 16 and 8 type DSA cells have been designed and put into production successfully.

The size of a single anode plate of these two types of cell is exactly the same as that of a 30-III type cell. The advantages of such design are (1) the standardization of anodes is convenient for mass production by adopting precise tool machines and (2) to provide appropriate conditions for enlargement of electrolysers.

2.3 150-kA Mercury Cells (Table 4)

These cells are employed in a plant producing about half of the mercury cell caustic. The electrolyser is 13 m in length and 1·1 m in width. The cell bottom is a

TABLE 3

16 and 8 type DSA diaphragm cells: design and operating characteristics

Item	16 type	8 type
Number of anodes	26	13
Effective anode area (m²)	16·4	8·32
Current load (kA)	24–28	12–15
Current density (kA/m²)	14·5–16·8	14·5–18
Current efficiency (%)	95	95
Cell voltage (V)	3·20–3·38	3·16–3·34
Caustic production per electrolyser (t/day)	0·81–0·95	0·39–0·49
Power consumption (d.c.) (kWh/t NaOH)	2 260–2 380	2 230–2 360

TABLE 4

150-kA mercury cells: design and operating characteristics

Item	
Number of anodes	30
Active anode area (m^2)	14·3
Current load (kA)	140–150
Current density (kA/m^2)	9·8–10·5
Cell voltage (V)	4·7–4·8
Power consumption (d.c.) (kWh/t NaOH)	3 300–3 400
Sodium content in amalgam (%)	0·2–0·25
Caustic production per electrolyser (t/day)	4·75–5·0
The product concentration NaOH (%wt)	45
Quantity of mercury per cell (kg)	2 600

thick inclined steel plate, which is planed smoothly and worked to form slots along the bottom. The side wall is made of rubber-lined channel steel. The anodes are made of thin titanium wire grids welded onto titanium structures. The surface of the anode is coated with Ru–Ti active coating. The cell cover is made of pasted double layers of flexible rubber resistant to oxygen and chlorine. An automatic jumper switch and an anode assembly lifting device are fitted to each electrolyser. The flexible cell cover moves up and down simultaneously with the anode assembly to keep the seal hermetically. Both ends of the electrolyser are equipped with washing boxes, the inlet box for removing traces of alkali contained in the mercury and the outlet one for removing traces of depleted brine adhering from the amalgam.

Saturated brine flows through a rotameter into the electrolyser. The depleted brine passes via a gas separator to the dechlorination and the resaturation sections. Sodium amalgam is fed to the top of a vertical denuder packed with graphite granules and flows down through a distributor. Deionized water is fed to the bottom of the denuder. The finished caustic liquor overflows to storage. Hydrogen is cooled in a reflux condenser to remove water and mercury. The mercury from the denuder is pumped through the inlet washing box and returned to the electrolyser.

3 RECENT TECHNOLOGICAL DEVELOPMENTS

3.1 Diaphragm Cells

3.1.1 Cell Structure Improvements

(A) The accuracy of machining has been improved and as a consequence the flatness, smoothness and verticality of components has been considerably improved. Consequently gaps between electrodes are more uniform.

(B) Mathematical models have been adopted in the investigation of the optimum structures and geometrical shapes of cell components.

(C) The brine gap has been reduced from 9 to 6 mm. Some of the experimental cells have been put into production.

(D) Punched sheet has replaced woven reticular cathodes to make the surface smoother and flatter.

(E) Expanded anodes are now being tested.

3.1.2 Improvements of Anode Coatings

Ru–Ti coating has been used as the electro-active material for a long time. Coating methods, procedures and formulations, including the optimum amount of ruthenium to be used, have been tested and improved through experiment and in practice. Recently, new coating constituents, e.g. tin, iridium, etc., have been introduced into the formulation to form a three-component system. The addition of a medium layer is also being tested. These subjects are now in pilot-plant scale. Universities and research institutes have played an active and important part in this research work.

In the meantime, the stripping of ruthenium from used coatings by electrolysis has achieved success and has already been adopted by some plants. Another method using a surface active agent to strip the ruthenium has also been proved to be successful in industrial cells. The recovery yield is high and no corrosion to the titanium substrate occurs. Several plants are interested in this method and are preparing to put it into operation.

3.1.3 Research and Application of Modified Diaphragms

Ever since DSA cells were put into production a number of plants and research institutes have concentrated their efforts on the development of modified diaphragms. They are engaged in experiments using various kinds of fluorine compound modifiers. Among these, PTFE emulsion and fibre are relatively satisfactory. The former is employed in a large plant and the latter has been applied in many single cell experiments and fair results were obtained. A factory has been built up to provide the modifier which, in combination with short gap cells, will achieve a significant reduction in cell voltage.

In order to bring about the application of modified diaphragms in production successfully, we have taken the following measures:

(A) To select chemically stable, high strengthened and uniformly sized asbestos, in addition to careful pretreatment. For this purpose we are working out a standard of asbestos suitable for diaphragm cells.

(B) The impurities in the feed brine must be minimized to avoid blocking. The producers are taking measures in this respect both in technology and administration. Some of the producers have adopted secondary filtration to ensure the brine quality.

(C) To keep the operation steady, operating instructions must be strictly observed; current changes, shutdowns and the changes of all the operating conditions should be avoided as far as possible.

3.1.4 Research on Active Cathode Coatings

Recently, efforts have been made to develop active cathode coatings, among which the electroplating of Ni–Zn, the chemical plating of Co–W–P and the plasma spraying of Ni–Al methods gave fair results and have been adopted to make trials on many technical electrolysers for years. Today discussions and evaluations on concrete measures and economic aspects of mass applications are held so as to practice these achievements.

3.2 Mercury Cells

3.2.1 Improvements in Anode Coatings

Owing to the higher current density in mercury cells, the compositions and formulations, as well as the quantities of active materials in coatings, are all different from those used for diaphragm cells. A series of experiments have been performed in which a Ru–Ti–Sn ternary composition was used on industrial cells. Another new formulation aimed at lengthening the coating life has been applied on an industrial cell to confirm the result.

3.2.2 Development of an Anode Protection Device

At present the automatic ascending device of the anode assembly does not work satisfactorily because the computer control unit has not yet been employed to regulate the operation conditions and consequently the gap between the electrodes is much larger than that in foreign advanced mercury cells. This is the most important problem we are trying to solve.

3.2.3 Mercury Pollution Control

After persistent and continuous research work for years, we have worked out a series of methods for the elimination of the major emissions of mercury.

(A) Removal of mercury vapour in hydrogen and in mercury contaminated air by a hypochlorite scrubbing method is already employed in production.

(B) Prevention of mercury sedimentation in brine resaturation and purification will be put into operation in the near future.

(C) Filtration by active charcoal to decrease mercury content in caustic liquor. The design work has been finished.

(D) An ion exchange method to treat the mercury contaminated waste water is about to begin construction.

After the realization of the above items, the mercury emission will be reduced to less than 50 g/t caustic soda (100%).

In summarizing the above descriptions it can be seen that, although many aspects of the research work on chlorine cell technology have been performed with a fair degree of success, the utilization and application of these achievements on a commercial scale has not been realized. The predominant factor is due to the comparatively low cost of electric power in China. In some cases, as a newly developed technique is planned to be adopted in production, the value of output can scarcely counterbalance the value of input, thereby making the overall economics unprofitable and naturally restricting the enthusiasm of the caustic soda producers to make the innovations and reforms. This will inevitably change along with the deepening of the reform of the economic system. The price of energy is being and will continually be readjusted to a greater extent. This trend will no doubt encourage and accelerate the developments and applications of new technologies on chlorine cells.

4 A LOOK TO THE FUTURE

On the basis of the present situation in China, the expansion of the chlor-alkali industry in future should be according to the following principles.

(A) In the coming decades diaphragm cells will still play the most important part in the chlor-alkali industry. The technology of existing commercial DSA cells is still in its primary stage, and there is much to be done to adopt advanced technology. It should be rapidly improved within a few years so as to reduce cell voltage, to raise current efficiency, and allow cell capacity to be increased without a corresponding rise in power consumption, and finally, to replace graphite anode cells by steps. The adoption of new techniques should adhere to the following principles: (1) to attain a remarkable effect and (2) to expend reasonable cost for the purpose of saving power to enhance economic benefit. Some of the key techniques may be imported from abroad with economic benefit.

Small DSA cells are suitable for extension and transformation of small plants. Moreover, the manufacture of small cells is provided with better characteristics more easily. It is convenient to the direction of development.

(B) The existing graphite anode cells will continue to produce a significant quantity of caustic soda. We should strengthen the technical administration of the plants to realize optimized operations according to the respective situation of the plants, thus giving play to their benefits.

(C) Mercury cell technology will not be increased as mentioned before. The existing equipment still possesses fair economic benefits and is able to furnish high quality products. A number of cells are maintained fairly well and are able to operate normally. On condition that mercury pollution can be overcome, these cells will continue to be utilized.

(D) Membrane cells of about 100 000 t/year capacity have been imported and put into production. In the last decade of this century other sets of equipment with the capacity of 200 000 t/year will be imported successively. However, at the present time, the price of importing the technology and the equipment is expensive, so that the cost of production is much higher than that from diaphragm or mercury cells, hence it is impossible to expand membrane cell capacity by importing in very large quantities. Today a branch of the Chinese Academy of Sciences cooperating with other research units is concentrating its efforts on developing ion exchange membranes. Those plants having imported membrane cells are now digesting and assimilating the foreign advanced technologies, striving to collect the merits of each type of cell, and working to design a new type of membrane cell. Other auxiliary procedures—the precise filtration of brine, secondary brine purification by chelating resin, dechlorination of depleted brine, etc.—can be resolved by ourselves. Only by basing on domestic technologies and facilities can membrane cells be rapidly developed.

5

THE NEED TO CARRY OUT HAZARD AND OPERABILITY STUDIES ON EVERY NEW PLANT

R. Ellis Knowlton

Chemetics International Company, Canada

1 INTRODUCTION

Chemetics International Company is part of the Canadian company C-I-L Inc., which in turn is a subsidiary of the British company ICI. Corporate headquarters and principal design offices are in Vancouver, BC, Canada. The Company has a design office in Toronto, Ontario, Canada, and has a 90% ownership of Modo-Chemetics based in Ornskoldsvik, Sweden.

The Company designs and builds certain types of chemical plants worldwide. The main types of chemical plants and processes are electrochemical (chlor-alkali and sodium chlorate), sulphuric acid, nitration plants, certain other bleach processes and chemicals, for example oxygen bleaching systems and calcium hypochlorite. The Company is also involved in related technologies such as chlorine dioxide generators, nitric and sulphuric acid concentrators, and certain types of computer control systems.

The Company has always stressed the importance of Process Safety. Since 1975 it has been policy to carry out a Guide Word Hazard and Operability Study, led by properly trained Study Leaders, on every design. In addition, since 1980 Chemetics has provided a training service in Hazard & Operability Studies worldwide.

This chapter will discuss the technical and commercial advantages of executing Hazard & Operability Studies on every design and of providing training in Hazard & Operability Studies. These general advantages are illustrated with a specific example, namely a new chlor-alkali plant where, on first inspection, the results of a Hazard & Operability Study could have been predicted to be negligible. However, as this chapter will show, this was not the case.

2 TECHNICAL AND COMMERCIAL ADVANTAGES OF HAZARD & OPERABILITY STUDIES

The most fundamental of all approaches to Process Safety is experience. If a unit had been operated, without change, for an infinite period of time, then all possible hazards would have been identified and rectified. However, experience alone, although valuable, is not sufficient for two reasons:

1. Experience is costly in terms of human suffering and financial loss.
2. Hazards, once identified as a consequence of experience, will require changes; some of these changes in turn could result in new and unidentified hazards.

Therefore some additional method of identifying hazards is required to supplement experience. Guide Word Hazard & Operability Studies provide a form of 'Synthetic Experience'. The function of each part of a design is described—the so-called 'design intention'. Notional or hypothetical deviations are generated by applying a series of distinctive words or phrases to the design intention. These hypothetical deviations are used as the basis for a search for the causes of and consequences of corresponding real deviations which could occur in practice. A note is made of any potential hazards or operational problems so that remedial action can be taken. The method can be used at the Piping & Instrumentation (P&I) stage of design when rectification is relatively cheap or at any subsequent stage when rectification could be more expensive. In skilled hands about 99% of hazards can be detected and rectified.

One technical advantage of Hazard & Operability Studies emerges immediately. Changes at the design stage are much cheaper than field modifications and many orders of magnitude cheaper than dealing with the consequences of accidents once they have happened. A second advantage was found almost immediately after Hazard & Operability Studies were first applied. The technique identified 'Operability' problems as well as hazards. Consequently plants were easier to start up and to operate, in addition to being safer. A third technical advantage also became apparent. This advantage is concerned with 'information sharing'. Hazard & Operability Studies are conducted with a multidisciplinary team (e.g. a 'design team'). The requirement to specify the exact 'design intention' in order to be able to generate deviations also results in all the team members becoming aware of the detailed purposes of every part of the unit and every part of the operating procedure. Any ambiguities are removed, leading to greater mutual understanding and commitment. This factor also leads to better designs, better operating procedures, faster start-ups and smoother running.

The first commercial advantage arises directly from the technical advantages mentioned above, namely further contracts from satisfied clients. However, this powerful advantage can only arise with existing customers. Company policy of applying Hazard & Operability Studies to every new design is also advantageous when dealing with new customers. Contracts are awarded for a wide variety of reasons but Process Safety is one of the attributes used by clients in making a decision. This is a consideration not only from the point of view of clients themselves but, to an increasing extent, it is also a consideration by the financial backers of a project if these are separate from the clients.

Chemetics offers a training service in Hazard & Operability Studies in general and for Study Leaders in particular on a worldwide basis. The main aim is to improve the effectiveness of Hazard & Operability Studies carried out within the Process Industries, however such training also increases the awareness of the company as a provider of chemical plants to potential clients. This awareness also extends to those who approve permits for the construction of plants.

The final commercial advantage arises with certain contracts in which the Company not only provides the plant (which will have had a Hazard & Operability Study led by a properly trained Study Leader) but may also provide training so that the client company can conduct its own Hazard & Operability Studies (Hazop) in the future. The client can even execute its own Hazop Study on the Chemetics design as a further check. In these circumstances there will be very few residual hazards to be detected but the 'technology transfer' will be improved.

3 WHY EVERY DESIGN?

Chemetics has designed and built fourteen chlor-alkali units, mostly within the last few years. The Company has also built many other similar plants—over thirty sodium chlorate plants, for example. It would therefore be a reasonable question to ask why is it necessary to carry out a Hazop Study on every new plant design? Why not execute a Hazop Study on the first one or two, identify all the hazards, remove them, and thereafter freeze the design and forget about Hazop Studies?

There are three answers to this question. The first is that it is Company policy and our customers expect it. If this was the only reason the result could be a superficial study. The Hazop Examination would amount to little more than uttering incantations (in the form of Guide Words) over the flowsheet in the hope of a successful design. Hazop Studies must 'make sense' to those who participate and therefore must be productive in terms of the identification of residual hazards and operability problems.

The second reason is that no two units are identical. There are always changes. These changes may be due to differing customer requirements. Further changes may be due to the ongoing technology improvements; the 'learning curve' changes. In addition, standards, for example for environmental protection, are becoming more stringent.

The final reason is that the composition of the design team may change. The 'information transfer' aspect of a Hazop Study will help clarify the design intention for those team members who are relatively new to the technology.

4 THE SUBJECT OF A SPECIFIC HAZOP STUDY

The unit was to produce 20 t/day of chlorine using ICI's FM21 membrane cells. This was to be used for pulp bleaching for a paper mill.

In order to reduce potential hazards, the unit would operate with the minimum amount of intermediate storage. There would be no inventory of liquid chlorine. The intermediate storage consisted of 4 h production with the chlorine stored in chilled water. Because of the low inventory the unit needed a highly reliable production of at least 50% of capacity. In addition, the unit had to be designed to have a 'turndown' to 25% of capacity. Finally, an increasing demand was foreseen and the unit had to be capable of an expansion to 30 t/day with the minimum interruption to production.

In order to meet the above requirements, it was decided to design the unit with two parallel banks of cells, each bank having its own electrical circuit. The unit before and after the cells was to be the normal single-stream design. This combination of parallel cell banks joined to an otherwise conventional single-stream unit was a major difference from any previous design.

There were three other sources of differences. A significant source of difference between one unit and the next may be in the utilities and this particular unit was no exception. The pressure of the steam service was much lower than usual. There were other minor client requirements which were non-standard and, finally, there were the usual 'learning curve' changes mentioned previously.

5 THE HAZARD & OPERABILITY STUDY OF THE UNIT

The Hazard & Operability Study was conducted as usual by a trained and experienced Study Leader. The Hazop Procedure consists of the following steps:

Definition
Preparation
Examination
Follow-up

5.1 Definition

At the Definition stage it was decided by the Study Leader and Project Manager that the examination would cover only those parts of the design which were different from the previous design. The 'standard' part had already received many Hazop examinations. This decision was based on the knowledge that with experienced Hazop Study Leaders and team members at least 99% of any residual hazards would have already been identified and corrected by the Hazop of earlier designs. Any attempt to carry out a further Hazop examination would not only be an undesirable expense but would lead to such boredom, that if there were any residual hazards they would probably remain undetected. Furthermore, the design team for this particular design had been involved in so many previous Hazop examinations of the 'standard' features that there would be no 'information transfer' element to the Hazop. Finally, there were no unusual or enhanced environmental requirements. The Hazop Team consisted of the following:

Project & Mechanical Engineer
Senior Process Engineer
Process Engineer
Instrument Engineer
Electrical Engineer
Study Secretary
Study Leader

5.2 Preparation

At the 'Preparation' stage the Study Leader collected the information which enabled him to plan which sections of the unit should receive a Hazop examination and which would be excluded as 'standard'. The aim was to focus the Hazop in those sectors which were different. In addition to the P&I drawings, the Study Leader obtained the operating instructions for a standard chlor-alkali unit. The Process Engineer had mentally decided how the parallel sections would be operated but had not produced written procedures. The Study Leader estimated the examination would require 18 team hours.

5.3 Examination

The Hazop Examination was carried out in ten sessions lasting a total of 27 team hours. This was considerably longer than estimated from the amount of plant to be examined. The reason for the extra time was due to the added complexity introduced by having two parallel systems. This complexity resulted in significant revisions to the operating instructions. Much of this revision was carried out during the Examination sessions. In retrospect, it would have been more cost effective if new operating instructions for the parallel section had been written prior to the start of the Hazop Examination.

As mentioned earlier, there were four sources of differences between a standard chlor-alkali unit and the subject of this study. These were:

Parallel operation
Utility differences
Other client needs
Learning curve changes

Each of these differences contributed to the potential hazards and operability problems.

5.4 Follow-up

The follow-up to the hazards and operational difficulties identified during the Hazop Examination can be classified as follows:

Hardware changes
Changes in operating methods, i.e. 'software' changes
No change required

Altogether 99 questions were recorded. The sources of differences and the follow-up decisions are shown below in Table 1.

The proportion of 'no change' items, 21%, is within the normal range of 10–30%. However, the rate of production of questions—about 4 per hour—was much lower than the usual rate of about 7 per hour. This was due to the amount of examination time devoted to revising the operating procedures.

All the follow-up decisions were implemented and the plant was started up on time. Had a Hazard & Operability Study not been carried out there would have been delays and expensive field modifications.

TABLE 1
Classification of results

Sources of differences	Follow-up action			
	Hardware changes	Software changes	No change	Totals
Parallel operation	17	8	0	25
Utility differences	5	0	3	8
Other client needs	7	5	1	13
Learning curve changes	27	9	17	53
Totals	56	22	21	99

6 CONCLUSIONS

The policy of requiring a Hazard & Operability Study to be carried out in every new design is sound on both technical and commercial grounds. There are always differences in design or in requirements and these differences will be a source of potential hazards and operability problems. Although in prospect it may be difficult to justify carrying out a Hazard & Operability Study, in retrospect such studies are invariably justified.

6

TOXIC EFFECTS OF CHLORINE—AN UPDATE FROM THE MHAP TOXICITY WORKING PARTY*

R. M. J. Withers

Loughborough University of Technology, UK

and

J. P. H. Shaw

ICI, UK

INTRODUCTION

Subsequent to the publication of the 1st Report of the Toxicity Working Party at the International Chlorine Symposium organised by the SCI in 1985 in London, and its publication in book form [1], CEFIC commissioned a special study [2] at the TNO–CIVO laboratories in Zeist, The Netherlands. Its primary aim was to obtain data using rats which would establish an LC_{low} or threshold LC. The protocol also included sufficient work with mice to provide a comparison with earlier work described in the literature. It will be recalled that the Working Party, in its earlier report, had been unable to draw any conclusions about the concentrations likely to cause low percentage fatalities, although it did make the positive statement that no lethallties have been demonstrated at 30 min exposure time at or below 50 ppm for any species.

A secondary aim of the CEFIC-sponsored work at TNO–CIVO was to try to shed more light on the effect of exposure time upon the concentration/lethality relationship. From the restricted range of experimental data that had been available to the Working Party values seemed to be roughly satisfied over a restricted time range by the simple relationship; fatal dose is proportional to concentration multiplied by the square root of time. However the Working Party was unable to commit itself to a precise formula on this aspect whilst fully recognising its importance.

In addition to these main aims, the TNO work involved studies on pathology, including the measurement of breathing rates.

* A preliminary report. The final review will be published by the Institution of Chemical Engineers as a Loss Prevention Bulletin in 1989.

The MHAP Toxicity Working Party were asked to comment upon the draft TNO report, with particular reference to:

—scope and consistency,
—mouse–rat–human response relationships, and
—relevance to other work and our previous findings.

In this chapter we now discuss each of these aspects in turn.

SCOPE AND CONSISTENCY

The TNO–CIVO study is an important piece of scientific investigation and members of the Working Party have spent some time in discussing and evaluating its findings. Accordingly this aspect forms the major part of our paper.

The study involved a total of 22 groups of 10 rats, with these groups distributed over exposure times of 5, 10, 30 and 60 min, and at concentrations over the range 321–5785 ppm. Twelve groups of 10 mice were studied at exposure times of 10 and 30 min over the range 457–1652 ppm. In addition, there were groups of rats and mice for control purposes, and for separate studies on pathology, including breathing rates.

The handling and pathological examination of the animals by TNO clearly conformed to a high level of scientific investigation, as did the production and analysis of the test atmospheres. The change in the behavioural pattern of the animals exposed for 5 or 10 min (breathing rates) is well documented, but makes extrapolation of the results more difficult. It would have helped to have either done more tests at 15–20 min or with animals under induced exertion at the shorter periods of exposure. Taken by itself for the animal species under study, it provides a statistically significant threshold LC and a clear indication of the effect on lethality of varying concentration and time, conclusions that the Working Party had been unable to draw from any previously published animal studies. However, the work should be regarded as complementing earlier work which the Working Party had judged to be soundly based, and to be reasonably consistent with comparable animal studies, e.g. Silver & McGrath or of Underhill, rather than replacing it, and many factors need to be considered when assessing the value of the TNO–CIVO study and relevance to an understanding of the lethal toxicity of chlorine to man.

Results by TNO–CIVO indicate a significantly higher level of dangerous concentrations than the average figures calculated by the Working Party from its study of earlier work. Thus for the rats the TNO LC_{50} at 30 min is c. 640 ppm whereas the Working Party's figure was 414 ppm, largely based on Vernot's work. For mice the TNO figure is c. 500 ppm at 30 min whereas the earlier assessment was 237 ppm derived from a variety of published findings.

The Working Party in 1985 postulated that the average LC_{50} for humans should be assumed to lie between 300 and 400 ppm at 30 min.

If this new TNO work had been available at the time of the 1985 review, it would have raised the Working Party's average figures of 414 and 237 to 561 and 274 for the LC_{50} in ppm of rats and mice respectively. This new data, therefore, could be used to suggest that the 30 min LC_{50} for humans is more realistically set at 400 ppm.

Table 1 shows earlier published LC_{50} values, together with those now available from the TNO work, to support these revised estimates. Of course it is a matter for debate as to whether the various determinations should be given equal weight in this way. However, it may be remarked that the higher values seem to be given by those workers (TNO–CIVO, Silver & McGrath) whose protocol reflected particularly high standards and the average obtained in this way seems more likely to be on the low side than too high in so far as fit animals in a healthy environment are concerned.

Because of its apparent inconsistency with other apparently highly credible animal data, the 1985 review queried the Schlagbauer mouse data, which lay below the lower band of all the other results. Conversely, however, care must be used when interpreting the fact that the TNO LC_{50} numbers are significantly higher than the earlier calculated average.

The high scatter or variability between the various studies, which is also shown in Fig. 1 (referred to later), was remarked upon in the previous report of the Working Party and will not be enlarged upon here. Suffice it to say that the TNO findings are clearly at the top of the range we have accepted, but that the Working Party do not consider them to be anomalous. They represent the results obtained from carefully controlled experiments with a healthy strain of rats/mice under modern laboratory conditions. More detailed notes on the protocols of the studies, whose results are listed in Table 1, were given in our previous report.

The 1985 review pointed out that the data available on the time/concentration relationship to lethality was not sufficiently consistent to draw any firm conclusions, but that lethality is proportional to ct^n with the $n = \frac{1}{2}$ seeming to provide the best fit.

In converting the various results to a common LC_{50} at 30 min the simple load formula given previously has been used. The table shows the effect of changing the

TABLE 1

Effect of adding TNO data to earlier publications

Author	Species	LC_{50} (ppm)	Exposure time (min)	LC_{50} at 30 min exp. time	
				$n = \frac{1}{2}$	$n = 0.67$
Alarie	Mice	302	10	174	134
Bitron	Mice	290	10	203	199
Lipton	Mice	628	10	362	280
Silver	Mice	618	10	357	275
TNO	Mice	503	30	503	503
Schlagbauer	Mice	131	30	131	131
Bitron	Mice	170	60		
Vernot	Mice	137	60	193	218
Average	Mice			274	248
Vernot	Rats	293	60	414	466
TNO	Rats	708	30	708	708
Average	Rats			561	587
Underhill	Dogs	650	30	650	650
Average	All species			495	495

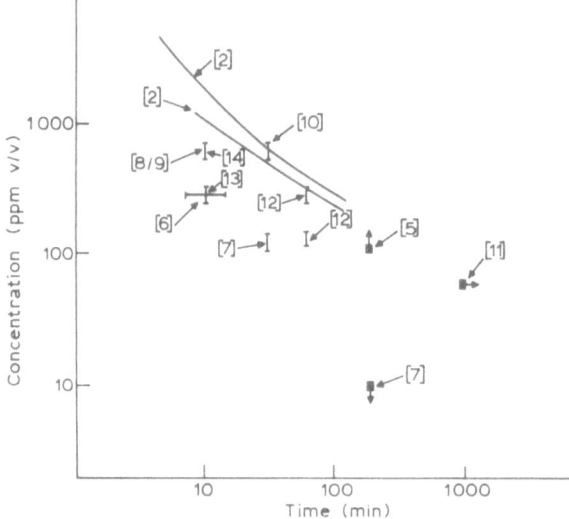

Fig. 1.   A display from the first Working Party report. Various LC_{50} levels for inhaled chlorine with added TNO data lines. [see Refs.]

exponent n to t in the load formula from $\frac{1}{2}$ to 0.67; the reason for this choice will be apparent later.

All of the above dwells on the LC_{50} interpretations and the internal consistency of the results published to date. However, the main aim of the TNO work was to cast some light on the values of lethal concentration likely to apply to very small percentages of the population (i.e. the threshold LC values). For this purpose we are interested in the 'slope' of the LC_{10}/LC_{90} line and the confidence limits that can be placed around it.

A more homogeneous animal population, and better controlled test conditions, are likely to yield a smaller difference between the threshold LC and the LC_{50}. The TNO 'slope' is in fact less than that suggested from the previous work of Underhill and of Silver & McGrath, primarily on dogs and mice respectively, though not that of Schlagbauer & Henschler. Withers & Lees have published a direct comparison of these slopes [3] and Fig. 2 is a development of a figure given in their paper with the addition of the TNO slopes for rats and mice. In this data display the concentrations have been 'normalised' to allow the 'slopes' to pass through a common 50% lethality point so as to facilitate a comparison. A more detailed description of the method is given in the Withers & Lees paper. It is inevitable that the difference between the upper and lower levels of a given confidence limit widens as the position moves away from the 50% mortality point. The TNO-derived LC_{01} for mice at 30 min at the lower level of 50% confidence is $c.\ 300$ ppm. Using the suggested revised Working Party LC_{50} for mice of 274 ppm in this approach would give a calculated LC_{01} at 30 min of $c.\ 165$ ppm.

All that the Working Party could commit itself to in 1985, on the basis of the information available to it, was a lower lethality bound of 50 ppm for all species in animal tests. Thus the TNO work leads to a supposition that a more realistic

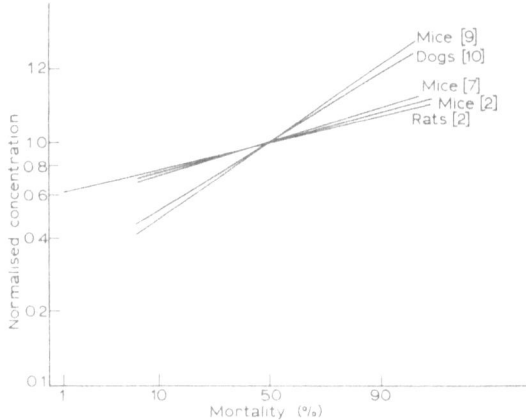

Fig. 2.   Concentration of chlorine lethal to various animal species. [see Refs.]

threshold *LC* for all animal tests might be appreciably higher than the previous perception of the lower bound. The relevance of this to the regulation of major hazard installations is discussed later.

It should, however, be emphasised that although the TNO work on rats has enabled the concentration/mortality relationship to be extrapolated down to the LC_{01} position with much greater confidence than was the case with any previously available data sets, it might well have been that had more work been done at the lower concentrations by, say, Silver & McGrath a somewhat different conclusion might have been drawn. It only requires a small change in the 'slope' constructed from 20–80% mortality data to make quite a difference to a point obtained by extrapolation at 1% mortality.

The TNO work also shows more clearly the time dependence of the toxic effect. With rats use of a simple load formula of the type $L = c \times t^n$ gave significantly less satisfactory reconciliation of the experimental data than more complicated formulae using different indices and constants for c and t. In other words, the value of n seems to depend on the value of both c and t, and the TNO–CIVO staff have gone to some lengths to validate various forms of probit equations with an added cross term.

The probit equation given by TNO with the best fit to their data from the rat experiments (lowest $X^2 = 15.56$) was

$$P = -16.67 + 1.33 \, InC_{exp} - 4.31 \, Int + 1.01 \, InC_{exp} \, Int$$

with P as probit, C_{exp} the exposure (concentration in mg/m^3) and t the exposure time in minutes.

Other equations were given in the report, including:

$$P = 32.29 + 3.69 \, InC_{exp} + 2.27 \, In(t - 3) \quad (X^2 = 18.87)$$

and a simple load equation version:

$$P = -26.84 + 2.89 \, InC_{exp} + 2.78 \, Int \quad (X^2 = 24.10)$$

but these provided a less good fit with the test results.

For mice the simple load equation was found to be satisfactory with the version

$$P = -33.74 + 4.05 \, InC_{exp} + 2.72 \, Int \qquad (X^2 = 12.97)$$

Thus, when sticking with the simple load formula as previously used, the best fit to the data for rats was obtained with a value on n equal to 0·96 and for mice with a value equal to 0·67. The value obtained previously by the Working Party from the Bitron & Aharonson work on mice was 0·4, from the Weedon work on rats was 0·6,

TABLE 2

Group code, concentration of chlorine in test atmosphere, exposure time corresponding to mortality rates and day of death

Group code		Concentration $(mg/m^3) \pm SD \, (n)$	Exposure time (min)	Mortality rate (%)	Day of death (number)			
Rats	A	1 654 ± — (1)	5	0				
	G	2 201 ± 92 (2)	5	0				
	I	2 399 ± 82 (3)	5	0				
	P	3 485 ± — (1)	5	0				
	S	4 798 ± — (1)	5	0				
	U	8 241 ± — (1)	5	0	a			
	W	16 801 ± 72 (6)	5	70	0(7)			
	B	1 680 ± 0 (2)	10	0				
	H	2 186 ± 81 (4)	10	0				
	J	2 363 ± 81 (6)	10	0				
	N	3 485 ± 0 (2)	10	0				
	T	4 798 ± 0 (2)	10	10	0(1)			
	X	6 519 ± 30 (6)	10	60	0(6)			
	E	1 586 ± 47 (7)	30	0				
	K	1 665 ± 32 (7)	30	30	0(1)	1(1)	6(1)	
	C	1 757 ± 35 (4)	30	50	0(2)	1(1)	3(2)	
	Z	1 870 ± 21 (16)	30	60	0(2)	1(3)	2(1)	
	Y	935 ± 32 (31)	60	0				
	Q	1 325 ± 35 (16)	60	40	1(2)	2(1)	11(1)	
	O	1 473 ± 19 (11)	60	60	0(3)	1(1)	2(1)	4(1)
	L	1 651 ± 29 (14)	60	80	0(6)	1(1)	4(1)	
	D	1 725 ± 57 (6)	60	100	0(6)	1(3)	5(1)	
Mice	B	1 680 ± 0 (2)	10	0				
	H	2 186 ± 81 (4)	10	0				
	J	2 363 ± 81 (6)	10	30	6(1)	13(2)		
	N	3 485 ± 0 (2)	10	40	0(2)	10(1)	13(1)	
	V	3 826 ± 38 (3)	10	100	0(8)	1(2)		
	T	4 798 ± 0 (2)	10	100	0(10)			
	R	1 328 ± 55 (8)	30	40	0(1)	8(2)	13(1)	
	E	1 586 ± 47 (7)	30	70	0(4)	8(1)	9(1)	11(1)
	K	1 665 ± 32 (7)	30	60	0(2)	1(2)	9(1)	11(1)
	C	1 757 ± 35 (4)	30	90	0(5)	10(3)	13(1)	
	Z	1 870 ± 21 (16)	30	70	0(3)	1(2)	8(2)	

Groups F and M have been removed from file due to incorrect exposure.
a 0 = day of exposure.
(n) = number of animals in trial.

and from the Weedon work on mice was 0·5. The Working Party did have some reservations about both of these earlier studies, however; they previously came to the conclusion that whilst the value of n in a simple load equation could be put approximately at $\frac{1}{2}$ there was much uncertainty. It could not accept the figure of 0·39 attributed to Eisenberg *et al.* used in the US Coast Guard model, either as to its absolute value or to its apparent precision.

The TNO work has therefore, to a large extent, reinforced the previous Working Party position. In the light of this new work a revised average n figure for mice could be 0·59. However, the rat data, although of excellent quality, raise a number of issues and it seems unwise to depart from the value of $\frac{1}{2}$ previously given on the grounds of uncertainty and doubt about the validity of the simple load formula. It may be remarked, however, that the average value of n is more likely to lie above $\frac{1}{2}$ than below it.

This uncertainty in the value of n does not affect greatly the compilation of an average LC_{50} value for all species at 30 min exposure time as Table 1 makes clear. Here the same final result is obtained whether the TNO value for mice is used or whether the Working Party assessment of $\frac{1}{2}$ is used. It obviously has more relevance when an average LC_{50} value at 30 min is used to estimate the LC figure at an exposure time of, say, 5 min.

Much useful physiological data is provided by the TNO test at short exposure times of 5 and 10 min, where earlier work tended to be less informative. The TNO work provides strong evidence to show the marked effect of chlorine upon the breathing rate of rats. The marked slowing of the breathing rate in some cases down to a seventh of the normal provides some explanation of the rats' astonishing ability to cope with very high concentrations for short exposure times. It has been estimated from visual observation that for rats an apparent 5-min exposure time may be equivalent to a 3-min exposure time in reality due to their ability to hold on to their breath. Unfortunately no corresponding data were taken by TNO with mice. Earlier workers, such as Lehmann, also demonstrated marked reductions in the breathing rates of rats exposed to chlorine (who, like dogs, then become very subdued) but have reported enhanced breathing rates with animals such as cats and mice, who may become very excited when exposed to chlorine.

Table 2, reproduced from the TNO report, provides a succinct summary of the data compiled on mortality. Figure 1 shows the TNO data positioned on the display given with our first report.

MOUSE–RAT–HUMAN RESPONSE RELATIONSHIP

It will be apparent from the observations made previously that whilst the TNO–CIVO study has provided fresh insights into a number of significant aspects relating to the lethal toxicity of chlorine, it has also re-emphasised many of the difficulties in the way of extrapolating the findings of animal studies to man in this regard. The earlier review by the Working Party, and by Withers & Lees [3], highlight the difficulties of basing an interpretation on a larger, physiologically similar species like the dog or a smaller obligate nose breather like the rat. The average human will also be in a less homogeneous population than the laboratory animal. Questions still remain about the effects of reflex reactions and toxicokinetics in man compared to animals.

Whilst it seems likely that the different responses of mice and rats can to a large measure be explained in terms of breathing rate changes, it is not at all clear how man would respond to such lethal levels of concentrations. It is known from observation that at relatively low levels of concentrations man is likely to hold his breath, but this is not to say that he would behave in the same way at higher concentrations.

It also seems likely that the high statistical significance of the TNO data is a direct reflection upon their excellent protocol and in particular the care taken to work with disease-free animals, to avoid secondary infection, and to select a uniform set from the species. The Working Party has noted a tendency for all more recent work to exhibit improvements in these directions and this may account for many of the differences between the older and the more recent reports. This after all is only to be expected. However, when we turn to the human situation in a real every-day environment there remains a problem as to how best to assess the vulnerability of a mixed population taken at random.

It could well be that improvements in the protocol and control of the animal experiments not only narrow the difference between the levels of confidence limits but might well reduce the 'slope' of the mortality/concentration relationship due to the removal of the small more vulnerable element from the population under test. The work of Bell & Elmes previously reported by the Working Party has demonstrated the greater resistance of disease-free rats to chlorine, for example.

Because of such uncertainties and the lack of knowledge concerning man's likely behaviour in high concentrations, the Working Party remains cautious when extrapolating the animal results to humans. Nevertheless, the TNO–CIVO work does suggest that for fit and healthy persons at least the lower level of lethal toxicity in percentage terms as well as the LC_{50} will be somewhat higher than suggested by the Working Party in 1985.

RELEVANCE TO OTHER WORK AND TO OUR PREVIOUS FINDINGS

The TNO–CIVO study represents the first large-scale attempt with proper consideration given to the statistical design of the experiments, so as to establish the inter-relationship of time and concentration in the lethal toxicity of chlorine. It is the first significant attempt to address the problems of establishing a threshold LC. The use of mice as well as rats has enabled a comparison to be made with earlier work which the TNO study complements rather than replaces.

In many ways the TNO study reinforces the position of the Working Party's previous findings. It confirms that we were correct in rejecting much inadequately referenced data, and in suggesting an LC_{50} for animals somewhat above that suggested in other reviews. It is also generally supportive of our earlier assessment of the concentration/time effects. In respect of concentration/mortality relationships as determined in animal experiments it has, however, broken fresh ground in the area of cross terms in probit equations. The Working Party was previously reluctant to specify a generally applicable probit equation, because no one set of data covered a wide enough span, and the mixing of otherwise reputable published figures could not be accepted as providing a statistically acceptable result.

TNO have now completed a well controlled and documented piece of work, where the results are statistically consistent and which provides a valid probit equation relating to their test conditions. It is, however, not possible to reject other apparently sound work, such as by Silver & McGrath, which gives a slightly different result (possibly reflecting differences in experimental conditions or strains of animal rather than variability in a single species). Reluctantly it is still not possible to agree a generally applicable probit for rats or mice which is consistent with all the better published data. However, for the purposes of comparative risk assessment, the use of an averaged probit may be contemplated as a matter of convenience so long as the limitations on its significance as an artefact are understood. The human dose–response curve derived from such a probit should be compatible with the slope around and the magnitude of the LC_{50} point now suggested by the Working Party in the light of the TNO data.

THE IMPLICATION FOR POSSIBLE HUMAN EXPOSURE AND THE ASSESSMENT OF MAJOR HAZARDS

The consideration of major hazard installations and the setting of standards for the minimising of risks to the actual human population can involve two discrete approaches. Overall societal risk evaluation requires a concentration/time probit-type equation, in order to judge the overall magnitude of the risks per foreseeable event. The LC_{50} data can then be used as a guide to relative levels of toxic effect. Alternatively one may seek to set a standard for exposure limits which is unlikely to cause fatalities amongst a healthy population and therefore should provide a good measure of protection for older or infirm individuals. This approach has been used [4] by the HSE for consideration of certain major hazard situations. This reference calculated the LC_{01} for mice at the lower 95% confidence limits from Silver & McGrath data and suggested this as the advisory threshold LC for a typical human population.

The TNO–CIVO results relating to threshold LC data are well above the lower lethality bound which was identified by the Working Party in its earlier report (50 ppm for 30 min for all species in any animal tests). This may have significant implications where dose/effect relationships tied to a concept of threshold or low lethality are used in assessment of risk. Taken together with other mice and rat data, the latest TNO results could be used to suggest an animal LC_{01} of 150–200 ppm for exposures ranging between 10 min and half an hour.

The Working Party is more reluctant to postulate a general probit for the purpose of hazard assessment for humans of varying ages and fitness, not least for the reason that such an equation is likely to provide an impression of precision in an area where none exists. In so far as the simple load equation is concerned, this may well be useful for rough approximations and recommends for chlorine an average approximate value of $n = \frac{1}{2}$ for the exponent of time but agrees that in many instances it is more likely to be greater rather than less. The Working Party now feels that a concentration of 400 ppm is the best approximation to an average LC_{50} at 30 min for fit and healthy humans rather than the range 300–400 ppm previously given. It is to be hoped that the TNO–CIVO work and the Working Party discussions will stimulate a review of toxicity criteria for chlorine in real situations, as the Working Party considers that there is a need for further consideration of the appropriate use of animal data for predictions of human response.

In its first report the Working Party discussed many other issues arising from its understanding of the lethal toxicity of chlorine. These included aspects of remedial medical treatment, the advice to be given when preparing emergency plans, and the assessment of case histories of major spillages causing fatalities. We see no reason for revising our views on any of these issues.

MEMBERSHIP OF THE WORKING PARTY (AUGUST 1987)

The MHAP Toxicity Working Party, as in 1984/85, was made up of individuals drawn from industry, industrial toxicologists, external consultants, the medical profession, and the UK Health and Safety Executive. All were selected on the basis of their relevant experience in major hazards assessment and the industrial handling of chlorine. The members were:

R. M. J. Withers (Chairman)
D. G. Farrar
P. M. Hext
I. W. Hymes
J. P. H. Shaw
R. P. Pape
D. Clarke

REFERENCES

1. Withers, R. M. J., 1st Report of MHAP Toxicity Panel, *Modern Chlor-Alkali Technology*, Vol. 3. Ellis Horwood, Chichester, 1986. *Loss Prevention Bulletin*, Chlorine Toxicity Monograph published by the Institute of Chemical Engineers (1987).
2. Zwart, A., TNO Report No. V87.089/260851, May 1987, TNO–CIVO, Zeist. Published in *J. Haz. Mater.* **19** (1988) 195–208.
3. Withers, R. M. J. & Lees, F. P., The assessment of major hazards: the lethal toxicity of chlorine. Part 1: Review of information on toxicity. *J. Haz. Mater.*, **12** (1985) 231.
4. Davies, P. C. & Hymes, I. W., *The Chem. Engnr* (June 1985) 30–3.
5. Bell, D. & Elmes, P., The effects of chlorine gas on the lungs of rats. *J. Path. Bact.*, **84** (1965) 307–16.
6. Bitron, M. & Aharonson, E., Delayed mortality of mice following inhalation of CH_2O, SO_2, Cl_2, and Br_2. *Am. Ind. Hyg. Ass. J.*, **39**(2) (1978) 129.
7. Schlagbauer, M. & Henschler, D., Toxicitat von Chlor und Brom. *Int. Arch Gewerbepath. u. Gewerbehyg.*, **23** (1967) 91–8.
8. Silver, S. D. & McGrath, F. P., Chlorine median lethal concentrations for mice. Edgewood Arsenal Technical Reports 351, May 1942.

9. Silver, D. D., McGrath, F. P. & Ferguson, R. L., Chlorine median lethal concentration for mice. Edgewood Arsenal Technical Report 373, 17 July 1942.

10. Underhill, F. P., *The Lethal War Gases.* Yale University Press, 1920.

11. Weedon, F., Hartzell, A. & Setterstrom, C., *McCallen and Setterstrom: Contributions to the Boyce Thompson Institute,* **11** (1940) 325–65.

12. MacEwen, J. & Vernot, E., Acute toxicity of chlorine. In SysteMed Corporation Toxic Hazards Research Unit Annual Technical Report: 1972, Report No. AD 755-358 (1972).

13. Alarie, Y., Toxicological evaluation of airborne chemical irritants and allergens using respiratory reflex reactions. In *Inhalation Toxicology and Technology*, ed. B. K. J. Leong. Ann Arbor Science Publishers, 1981, pp. 207–31.

14. Lipton, M. & Rotariu, G., The toxicity of chlorine gas for mice. University of Chicago Toxicity Report No. 6 (1941).

7

INFLUENCE OF SYSTEMS DESIGN AND OPERATION PARAMETERS ON MEMBRANE PERFORMANCE

Klaus-Jochen Behling

Du Pont de Nemours (Deutschland) GmbH, FRG

and

David L. Peet

EI Du Pont de Nemours and Company, USA

INTRODUCTION

During the past 10 years membrane technology has made rapid advances in the chlor-alkali industry both in terms of the development of the technology and penetration into commercial plants. Figures 1–3 highlight these developments.

Electrolyser power consumption dropped from about 3100 kWh/t NaOH in 1975 to 2200 kWh/t NaOH today. During the same period membrane current efficiency increased from 90% to 96% and above, and caustic concentration went from 10–20% to 35% (Table 1).

Parallel to the technical development there was a rapid increase in the commercial acceptance of this new technology, especially since the early 1980s. Today more than 12% of the worldwide chlor-alkali capacity is run on membrane technology. Seventy per cent of current membrane plant capacity is from replacement of diaphragm or mercury plants and 30% of the capacity are new grass roots plants (Fig. 1).

TABLE 1
Membrane technology development

	1975	1987
Cell power consumption (kWh/t NaOH)	3 100	2 200
Current efficiency (%)	90	96
Caustic concentration (%)	10–20	35

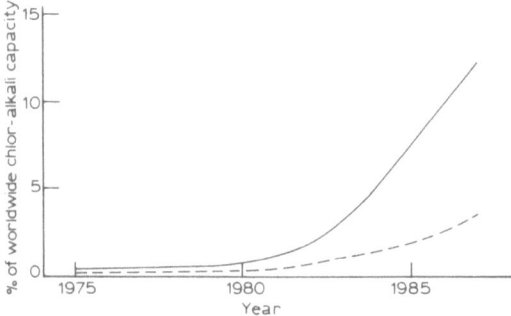

Fig. 1.   Membrane technology penetration.

Membrane plant sizes range from small on-site plants to major units with up to 850 t/day caustic capacity. One hundred and four membrane plants with a total capacity close to 14 000 t/day were operating at the end of 1987 (Table 2).

During the early development phases of the technology the focus of membrane producers such as Du Pont was on membranes with lower power consumption, higher caustic concentration and defining a competitive position versus diaphragm and mercury technology.

Now we look at a broadly accepted technology with a variety of membrane types and electrolyser designs operating in plants under a wide range of conditions.

Low power consumption and high current efficiency over long periods of time have been demonstrated.

There is now a solid base of commercial experience to correlate laboratory and plant data, and to relate membrane performance to plant design and operating conditions.

OBJECTIVE

Before the background of this experience it is the objective of this chapter to discuss membrane performance with regard to the effects of brine impurities, systems and membrane design, and to analyse the effects of membrane life and performance on plant operating costs.

TABLE 2
Membrane plants

Plant capacity (t/day NaOH)	Number of plants	Total capacity (t/day NaOH)
0–50	42	990
51–200	41	4 660
201–850	21	8 130
	104	13 780

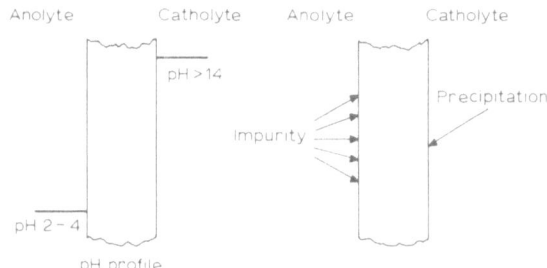

Fig. 2. Brine impurity precipitation.

EFFECT OF BRINE IMPURITIES ON MEMBRANE PERFORMANCE

One of the challenges that came along with the new membrane technology was the need for higher quality brine as compared to the diaphragm or mercury processes. It was noted already in the early stage of membrane technology development that certain cations, such as calcium and magnesium, can cause a reduction of current efficiency through precipitation in the barrier layer of the membrane. Similar effects were also observed with anionic impurities such as sulphates and neutral compounds such as silica.

The mechanism for brine impurity precipitation is now well understood. Calcium or magnesium cations move into the membrane just as do the desired sodium ions. As the pH increases within the membrane metal ions that form insoluble hydroxides precipitate in the membrane whilst the sodium ions continue to move through.

Precipitates of calcium and magnesium hydroxide physically disrupt the barrier layer of the membrane. This cathode side layer governs the back migration of hydroxyl ions and thus the current efficiency of the membrane (Fig. 2).

The damage from brine impurity precipitates is irreversible; voids created by the precipitates will remain even after the precipitates might have been washed out.

Besides the electrically-induced flow of cationic brine, neutral or anionic impurities such as silica or sulphate ions can enter the membrane through diffusion and electro-osmotic water flow from the anolyte to the catholyte. Again such impurities can form precipitates in the high pH region of the membrane, which is also more dehydrated than the anolyte side due to the higher ionic strength of the catholyte.

An important factor in the success of commercial membrane cell plants has been the development of high-performance ion-exchange resins that remove calcium, magnesium, barium or strontium from the feed brine. These resins have proven to be effective in removing, for instance, calcium and magnesium to below 50 ppb, which is required as a maximum for NAFION* membranes.

With growing commercial experience and data from operating plants plus extensive research it became evident that there are synergistic effects of certain impurities.

Du Pont researchers identified complex sodium–aluminium silicates in precipitates—specific examples are Faujasite and Sodalite. These complex salts are

---

* Du Pont's trademark for its perfluorinated ion-exchange membranes.

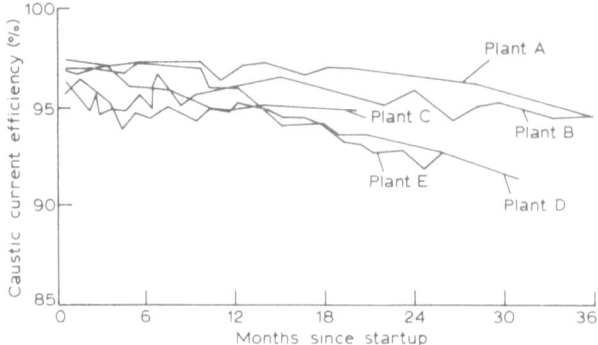

Fig. 3.   Long-term plant performance with NAFION 90209 membrane.

formed when silica reacts with other elements under the basic conditions within the membrane.

One of the keys to understand and prevent excessive precipitation of complex silicates was detailed work by Bissot of Du Pont [1] about the mechanism of silica transport through the membrane. It showed that the rate of silicate precipitation is proportional to current density. It also showed—somewhat surprisingly—that a thinner membrane gave less silica transport and therefore less precipitation of complex silicates.

When predicting membrane performance one must therefore not only look at individual brine impurities but also combinations of the different elements. Plant and membrane design factors such as current density and choice of membrane type are also critical.

The large number of operating data from commercial plants and laboratory studies mentioned a moment ago have allowed us to develop a correlation between membrane design, plant operating conditions, brine impurities and the predicted membrane durability, taking into account both individual and synergistic brine impurity effects.

With this correlation we are now more flexible with regard to brine specifications and can give our partners broader options both in the planning stages for new plants and for the evaluation of options in existing plants.

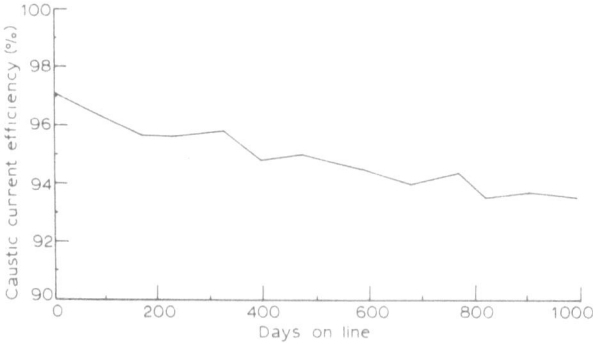

Fig. 4.   Long-term plant performance with NAFION 954 membrane.

Examples of commercial plant data are given in the following figures to demonstrate this predictive capability of membrane performance in finite and narrow gap cells.

Figure 3 shows plots of current efficiency versus time for five plants using NAFION 90209 membrane. All plants started up at 96–97% current efficiency and achieved membrane life of longer than 2 years.

Differences in brine quality become apparent when comparing plants A and B, which show very little performance decline and run at 95% current efficiency after more than 3 years on line with other plants which, with poorer brine quality, show a faster decline of performance and current efficiencies of around 91% after 30 months on load. These five plants all operate at about the same current density of 3·0 kA/m².

Figure 4 shows a graph from a plant operating with a narrow gap membrane—NAFION 954. Running also at 3 kA/m² with good brine quality, current efficiency is still close to 94% after 1000 days on load.

EFFECT OF CURRENT DENSITY ON MEMBRANE PERFORMANCE

As indicated before in connection with brine impurity studies, another important consideration for plant design and its impact on membrane performance is current density. Higher current density means higher impurity transport and thus a higher rate of impurity precipitation in the membrane. This leads directly to a faster rate of current efficiency decline.

Taking an actual plant with a given brine quality operating at 3 kA/m² with a current efficiency of 95% after 36 months on load, we would project a current efficiency of about 93% after 3 years at 4 kA/m² (Fig. 5). Please note that the projected rate of current efficiency decline is not directly proportioned to current density. The rate of current efficiency decline in operating plants is the result of both gradual impurity damage and any physical damage that occurs due to handling or electrolyser design and fabrication. This physical damage should not be a function of current density and membrane life is therefore not reduced at a direct ratio of current densities.

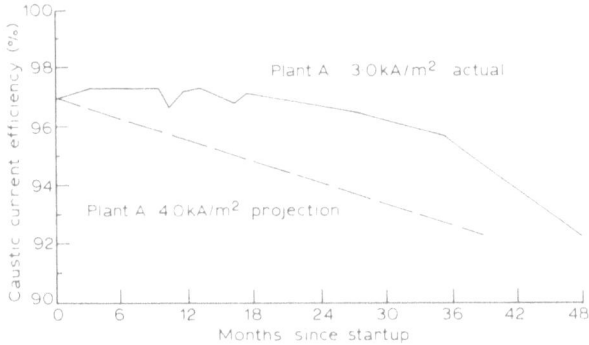

Fig. 5. Long-term plant performance with NAFION membrane.

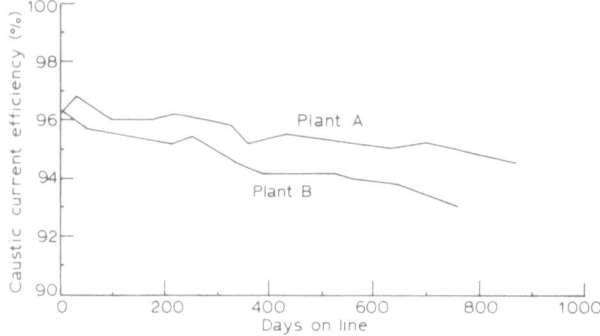

Fig. 6.   Long-term plant performance with NAFION 961 membrane.

Projected and observed current efficiency is in good agreement. Figure 6 shows two plants using NAFION 961. Both have about the same brine quality, and both show excellent performance after more than 2 years on line. Plant A runs at a current density of $3.0\,kA/m^2$ and plant B is operated at $3.5\,kA/m^2$.

HIGH CAUSTIC MEMBRANE

The majority of commercial membrane plants are designed for and operate at caustic concentrations in the range 32–35% NaOH. Whilst this range remains the focus for new plants in the foreseeable future, interest has been expressed by some chlor-alkali producers and technology sellers in a higher caustic concentration membrane. Our experience with 30–35% caustic membranes indicates that higher caustic concentrations will reduce membrane hydration. This will slightly reduce the tolerance to brine impurities.

Du Pont now has a developmental membrane for 40–45% caustic operation. In laboratory tests this membrane has given current efficiencies above 95% over a period of 220 days. These tests were done with very pure laboratory brine and were not designed to measure the effect of brine impurities.

Producing more concentrated caustic will have advantages for some plant sites since it may reduce investment and operating cost in grass roots or mercury plant conversions due to the elimination of caustic evaporation. Probable disadvantages will include a higher cell voltage versus 30–35% NaOH and a possible lack of high value alternative uses for an existing steam supply. Economic advantages to produce 40–45% caustic will therefore be very site-specific.

INFLUENCE OF MEMBRANE LIFE ON PLANT OPERATING COST

This chapter finishes with some considerations about the effect of membrane performance and life on production cost. Whilst economics are predominantly site-specific, there are certain benefits of membrane technology which are valid for almost all existing and potential new plants. Among these are:

—lower power consumption and investment than the alternative technologies;

—membrane technology is not associated with environmental concerns;
—it has greater operating flexibility;
—production cost considerations must involve membrane cost and life;
—electrolyser renewal cost—excluding membrane; and
—loss of production during renewal.

As commercial membrane plant operators experience membrane life beyond original expectation the importance of physical durability has increased. For example, if the electrolyser gaskets last for 2–3 years but the membrane is still performing well at the time of gasket renewal the membrane should be tough enough to withstand the handling during disassembly and reinstallation.

We have examples of commercial plants where NAFION 90209 went through an average of two gasket changes during 4 years of operation. Since membrane life is an integral part of production cost it has to be taken into consideration during the planning of new plants.

Table 3 shows the projected effect of current density on membrane life and the associated production cost per tonne of caustic.

Actual experience in a 200 t/day caustic soda plant with NAFION 90209 membrane is a 4-year life. With an installed membrane volume of 2200 m² the membrane costs are about 4·60 US dollars/t caustic at current membrane cost.

A plant with the same capacity operating at 4 kA/m² would require 25% less membrane. Life expectancy would, however, be only 3·25 years and the membrane cost 4·36 US dollars/t caustic. The membrane cost difference between these two alternatives is small. Note, however, that these cost figures do not include other electrolyser renewal costs, which will vary with electrolyser design.

When looking at electrolyser renewal costs excluding membrane, gasket versus membrane life-cycles need to be considered as well as cost of labor and other materials. Advancements have been made to coordinate gasket and membrane life-cycles to minimize these additional labor and materials costs.

Such additional costs add up to 10–20% of membrane cost for each replacement, that is about 50 cents to one US dollar/t caustic, which is substantial. Also, especially in times of high capacity utilization, loss of production during gasket or membrane

TABLE 3
Effect of membrane life on production cost

	Example: 200 t/day plant *NAFION 90209*	
	Actual experience *(3·0 kA/m²)*	*Projection based on experience (4·0 kA/m²)*
Membrane life	4 years	3·25 years
Membrane volume (m²)	2 200	1 700
Membrane cost	$600/m²	$600/m²
Membrane cost per tonne NaOH produced	$4·58	$4·36

renewal can be an important cost factor. We feel that one assumption is safe to make: in general, a longer membrane life will lead to lower renewal costs over longer periods of time and also to better capacity utilization.

SUMMARY

Experience gained over the past years in operating plants and laboratories, together with close cooperation of technology sellers, membrane suppliers and not the least the operators of commercial plants, have created a much better understanding of systems and membrane design factors that affect membrane performance. Laboratory studies of brine impurity effects on membrane performance show a good correlation with plant experience. This better understanding leads to longer membrane life in existing plants and helps in the planning of new plants.

Long membrane life has proven to be an important factor in achieving lower production costs.

REFERENCE

1. Bissot, T. C., ECS Meeting, May 1986, Boston, Massachusetts.

8

NEW DEVELOPMENTS IN ELECTRODE COATINGS FOR CHLOR-ALKALI PROCESSES

A. M. Couper, W. N. Brooks and D. A. Denton

ICI Chemicals and Polymers Ltd, UK

INTRODUCTION

Electrolytic cells have been used for the commercial production of chlorine for over a century. For the first 7–8 decades of this period all chlorine produced was liberated at graphite anodes until, in the 1950s, developments by Cotton and Beer led to the introduction of coated titanium anodes which offered greater dimensional stability, lower overpotentials and longer service life. A widespread replacement of graphite took place during the 1970s, but even now a few smaller operators still continue to use graphite anodes. Since the (relatively late!) arrival of the coated titanium anode enormous progress has been made, leading to products specifically designed for a wide range of electrolytic processes.

This chapter deals with two aspects of electrode coating technology. Firstly, the development of the membrane cell has led to a requirement for anode coatings to operate in an environment rather different from those in the traditional mercury and diaphragm cell technologies. In the first part of the chapter the operating environment in the membrane cell is reviewed and electrode coatings with improved lifetime and by-product distributions are described.

Secondly, the past decade has seen a marked increase in interest in the electrolytic generation of hypochlorite (electrochlorination) for the on-site generation of hypo for sterilisation purposes. Although superficially all electrochlorinators operate along similar lines (i.e. the electrolysis of brine in an undivided cell), in practice the conditions vary widely between different cells and from one location to another. The consequences of these different operating conditions upon electrode coatings are reviewed and a range of ICI coatings is described.

ANODE COATINGS FOR CHLOR-ALKALI MEMBRANE CELLS

Operating conditions differ widely between the three types of cell used for chlorine production in a number of respects:

—current density,
—feed and exit brine pH,
—inter-electrode gap,
—likelihood of inter-electrode shorting and current reversal,
—hydrodynamic (both liquor and gas) regimes, and
—brine impurity levels.

The implications of these differences on the requirements of electrode coatings in the various cell technologies have been described previously [1, 2]. As far as membrane cells are concerned, the key features are:

—moderate current densities (typically $3–4\,kA/m^2$),
—variable pH with caustic permeation through the membrane,
—a low (or zero) anode–membrane gap,
—some current reversal if cells are shorted off-load, and
—potentially high gas volume fractions in the anolyte.

A good anode coating is judged by its ability to operate under these conditions whilst retaining:

—a low overpotential for chlorine evolution,
—a good by-product distribution,
—good gas release properties,
—minimal abrasion of the membrane surface, and
—long life at reasonable cost.

One of the most important factors to consider in the membrane cell is anolyte pH. It is in this respect that the membrane cell differs most widely from mercury or diaphragm cells in terms of the demands placed on anode coatings. Even at a good membrane current efficiency, say 95%, the rate of permeation of hydroxide across the membrane is relatively high, equivalent to over $0.25\,kg$ of caustic soda per square metre per hour at a current density of $3.5\,kA/m^2$. This tendency towards an alkaline environment at the anode surface has two deleterious consequences. Firstly, it promotes the formation of a number of by-products, and secondly, it enhances the wear rate of anode coatings. Fortunately, by careful design of the anode coating, one can minimise the impact of these effects.

As the pH of the anolyte increases a number of undesirable side reactions become thermodynamically more favourable:

Hypochlorous acid formation:

(i) $$Cl_2 + H_2O \rightarrow HOCl + Cl^- + H^+$$

Chlorate formation:

(ii) $$2\,HOCl + OCl^- \rightarrow ClO_3^- + 2\,Cl^- + 2\,H^+$$

Oxygen evolution:

(iii) $$2\,HOCl \rightarrow O_2 + 2\,Cl^- + 2\,H^+$$

(iv) $$2\,H_2O \rightarrow O_2 + 4\,e^- + 4\,H^+$$

Dissolution of high valent ruthenium species from ruthenium-containing anode coatings:

(v) $$RuO_2 + 2H_2O \rightarrow RuO_4^{2-} + 4H^+ + 2e^-$$

Reaction (iii) has been considered by Kotowski & Busse [3] as a superposition of the reduction of hypochlorous acid to chloride:

(vi) $$HOCl + H^+ + 2e^- \rightarrow Cl^- + H_2O$$

and its oxidation to oxygen:

(vii) $$HOCl + H_2O \rightarrow O_2 + 3H^+ + Cl^- + 2e^-$$

However, reaction (vii) is itself really a superposition of two reactions, the oxidation of water (i.e. reaction (iv)) and the reduction of hypochlorite (reaction (vi)). The main sources of oxygen can thus be considered as the direct evolution reaction (iv), together with reaction (iii), which may occur either in solution or via heterogeneous catalysis, either on the electrode surface or on suitable particulate material in the anolyte liquor.

Side reaction (v)—the dissolution of RuO_4^{2-} or other (e.g. RuO_4) soluble/volatile high valent oxo/chloro species—leads not to significant by-products but to increased anode coating wear rate.

The extent of total by-product formation can only be minimised by:

—operating at high membrane current efficiency, or
—by acidifying the feed brine in order to compensate for membrane inefficiency, typically by HCl addition.

However, since one of the by-product formation reactions, reaction (iv), is electrochemical in nature, one can control the balance between the different by-products by suitable choice of anode coating. The evolution potentials for chlorine and oxygen are shown in Fig. 1 for different values of anolyte pH.

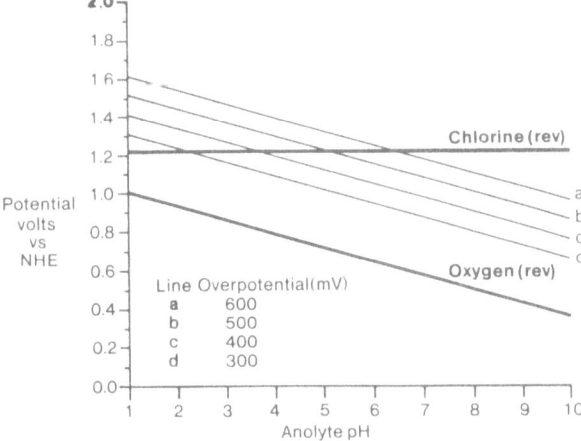

Fig. 1.   Potentials for oxygen and chlorine evolution at different anolyte pH values. 230 g/litre brine at 90°C.

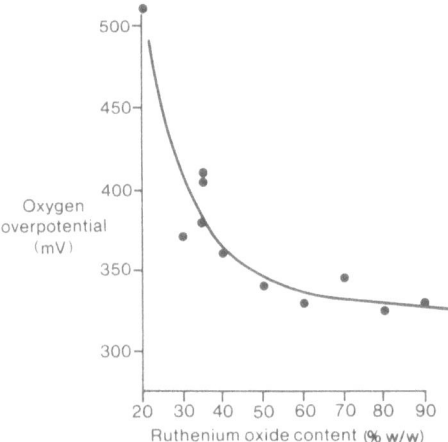

Fig. 2. Relationship between oxygen overpotential and ruthenium content for $Ru_xTi_{(1-x)}O_2$ anode coatings. 1M sulphuric acid, 30°C, 1·5 kA/m².

This graph shows clearly that the reversible potential for oxygen evolution is less than that for chlorine at all pH values. If an anode coating showed reversible behaviour towards both reactions very little, if any, chlorine would be evolved at the anode—only oxygen would be formed. Fortunately, whilst the evolution of chlorine is close to being reversible on RuO_2-based anode coatings, with overpotentials of only a few 10s of millivolts, that for oxygen is far less so. Figure 1 also shows the oxygen evolution line, shifted by different values of overpotential, over the range 300–600 mV. As the overvoltage for oxygen evolution increases the pH value at which the chlorine and oxygen lines cross also increases. For typical brine pHs of 3–5 one can therefore minimise oxygen evolution by designing the anode coating to have a high overpotential for oxygen evolution. It has been shown [4], for example, that oxygen evolution overpotential is related to the mole fraction of Ru in

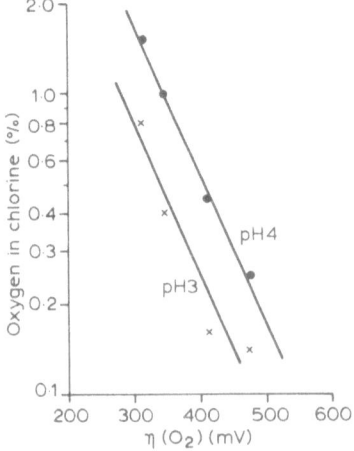

Fig. 3. Relationship between oxygen overpotential and oxygen in chlorine content. Laboratory conditions, 0·2 kA/m².

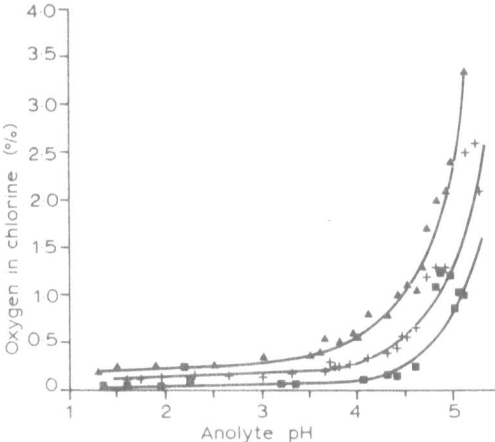

Fig. 4.   Oxygen in chlorine content for different MET coatings; standard (▲), low oxygen (+) and new low oxygen (■): effect of anolyte pH (after Denton *et al.* [4]).

$Ru_xTi_{(1-x)}O_2$ coatings (Fig. 2). Other modifications to composition and/or the conditions of preparation of anode coatings can also be made in order to modify oxygen overpotential.

Experimentally, one indeed observes a close relationship between oxygen overpotential and the measured content of oxygen in chlorine. Figure 3 shows data for a range of experimental coatings evaluated in laboratory cells. As one might expect, from a simple Tafel law argument, a logarithmic relationship is observed:

$$\log (\text{oxygen content}) = -A\eta(O_2) + BpH$$

where A and B are constants. A similar relationship has been reported by Kotowski & Busse [3].

ICI Chemicals and Polymers Ltd have used these principles to design low oxygen evolution anode coatings for use in membrane cells when this particular by-product distribution is required by the clorine user. The performance of some of these low oxygen anode coatings was reviewed at the last SCI Chlorine Symposium in 1985 [1]. Since that time ICI have been evaluating coatings with even lower oxygen evolution characteristics (Fig. 4). These coatings are now being evaluated in the FM21 membrane cell.

The second part of this chapter deals with coatings for electrochlorination. This is an area complementary to chlorine/caustic production in that the fundamental electrode reactions are similar, but one in which the operating conditions place very different constraints upon the electrode coatings.

ELECTRODE COATINGS FOR THE ELECTROLYTIC GENERATION OF HYPOCHLORITE

Over the last decade or so the direct electrolytic production of sodium hypochlorite, a process often referred to as electrochlorination, has received increasing attention. Hypochlorite is a powerful oxidant which is used for water treatment in a wide variety of situations, for example:

—for reducing marine growth (algae, molluscs, etc.) in seawater used for cooling in coastal power stations and on board ships,
—as a disinfectant/biocide in sewage treatment,
—for the sterilisation of water for drinking and food processing,
—for the disinfection of swimming pools,
—to sterilise water for injection into oil/gas-bearing rock strata during extraction processes, and
—for cyanide detoxification.

The Electrochlorination Process

The electrochlorination process is fairly straightforward; hypochlorite is produced in an undivided cell by a reaction between the products of the anodic and cathodic half-reactions:

Anode $\qquad 2Cl^- \rightarrow Cl_2 + 2e^-$
Cathode $\qquad 2H_2O + 2e^- \rightarrow 2OH^- + H_2$
Solution $\qquad Cl_2 + 2OH^- \rightarrow H_2O + ClO^- + Cl^-$

In practice a number of side reactions compete with these desired processes, leading to a loss of current efficiency:

Anode $\qquad 2H_2O \rightarrow O_2 + 4H^+ + 4e^-$
$\qquad\qquad ClO^- + 2H_2O \rightarrow ClO_3^- + 4H^+ + 4e^-$
Cathode $\qquad ClO^- + H_2O + 2e^- \rightarrow Cl^- + 2OH^-$
Solution $\qquad 2HOCl + OCl^- \rightarrow ClO_3^- + 2HCl$
$\qquad\qquad 2ClO^- \rightarrow O_2 + 2Cl^-$

In addition to these loss processes, the alkaline conditions prevalent at the cathode surface can lead to the formation of calcium and magnesium hydroxides, which can form a deleterious scale deposit on the cathode surface. The thermodynamics and kinetics of these processes are influenced by a number of variables:

—salt concentration,
—mass transport effects,
—temperature,
—current density, and
—the presence of impurities (principally hardness) in the brine.

In turn, these parameters have an important influence on the current efficiency and power performance of the cell. As in the case of membrane cell anodes, side reactions lead not only to a loss of current efficiency but also to decreased anode coating lifetime.

Electrochlorinator Cell Design and Operating Characteristics

A very large number of electrochlorinator cells are currently marketed, reflecting the diverse needs of the different application areas. The scale and conditions of operation vary widely between different cells. Although electrochlorinator cells, being undivided, are simpler than divided chlor-alkali cells, their mechanical designs vary widely in a number of aspects:

Electrical configuration	monopolar, bipolar
Electrode forms	plates, tubes, meshes
Anode material	coated titanium
Cathode material	titanium, nickel, steel
Inter-electrode gap	1–6 mm
Electrode orientation	horizontal or vertical, concentric tubes

Likewise, the operating parameters of different devices also vary:

Current density	0.2–$5\,kA/m^2$
Salinity	0.15–3% w/v NaCl
Temperature	3–$40°C$
Hypochlorite strength	2–$10\,000$ ppm w/v as Cl_2
Duty cycle	continuous or intermittent

Different cell manufacturers use different approaches to the problem of scale removal. Firstly, purified brine can be used to prevent the problem occurring. However, in many cases this is impractical and scale deposition can be minimised by ensuring a high Reynolds number at the electrode surface, either by using high linear flow velocities or by promoting turbulence, or both. Alternatively, periodic polarity reversal can be used to prevent scale build-up; in this regime scale formed during the time when an electrode is cathodic is redissolved during the next anodic half-cycle. Finally, scale can be removed by acid washing or by mechanical brushing or scraping.

The performance achieved by a particular cell depends both upon its design characteristics and the conditions under which it is operated. Since both of these vary widely, different cells offer very different performance characteristics:

Energy consumption	3.0–$6.0\,kWh/kg\,Cl_2$
Salt consumption	3.5–$4.5\,kg\,NaCl/kg\,Cl_2$
Current efficiency	$<60\%$ to $>90\%$

Because of this wide variation in cell design and operating characteristics, one cannot address the whole electrochlorinator anode coating market with a single coating type. A range of complementary coatings is needed, each coating being optimised for a particular set of operating conditions. A range of coatings offered by ICI Chemicals and Polymers Ltd will be described for the following application areas:

Brine feed	*Temperature* (°C)	*Coating*
Synthetic dilute brine	>10	ECB (precious metal oxide)
Synthetic dilute brine— low salinity	>10	K (Pt/Ir)
Seawater	>5	ECS (precious metal oxide)
Seawater	>3	ECSL (precious metal oxide)
		L (Pt/Ir)
Heavy-duty GP	>5	MA (Pt electroplate)

Coatings for Use in Synthetic Brine

In applications such as the sterilisation of swimming pools and of water for drinking or food processing, electrochlorinators are typically run with synthetic brine. Since salt has to be transported to make up the feed brine in these applications, it is important that a high current efficiency is maintained at high salt conversion. In such applications, ICI recommend the use of the 'ECB' precious metal oxide coating. This offers an excellent performance in terms of current efficiency over a range of salt concentrations and current densities. A lifetime in excess of 5 years is obtained for salt concentrations above 20 g/litre, and for current densities below 2·5 kA/m². These constraints on salt concentration and current density can be relaxed somewhat by opting for a Pt/Ir coating (designated type 'K'). This coating can be used at low salt concentration and at current densities up to 5 kA/m². Figures 5(a) and (b) compare the current efficiency of the ECB and K coatings at different salt concentrations and current densities.

Coatings for Use in Seawater

For applications such as coastal power stations, electrochlorinators are fed with seawater. In many parts of the world the seawater feed may be quite warm ($>10°C$);

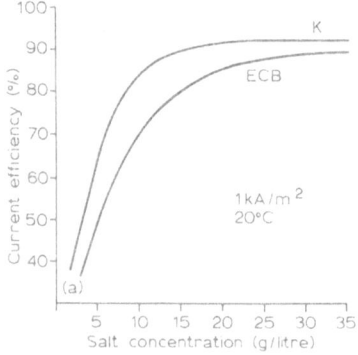

Fig. 5(a).   MET coatings for synthetic brine: relationship between current efficiency and salt concentration for type ECB and K coatings. 1 kA/m², 20°C.

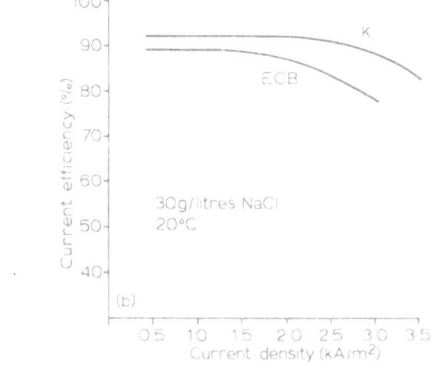

Fig. 5(b).   MET coatings for synthetic brine: relationship between current efficiency and current density. 30 g/litre NaCl, 20°C.

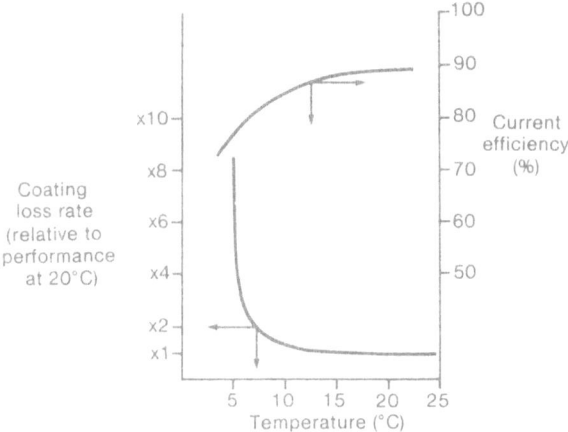

Fig. 6. The MET-type ECB coating: the effects of temperature upon current efficiency and anode coating wear rate. 30 g/litre NaCl, 1 kA/m².

this applies to the tropics at all times throughout the year and to more temperate zones during the summer months. In this case coatings of the type ECB or K can be used satisfactorily, with good service lifetimes. However, in many locations, the UK included, winter temperatures may drop as low as 4–5°C, with temperatures of less than 10°C for 4 or more months of the year. At these temperatures the lifetime of coatings such as the ECB or K types is dramatically reduced. Figure 6 shows the effect of decreasing temperature on the current efficiency and lifetime of the ECB coating.

At one time the only solution to this problem was to gradually reduce the operating current density, as the temperature fell during the winter months. Increasingly, operators have needed to operate at full capacity throughout the year, and ICI have developed new coatings to address this need. The reasons for decreased performance at low temperatures are several. Firstly, the observed

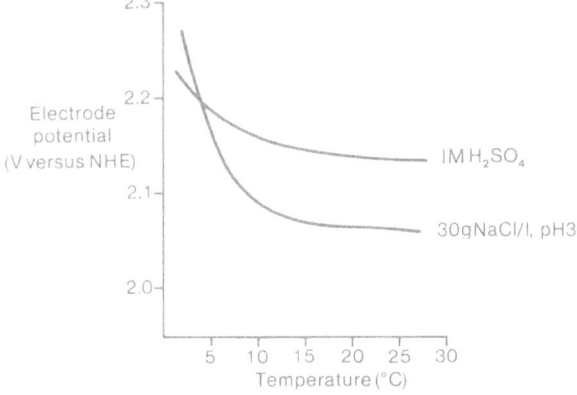

Fig. 7. Relationship between temperature and overpotentials for oxygen and chlorine evolution. MET-type MA Pt electroplate coating, 2·5 kA/m².

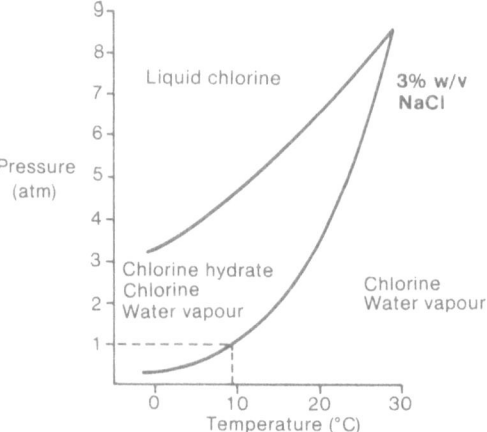

Fig. 8.   Phase diagram for the chlorine/water system showing *P–T* region of stability for chlorine hydrate. 30 g/litre NaCl.

electrode potentials for the evolution of chlorine and oxygen have very different temperature sensitivities (Fig. 7). Also, at low temperatures, chlorine is known to form a hydrate. At less than 10°C formation of hydrate at the surface of the anode may lead to a diffusion barrier [5] which in turn may lead to local chloride depletion (Fig. 8).

The ECS coating, a modified variant of the ECB precious metal oxide coating, has been developed to operate at low temperatures; this coating gives good lifetime and current efficiency performance at moderate current densities, as shown in Fig. 9.

For higher current density operation a more durable precious metal coating has been developed, designated ECSL. This coating, whilst not quite as good as ECS in

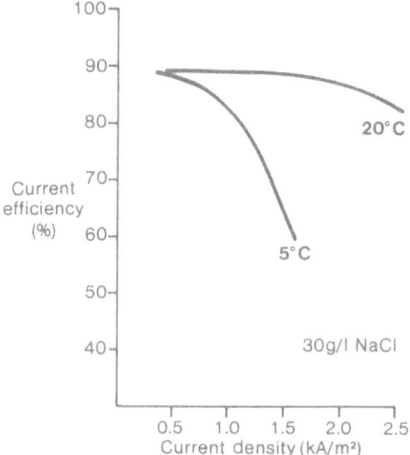

Fig. 9.   The MET-type ECS coating for seawater: relationship between current efficiency and current density at 5 and 20°C. 30 g/litre NaCl.

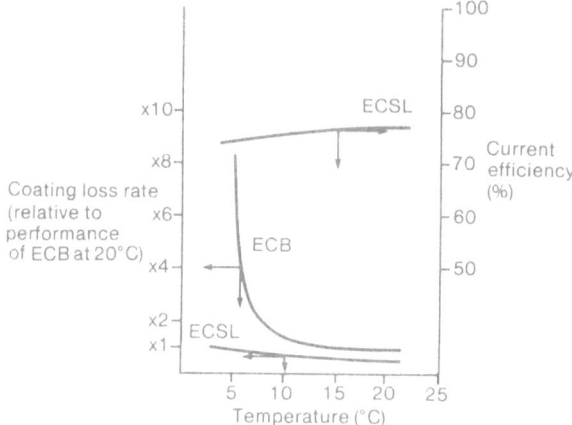

Fig. 10. The MET-type ECSL coating: effects of temperature upon current efficiency and wear rate. 30 g/litre NaCl, 1 kA/m².

terms of its current efficiency, offers excellent lifetimes in excess of 5 years, even at low-temperature, high current density operation. Some of the characteristics of ECSL are shown in Fig. 10. For operation at the highest current densities the type L coating gives improved performance.

Coatings for Use with Scale Removal Procedures

As described earlier, the operation of electrochlorinators in the presence of calcium or magnesium impurities in the feed brine can lead to scale formation on the cathode. Furthermore, in seawater containing appreciable amounts of Fe or Mn, oxide deposits can be formed on the anode surface. Different electrochlorinator cell manufacturers adopt different methods of scale prevention and removal. It is found that units operating on synthetic dilute brine made from softened water and purified salt are virtually free from scaling. In those where the water feed is harder descaling may be required. In such circumstances periodic (sometimes every 2–3 weeks) cleaning of the electrodes with dilute mineral acids is commonly practised. All the coatings described above have been developed such that they are immune from damage during such cleaning processes. ICI recommend a number of particular cleaning compositions and procedures to meet particular customer operating conditions.

In some devices high linear flow velocities or turbulence promotion is used to prevent scaling. This can, in some circumstances, lead to abrasion of the anode surface, and under these circumstances precious metal oxide coatings are not sufficiently mechanically robust; instead a platinum electroplate coating (type MA) is recommended. Whilst the current efficiency performance of the MA coating is not as good as the oxide coatings, excellent lifetimes can be achieved, even under the most demanding conditions (Fig. 11).

Finally, periodic current reversal can be used to prevent scale formation. In this case the Pt electroplate coating, MA, is again recommended. Alternatively, coated

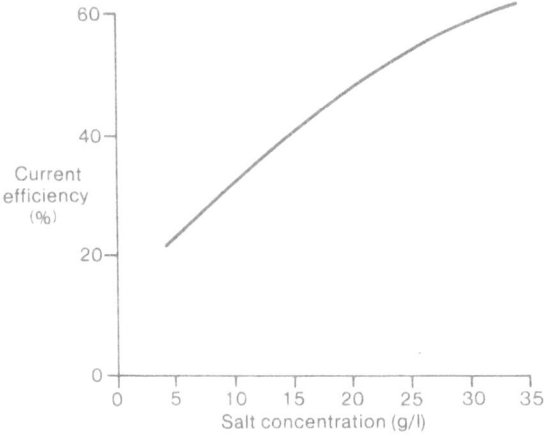

Fig. 11. MET-type MA Pt electroplate coating: relationship between current efficiency and salt concentration. 1 kA/m², 20°C.

electrodes based on a novel ceramic electrode material, EBONEX (marketed by Ebonex Technologies Inc., 5915 Hollis Street, Emeryville, CA 94608, USA), can be used with performance characteristics superior to coated titanium electrodes [6, 7].

The performance of the different coatings offered by ICI for different electrochlorination applications is summarised in Table 1. ICI has developed this extensive range of coatings for electrochlorination based on its many years of experience in electrocatalyst research and development activities, and also through its recent acquisition of the electrode business formerly belonging to Marston-Palmer. Several thousand kg/h total capacity are now produced using MET-coated electrodes. Electrodes can be coated on customers' own structures, or fabricated according to their own requirements, at ICI's plants in Winnington, Wolverhampton or Taiwan.

TABLE 1

Coating	ECB	K	ECS	L	ECSL	MA
Temperature (°C)						
Typical	10–30	10–30	10–30	5–30	5–20	5–40
Minimum	10	10	5	3	3	
Current density (kA/m²)						
Typical	1–2	2–4	1–2	2–4	1–2	0·5–4
Maximum	2·5	5	2·5	5	2·5	5
Salinity, typical (g/litre)	30	2–30	20–30	2–30	20–30	2–30
Cl_2 overpotential (mV)	70	70	70	70	110	600
Current efficiency (%)	90	90	90	90	75	65
Cleaning regimes						
Acid	OK	OK	OK	OK	OK	OK
Reversal	No	No	Limited	No	No	OK
High flow	No	No	No	No	No	OK

REFERENCES

1. Cairns, J. F., Couper, A. M. & Denton, D. A., ICI electrode coatings for membrane cells. In *Modern Chlor-Alkali Technology*, Vol. 3, Chap. 21, ed. K. Wall. Ellis Horwood, Chichester, 1986.
2. Brooks, W. N., Couper, A. M., Elson, I. H. & Cairns, J. F., ICI electrode coating technology—a brief history and recent developments. Paper presented to the 9th Japan Soda Industry Association Meeting, Kyoto, 21–22 Nov. 1985.
3. Kotowski, S. & Busse, B., The oxygen side reaction in the membrane cell. In *Modern Chlor-Alkali Technology*, Vol. 3, Chap. 22, ed. K. Wall. Ellis Horwood, Chichester, 1986.
4. Denton, D. A., De Souza, J. T., Entwisle, J. H., Lee, D. & Wilson, H. G., Developments in coatings for metal anodes. In *Modern Chlor-Alkali Technology*, Vol. 2, Chap. 13, ed. C. Jackson. Ellis Horwood, Chichester, 1983.
5. Kawashima, Y. & Ohe, K., A study of the anode material for sea water electrolysis. Paper presented at the 50th Meeting of the Electrochemical Society of Japan, Tokyo, May 1983.
6. Hayfield, P. C. S., US Patent 4 442 917, 1983.
7. Clarke, R. L. & Brooks, W. N., Some electrochemical applications for Ebonex ceramic electrodes. *Electrochem. Soc. Ext. Abs.*, 88-1 (1988) 579.

9

AN ELECTROCHEMIST'S VIEW OF CHLORINE MANUFACTURE

J. A. Harrison

University of Newcastle upon Tyne, UK

NOTATION

ba	Tafel slope for reaction of eqn (1)
bc	Tafel slope for reverse of reaction of eqn (1)
c_{dl}	Double layer capacity
$c_{Cl^-}^b$	Bulk concentration of chloride ions
$c_{Cl_2}^s$	Surface concentration of chlorine
$c_{Cl^-}^s$	Surface concentration of chloride ions
$c_{OH^-}^s$	Surface concentration of hydroxide ions
D_{Cl_2}	Diffusion coefficient for chlorine
D_{Cl^-}	Diffusion coefficient for chloride ions
E_{cell}	Cell voltage
E_1	Potential driving electrode reaction of eqn (1)
E_1^o	Standard potential for electrode reaction of eqn (1) against SCE and for $1\,mol\,cm^{-3}$ reactant and product
fl	Flux
F	Faraday constant
I_{cell}	Cell current
$(kSH)_1, k_1^o$	Standard rate constant for electrode reaction of eqn (1)
R_1	Internal resistance of a cell
δ	Diffusion layer thickness at an electrode

INTRODUCTION

As a starting point for this chapter the author looked back at the Proceedings of the previous Chlorine Symposia [1–3]. Some very impressive technical developments are reported, culminating in the latest cells based upon membrane technology. However, the role of the kinetics of the individual electrochemical reactions which form the basis of the industrial cells is not very clear in the performance figures. It is therefore not clear how much effort should go into trying to increase the rate of the anodic and cathodic reactions. However, isolated solutions to this problem have been published. Under particular conditions of current it is possible to insert probes containing an electrolyte and a reference electrode into a working cell and to measure the potential drops across parts of the cell, for example between an electrode and the electrolyte near to the electrode or across the membrane, etc. Examples of such data have been published a number of times (see, for example, Ref. 2, Chapter 10).

However, data at one current density may not be as satisfactory as the whole current–voltage characteristic. In order to set down a general answer to this problem we will proceed from an electrochemist's viewpoint. The simplest measurements that can be made on an industrial cell include cell current and voltage, and, by analogy with single electrode electrochemical investigations, the basic performance indicator can be considered to be the cell current (I_{cell}) against voltage (E_{cell}) curve. This contains information about all the potential drops in the cell; that is due to the two electrode reactions, and the ohmic losses in the electrical connectors, the electrolyte and the membrane, gathered into one ohmic resistor (R_l). Analyses of this problem have been presented [4, 5], however these have emphasised mainly the engineering aspects. Figure 1 shows a sketch of the way that the concentrations of the hydroxide ion, the chloride ion and the chlorine molecules are expected to behave in a chlor-alkali cell. In the steady state, for a simple diffusion layer the concentration gradients are linear. The individual current–potential curves

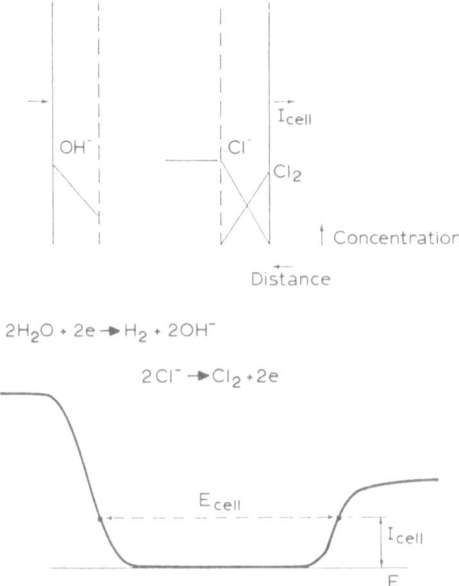

Fig. 1.   Simplified representation (top) of the steady-state concentration gradients of chloride ions, hydroxide ions and chlorine gas in the chlor-alkali cell. The corresponding current–potential curves at the anode and the cathode, and their relation to the cell current and voltage, are shown diagrammatically (bottom).

should also add, as illustrated, to form the electrical characteristic of the whole cell. The effect of the most important additional component, the ohmic resistances, must also be added. The form of the I_{cell}–E_{cell} curve is not immediately obvious due to the interaction of the various effects and can only be investigated by calculation, as shown in the next section.

THE CURRENT–VOLTAGE CHARACTERISTIC OF A COMPLETE CELL

The theory presented here is based on the theory for electrode reactions at single electrodes, which is familiar to electrochemists. The features of the single electrode reactions which are used represent only those elements which seem to be important in the chlor-alkali cell (see below). In particular, it is assumed that the electrochemical reaction is irreversible at both electrodes, and that the transport of all species is characterised by a transport coefficient, D, which is independent of potential and concentration. If fl is the flux of reactant to the electrodes (determined by the constant current flowing), then three similar equations for the flux can be set up for each of the electrodes. For example, for the chlorine electrode with E_1^o(SCE, $1 \, mol \, cm^{-3}$) $= 936 \, mV$

$$2Cl^- \rightarrow Cl_2 + 2e \tag{1}$$

The flux is given by

$$fl = c_{Cl}^s \cdot k_1^o \exp[2 \cdot 303(E - E_1^o)/ba] \tag{2}$$

$$fl = D_{Cl^-}(c_{Cl^-}^b - c_{Cl^-}^s)/\delta \tag{3}$$

$$fl = D_{Cl_2} c_{Cl_2}^s/\delta \tag{4}$$

It will be assumed that the area of each electrode is unity. In practice it will be useful to include the two electrode areas. Given ba, D, δ, bulk concentration of Cl^-, and fl, the three equations can be solved for the surface concentration of Cl^- $(E - E_1^o)$ and the surface concentration of Cl_2. The results are

$$c_{Cl}^s = \frac{D_{Cl^-} c_{Cl^-}^b/\delta}{k_1^o \exp[2 \cdot 303(E - E_1^o)/ba] + D_{Cl^-}/\delta} \tag{5}$$

$$(E_1 - E_1^o) = \frac{ba}{2 \cdot 303} \ln\left\{\frac{fl D_{Cl^-}/\delta}{(D_{Cl^-} c_{Cl^-}^b \cdot K_1^o)/\delta - fl k_1^o}\right\} \tag{6}$$

$$c_{Cl_2}^s = fl\delta/D_{Cl_2} \tag{7}$$

The current–voltage curve can now be constructed, given I_{cell}, by

$$I_{cell} = 2Ffl \tag{8}$$

$$E_{cell} = (E_1 - E_1^o) - (E_2 - E_2^o) + E_1^o + E_2^o + I_{cell}R_1 \tag{9}$$

A feature of the above equations is the value of the potential at $I_{cell} = 0$:

$$E_{cell} = \frac{ba}{2 \cdot 303} + \frac{bc}{2 \cdot 303} + E_1^o - E_2^o \tag{10}$$

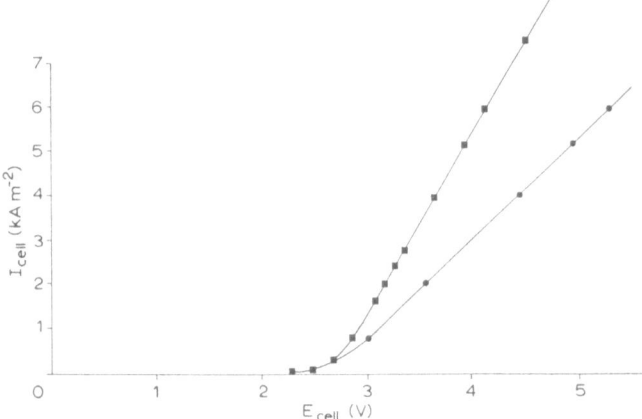

Fig. 2. Calculated cell current–voltage curve for a whole cell. The individual electrode reactions are $2H_2O + 2e \rightarrow H_2 + 2OH^-$ with $E_2^o(SCE, 1\,mol\,cm^{-3}) = -1250\,mV$ and $2Cl^- \rightarrow Cl_2 + 2e$ with $E_1^o(SCE, 1\,mol\,cm^{-3}) = 936\,mV$, and the parameters $k_1^o = 10^{-6}$, $k_2^o = 10^{-10}\,cm\,s^{-1}$, $ba = 40$, $bc = 120\,mV$, $D = 10^{-5}$ $cm^2\,s^{-1}$, $\delta = 10^{-2}\,cm$ and $c_{Cl}^b = 4M$. ■, $R_I = 2.1 \times 10^{-4}\,ohm\,cm^2$; ●, $R_I = 4.1 \times 10^{-4}\,ohm\,cm^2$.

Figure 2 demonstrates the typical behaviour of the above equations when $I_{cell}-E_{cell}$ curves are plotted, for trial values of the parameters. The trial parameters have been made reasonably realistic. However, they have been simplified. A more detailed consideration of the kinetics of the chlorine evolution reaction is given below. As can be seen the ohmic effects appear mainly in the slope of the curves and the electrochemical kinetic effects in the position along the potential axis. In Fig. 3 the electrochemical rate constants have been increased until the electrochemical rates no longer effect the $I_{cell}-E_{cell}$ curve. Under these conditions transport effects are the limiting factor which determines the reaction rate at each electrode. The surface concentrations are shown in Fig. 4. It is interesting that the current–voltage curves given in the last Symposium [3] for the Hoechst–Uhde (Chapter 13, Fig. 10) and Eltech membrane gap (Chapter 17, Fig. 3) industrial cells have characteristics which

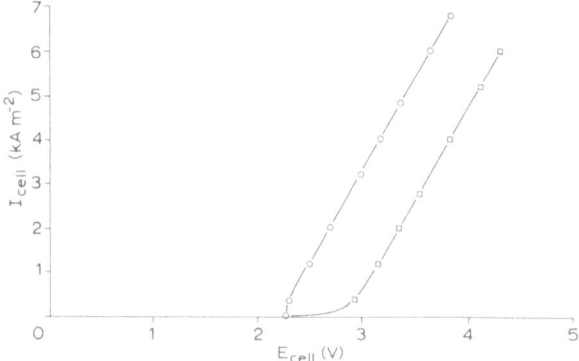

Fig. 3. Calculated cell current–voltage curve for a whole cell. The parameters are the same as for Fig. 2 but with ○, $k_1^o = 10^{-6}$, $k_2^o = 10^{-6}\,cm\,s^{-1}$, $R_I = 2.1 \times 10^{-4}\,ohm\,cm^2$; □, $k_1^o = 10^{-10}$, $k_2^o = 10^{-10}\,cm\,s^{-1}$, $R_I = 2.1 \times 10^{-4}\,ohm\,cm^2$.

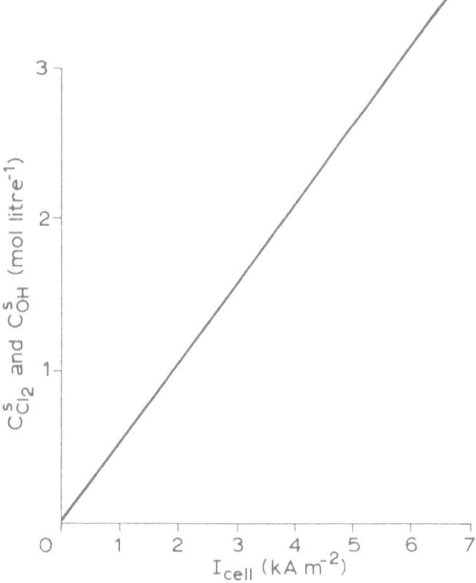

Fig. 4. Theoretical surface concentration of hydroxide ion and chlorine gas generated by the electrode reactions.

are similar to the theory presented above. In particular, the potential at zero current and the slope and position of the lines must have the significance given in the equations above. The potential at zero current is given by eqn (10). For the simplest expected Tafel slopes of 40 mV for the chlorine evolution reaction and 120 mV for the hydrogen evolution reaction it should have a value of 2346 mV. In addition, the lines are not straight, which must be attributed to the properties of the individual

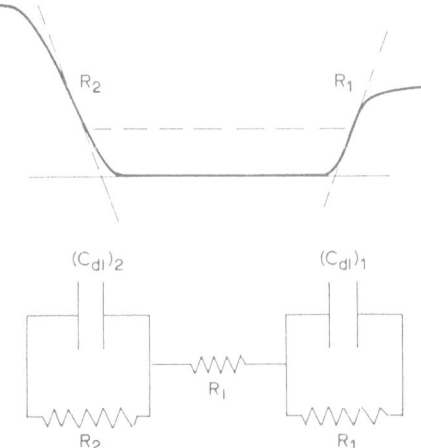

Fig. 5. Representation of the response of a cell to small amplitude alternating current potential. It is expected that the response to large amplitude alternating current will have similar characteristics. If the circuit elements are known the electrical efficiency of the cell to alternating current and to fluctuations in the current can be estimated.

electrodes and to the effects of bubbles discussed below. It would be an interesting exercise, using a computer, to split the industrial cell current–voltage curve into the two electrode reactions, given, for example, information about one of the reactions (see below).

A further property of industrial cells is the response to alternating current. A model based upon experience in electrochemistry with small amplitude signals is illustrated in Fig. 5. As shown in the next section measurements have been made of the small amplitude double layer capacity for the single chlorine electrode, so the order of magnitude of the circuit elements are known. The relevance of this model to industrial performance is that it would allow an estimate of the electrical efficiency of the cell to alternating current ripple and fluctuations in the supply of power.

THE CURRENT–POTENTIAL CHARACTERISTIC OF THE CHLORINE EVOLUTION

Reaction at a Single Ruthenium Dioxide/Titanium Dioxide Electrode

This section will be a review, chiefly of our own work with selected references, on the kinetic characteristics of the overall reaction of eqn (1), which should really be entered into the above model. The kinetics have been investigated at single stationary and rotating disc ruthenium dioxide/titanium dioxide electrodes by steady-state current–potential curves and by impedance measurements. Only a combination of current and impedance measurements is capable of measuring the kinetics under the potential and current conditions which are present in a real cell. As an example, a current–potential curve is shown in Fig. 6. The data were measured at a stationary electrode (horizontal, pointing upwards) and the curve is corrected for the measured ohmic resistance, via the impedance. The curve is strongly affected by bubble formation. It is problematical how to quantitatively describe this effect. If eqns (2) and (8) are used as a model to describe the single electrode current–potential curve, then either a non-constant value of ba or of k could be invoked. It would seem more correct to keep ba fixed as it is a characteristic of the electron exchange reaction. An example of the resulting standard rate constant–potential curve is shown in Fig. 7, which, together with the double layer capacity curve (Fig. 8), completely characterises the electrochemistry of the chlorine evolution reaction on titanium dioxide/ruthenium dioxide electrodes at a particular chloride ion concentration. It can be surmised that up to a potential of 1125 mV SCE only the electron transfer reaction with its $ba = 40$ mV Tafel slope determines the rate. The first-order chloride ion concentration dependence of the electron transfer reaction suggests that the overall reaction of eqn (1) can be split into two kinetic steps (see the latest paper [6] and references therein). Thereafter, at more positive potentials, the bubble layer takes over as a major factor in the electrode kinetics. From the data in Figs 7 and 8 the experimental current or impedance can be reconstituted, or calculated for any conditions. Data of this type, particularly the double layer capacity, have been used to characterise a number of different preparations of titanium dioxide/ruthenium dioxide type electrodes [8–10]. The shape of the double layer capacity curve in Fig. 8 also has some significant features. A double layer

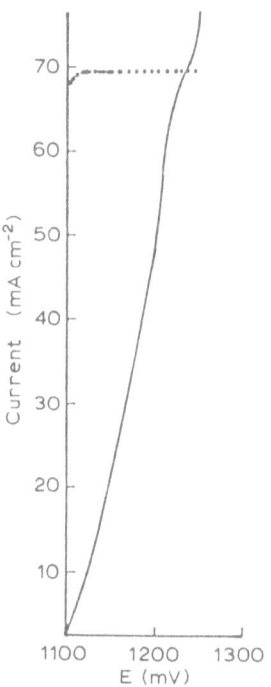

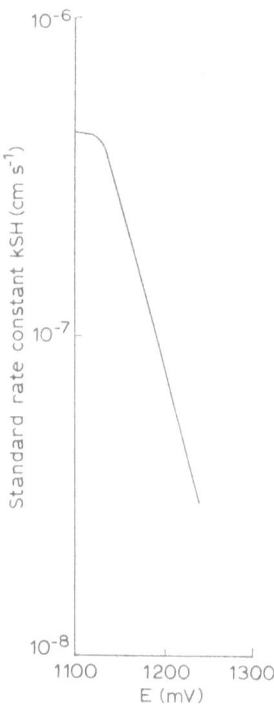

Fig. 6. A current–potential curve for the chlorine evolution reaction on a ruthenium dioxide/titanium dioxide electrode in 5·13M NaCl.

Fig. 7. The standard rate constant–potential curve (calculated with $E^\circ(SCE, 1\,mol\,cm^{-3}) = 1050\,mV$) for a ruthenium dioxide/titanium dioxide electrode in 5·13M NaCl, which, together with the double layer capacity curve shown in Fig. 8, completely characterises the electrochemistry of the chlorine evolution reaction on a ruthenium dioxide/titanium dioxide electrode.

capacity–potential curve contains information about electrode area and the change of the charge distribution in the solution with potential [11, 12]. These features can be interpreted in Fig. 8. At less positive potentials where no chlorine evolution is taking place the double layer capacity is constant and is proportional to the real electrode area. By comparison of Figs 7 and 8 it can be seen that when the reaction starts the double layer capacity rises. This can be interpreted as an increased change in the charge distribution due to chloride ion loss, and demonstrates that the reaction is behaving simply as an electron transfer reaction. At more positive potentials the double layer capacity falls due to the dominance of the chlorine bubble layer. The large values of the double layer capacity at the less positive potentials show that the real area is large. The reason for the apparent large catalytic effect of ruthenium dioxide/titanium dioxide electrodes is solely due to an area effect. Electrochemical measurements can give much information about the bubble layer [13]. Studies have only been carried out for the hydrogen evolution reaction on a nickel electrode in acid solution. No studies have yet been made on the chlorine evolution reaction itself. These could be made, but there is a narrow potential

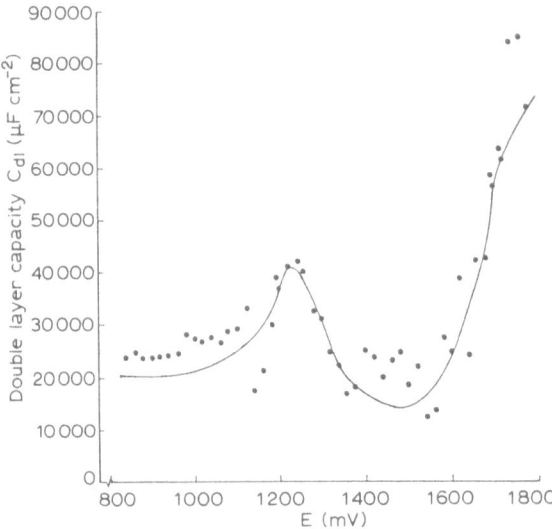

Fig. 8. A typical double layer capacity–potential curve for the chlorine evolution reaction on a ruthenium/titanium dioxide electrode. The curve was measured during the chlorine evolution reaction.

window available, due to the proximity of the oxygen evolution reaction. Double layer capacity measurements in sodium chloride solution and hydrochloric acid solution [14] (Fig. 9) suggest that these curves could be used to assess the extent of the bubble layer on different catalytic electrode materials. At a particular potential the ratio of the two capacities should be a measure of the average coverage with bubbles.

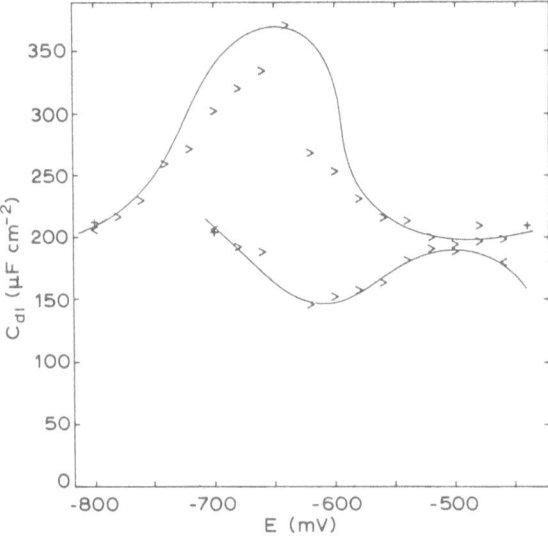

Fig. 9. An example of the effect of a bubble layer on the double layer capacity curve measured on a nickel electrode. The upper curve was measured in 1M NaCl and the lower curve in 1M HCl during the hydrogen evolution reaction.

REFERENCES

1. Coulter, M. O. (ed.), *Modern Chlor-Alkali Technology*. Ellis Horwood, Chichester, 1980.
2. Jackson, C. (ed.), *Modern Chlor-Alkali Technology*, Vol. 2. Ellis Horwood, Chichester, 1983.
3. Wall, K. (ed.), *Modern Chlor-Alkali Technology*, Vol. 3. Ellis Horwood, Chichester, 1986.
4. Pickett, D. J., *Electrochemical Reactor Design*. Elsevier, New York, 1977.
5. Hine, F., *Electrode Processes and Electrochemical Engineering*. Plenum Press, New York, 1985.
6. Harrison, J. A. & Hermijanto, S. D., *J. Electroanal. Chem.*, **225** (1987) 159.
7. Harrison, J. A., *J. Appl. Electrochem.*, **15** (1985) 495.
8. Harrison, J. A., Caldwell, D. L. & White, R. E., *Electrochim. Acta*, **28** (1983) 1561.
9. Harrison, J. A., Caldwell, D. L. & White, R. E., *Electrochim. Acta*, **29** (1984) 203.
10. Harrison, J. A., Caldwell, D. L. & White, R. E., *Electrochim. Acta*, **29** (1984) 1139.
11. Cooper, I. L. & Harrison, J. A., *Electrochim. Acta*, **29** (1984) 1147.
12. Cooper, I. L. & Harrison, J. A., *Electrochim. Acta*, **29** (1984) 1165.
13. Harrison, J. A., *J. Appl. Electrochem.*, **15** (1985) 495.
14. Harrison, J. A. & Kuhn, A. T., *J. Electroanal. Chem.*, **184** (1985) 347.

10

POISON TOLERANT PLATINUM CATALYSED CATHODES FOR MEMBRANE CELLS

D. S. Cameron, R. L. Phillips and P. M. Willis

Johnson Matthey Technology Centre, UK

1 INTRODUCTION

Chlorine and sodium hydroxide are manufactured almost·exclusively by the electrolysis of brine (aqueous sodium chloride), the overall reaction being represented by eqn (1):

$$2\,NaCl + H_2O \rightarrow 2\,NaOH + Cl_2\uparrow + H_2\uparrow \tag{1}$$

During the past decades, the chlor-alkali industry has been revolutionised by the advent of anodes catalysed by precious metal, which have almost entirely superseded the graphite previously used. Increasing public concern for environmental pollution has also led to pressure for a change from mercury cathodes, and from asbestos as diaphragm separators. This is encouraging a move towards cells incorporating ion exchange membrane separators, enabling greater efficiency in manufacturing, and making products of higher purity.

The chlor-alkali industry is one of the largest single consumers of electrical power, electricity being the major cost factor in production, with power constituting up to 60–70% of plant running costs. It is thus desirable to use electrodes giving the highest possible energy utilisation, and therefore the lowest overpotential, given that current efficiencies are typically in excess of 95%. With the various changes incorporated into the anode and separator technologies, one of the last remaining areas for significant cell voltage improvement is at the cathode.

In the past few years, Johnson Matthey has demonstrated that cathodes catalysed with precious metal can provide optimum savings in energy with consequent cost reductions.

2 ENERGY-SAVING CATHODES

In 1982 a breakdown of the factors contributing to an overall cell voltage of 3·46 V at 3 kA/m² was given by Thomas & Rudd [1]. This indicated that after the voltage drop in the membrane, the cathode was the second largest avoidable potential loss, accounting for some 10% of the overall power consumption of the cell. With improvements in membrane technology and cell design other voltage losses have been reduced over the intervening years, but the cathode losses still remain. Ferrous metal cathodes, traditionally used in diaphragm cells, are insufficiently stable in the more concentrated caustic routinely produced in membrane cells. There is, therefore, an incentive to move to stainless steel or nickel cathodes. However, all three of these materials have significant overpotentials for hydrogen evolution, and there is a considerable advantage to be gained by moving to electrodes which can minimise overpotential losses.

Cathodes catalysed with very small amounts of platinum group metals (that is a member of the platinum family and including palladium, rhodium, iridium, ruthenium and osmium) and exhibiting greatly reduced overpotentials for hydrogen evolution have been developed by Johnson Matthey over the past 9 years.

These platinum/ruthenium coatings, together with the plant constructed for their application on an industrial scale, were described in a previous paper [2]. They have been shown to give substantial power savings at full plant scale with good durability and robust operation, including cell dismantling and rebuilding. They have now been operating successfully in a production plant for over 4 years.

3 IRON CONCENTRATIONS IN CHLORINE CELLS

In common with any other catalytic process, precautions need to be taken to prevent contamination of the catalyst surface and its subsequent loss of activity. Any cathode is particularly susceptible to contamination, due to the tendency to collect cations from solution. In the majority of industrial membrane chlor-alkali plants, the most common poison in the catholyte is soluble iron, which arises from impure brine anolyte feed, and/or the slow dissolution of ferrous metal hardware in the caustic recycle loop.

Iron exists in the catholyte in a number of forms, but it is the removal of soluble iron, shown in Figs 1 and 2, which causes problems in membrane cells:

$$Fe^{2+} + 2e^{-} \leftrightarrow Fe \tag{2}$$

$$Fe^{2+} + 2OH^{-} \leftrightarrow Fe(OH)_2 \tag{3}$$

Equation 2 represents the plating out of metallic iron on the cathode, which hitherto has been assumed to reduce the effectiveness of the precious metal coating. Equation 3 represents the precipitation of ferrous hydroxide, which in practice occurs at pHs rather lower than those existing in the catholyte. The membrane which separates the anolyte from the catholyte has a very high pH gradient across its thickness, with pH 3–4 on the anolyte side and pH 11 on the catholyte side. If iron hydroxide precipitates out on or in the membrane, it can be the cause of resistance to liquor flow, and in extreme cases delamination and membrane rupture.

All chlor-alkali membrane plants use brine which is more or less impure, and processes are incorporated to remove heavy metals and multivalent cations, such as calcium and magnesium, before the liquor is admitted to the cell. Normally, iron is reduced to below 0·1 ppm. However, the catholyte is frequently recycled round pipework which can reintroduce impurities, particularly iron. There is, therefore, a need for cathodes to be tolerant of, for example, up to 5–10 ppm of iron while retaining their low overpotential. This would also increase the viability of incorporating such systems into existing plants as a retrofit.

4 POISON TOLERANT CATHODE COATINGS ON NICKEL ELECTRODES

A platinum-based coating capable of catalysing the evolution of hydrogen with high current efficiency and low overpotential has been developed by Johnson Matthey, details of this coating being given at a previous meeting of the International Chlorine Symposium [2]. Typical hydrogen evolution overpotentials of 50–70 mV are obtained at current densities of 3 kA/m² in 35% caustic soda solution at 90°C, with long-term stability, representing gains of some 200–300 mV over base metal cathodes. However, this stability is only achieved after an initial loss of some of the overpotential advantage, especially in the presence of iron.

It was appreciated that maximum benefit would accrue if these cathodes could be developed with in-built poison resistance to cope with levels of iron contamination normally encountered in production plants, which would usually employ steel piping and pump components. Although electrodes have been claimed with resistance to poisoning in up to 500 ppm of iron, in our experience the maximum solubility of iron is typically only 60 ppm in static, concentrated caustic solutions. According to plant operators, iron levels of up to 2 ppm are normally encountered in the catholyte in industrial-scale cells. To ensure adequate iron tolerance we have carried out measurements at 5 ppm as a standard test and, surprisingly, have found that this is a more difficult criterion to meet than resistance to higher iron levels, such as that of a saturated solution.

Work has been successfully carried out to make the original platinum group metal catalytic coatings resistant to poisoning by iron. This has been achieved by the use of small amounts of material which forms a partial coating over the original platinum/ruthenium layer on a nickel substrate. The resulting electrodes have been tested on a laboratory scale, and also generated considerable interest when evaluated by several plant operators.

5 EXPERIMENTAL TECHNIQUES

5.1 Electrode Coatings

Platinum/ruthenium catalysed electrodes based on nickel were prepared as described previously [2] using the respective precious metal chlorides. After simple pretreatment procedures, metal deposition was carried out by immersion plating in a single operation. An additional proprietary post-treatment process was subsequently employed to impart the required iron resistance. This latter process is designed to be fully compatible with the production plant for electrodes, also described previously [2]. A wide variety of techniques was evaluated in the course of developing this deposition process, entailing testing for both initial activity and durability.

5.2 Analytical Techniques

Electrode samples 'as prepared' were checked for composition by wet chemical analysis involving dissolution and spectroscopic determination of the solutions. Selected electrodes were also examined by X-ray photoelectron spectroscopy (XPS) to obtain an impression of the surface composition, and scanning electron microscopy (SEM) for surface topography. In addition, scanning transmission electron microscopy (STEM) was also used to examine fine detail of surfaces, and to carry out elemental analysis of small areas.

5.3 Iron Analysis Method

Iron levels in the caustic electrolyte were regularly monitored by a colorimetric technique using a Hach iron test kit, and (more accurately) using an SP600 ultraviolet visible spectrometer. Absorption of ultraviolet light through a 4 cm path length at a wavelength of 500 nm was determined, and the concentration of iron calculated in conjunction with a calibration curve.

6 ELECTROCHEMICAL TESTING

6.1 Half Cell Tests

Measurements of electrode overpotential for hydrogen evolution were carried out using an electrochemical half cell constructed from polypropylene, the electrolyte being 35% caustic soda solution at 90°C, with a dynamic hydrogen reference electrode, calibrated against a reversible hydrogen electrode.

The current/voltage characteristics were measured in this three-electrode cell using a potentiostat and a high impedance voltmeter, the measured potential being corrected for resistive losses in the electrolyte with a value of the solution resistance obtained by means of an alternating current impedance measurement. Applying the dynamic hydrogen/reversible hydrogen electrode correction allows the calculation of the hydrogen overpotential at a given current density. In order to speed up these measurements, a system incorporating an IBM microcomputer and a frequency response analyser (Solartron 1250) has been devised. With this apparatus it is also possible to measure the 'double layer capacitance' of the electrodes, and hence their 'roughness factor' compared to smooth surfaces. The term 'roughness factor' is used merely as a convenience, but represents real changes in electrode surfaces taking place during iron deposition, etc.

6.2 Activity Tests

Preliminary tests were carried out on each electrode, consisting of using the IBM PC programmed to linearly ramp up the voltage, and simultaneously measure the current and also to operate the frequency response analyser. By this means overpotentials, automatically corrected for resistance effects together with roughness factor measurements, could be determined at the same time.

6.3 Durability Tests

These tests were carried out in polypropylene electrochemical half cells maintained at 90°C by a heating element in a platinum sheath. Tests were continued for periods varying from 20 h to over 20 days. A modified reference electrode system was employed so that measurements could be made while using contaminated electrolytes, without the risk of affecting the reference electrode. Electrodes were operated at constant current densities, and voltages were monitored using a Digital Equipment Ltd PDP11/73 computer.

In experiments where no iron was added to the caustic electrolyte, the 35% NaOH solution was purified by pre-electrolysing overnight in a polypropylene beaker using a large Pt/Ru-coated cathode and a platinum sheet anode at a current density of $70\,mA/cm^2$ (total current 1 A). Iron additions were made using a calculated volume of 'saturated' iron solution (generally 60 ppm Fe) prepared by heating an excess of $FeCl_2 . 4H_2O$ in 2·5 litres of 35% NaOH solution at 80°C, followed by cooling and decantation.

7 LABORATORY TEST RESULTS

7.1 Short-term Poisoning Trials with Catalysed Nickel Electrodes

Figures 1–3 show the effects of iron concentrations varying from zero to 7·0 ppm on electrodes tested at $2\,kA/m^2$ over a 20 h period. Figure 1, on flat sheet nickel coated with Pt/Ru only, shows a rapid rise in overvoltage at iron concentrations of 3 and 7 ppm, with performance at 0·0 and 0·5 ppm largely unaffected, at least over this short timespan. Figures 2 and 3 show the effects of a similar range of iron contaminants on two types of mesh electrode of fine and coarse construction respectively, indicating that an increased number of sharp edge structures appears to exert a beneficial influence on poisoning characteristics.

Electrodes of several different configurations were tested, including flat sheets and

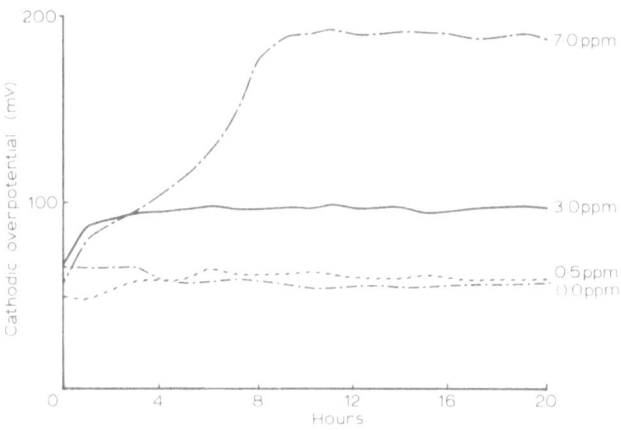

Fig. 1.   Effects of iron poisoning on potential for platinum/ruthenium catalysed nickel sheet electrode at $2\,kA/m^2$ current density.

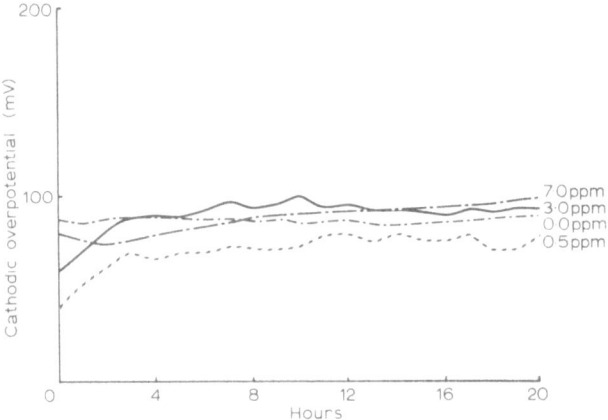

Fig. 2. Effects of iron poisoning on potential for platinum/ruthenium catalysed fine nickel mesh electrode at $2\,kA/m^2$ current density.

various expanded metal meshes. For reasons not wholly apparent, flat sheets invariably tended to poison more rapidly than meshes.

7.2 Medium-term Durability Trials

Electrodes were also tested for periods of 20 days in the half cell to compare the standard Pt/Ru coatings with the modified version, the results of these tests being given in Figs 4 and 5. Figure 4 shows the increase in overvoltage typically exhibited by unmodified electrodes, with a rapid increase during the first few days followed by stable operation at about 100–120 mV over the 20-day trial. Figure 5 shows the variation of the iron level in the electrolyte with time, the experiment being started with a nominal 8 ppm of iron present. In contrast, the modified coatings tend to remain below 80–100 mV (Fig. 6) with a slower rate of uptake of iron (Fig. 7) as evidenced by the more constant electrolyte iron content.

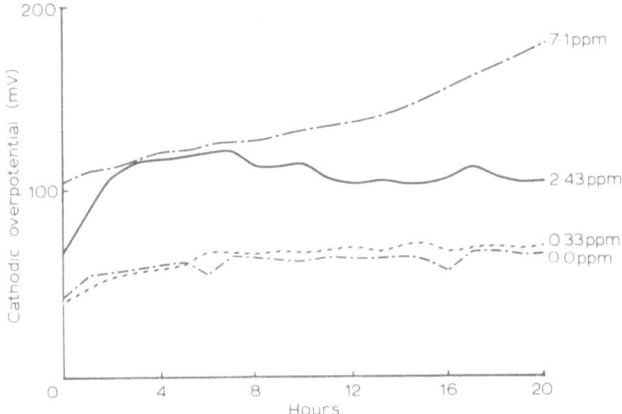

Fig. 3. Effects of iron poisoning on potential for platinum/ruthenium catalysed coarse nickel mesh electrode at $2\,kA/m^2$ current density.

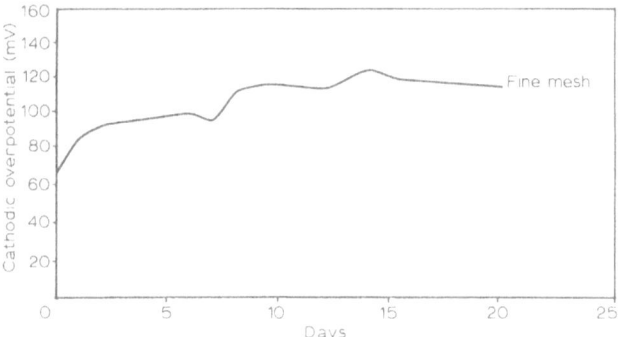

Fig. 4. Medium-term trials (20 days) on platinum/ruthenium catalysed nickel mesh electrodes at $2 kA/m^2$ current density.

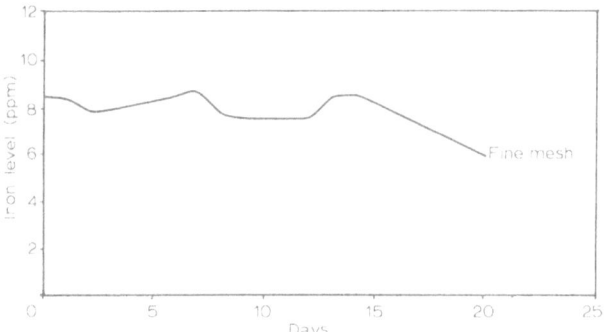

Fig. 5. Iron levels measured during medium-term trials.

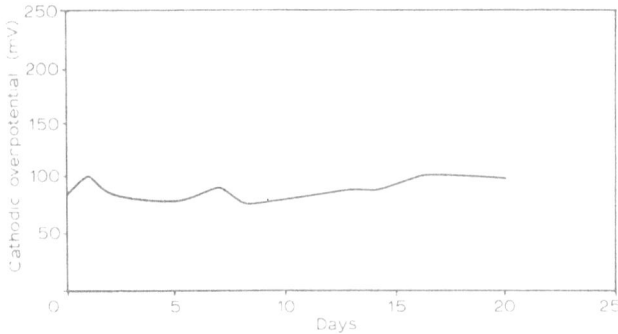

Fig. 6. Medium-term trials (20 days) on platinum/ruthenium modified coatings on nickel mesh electrodes at $2 kA/m^2$ current density.

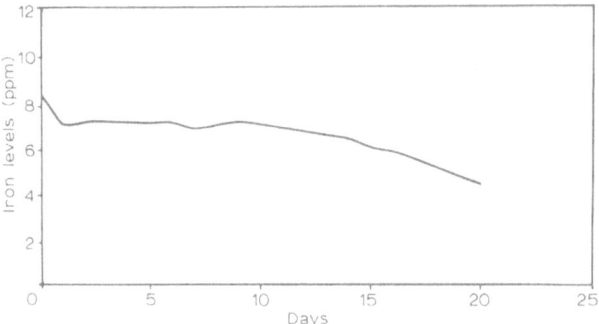

Fig. 7.   Iron levels measured during medium-term trials.

7.3 Electrodes on Stainless Steel Substrates

Since the catalytic coatings were found to tolerate iron in the electrolyte, it was decided to evaluate them on a stainless steel support, as well as the pure nickel materials previously used. Not only did it prove feasible to deposit the precious metal and the protective coatings on to the stainless steel substrates, but preliminary results indicate that these give overpotentials equally as low as nickel-based, catalysed electrodes, and are also resistant to iron poisoning. However, reversal of current flow, which is frequently encountered in real plants at periods of plant shutdown or interruptions in operation, can lead to some iron dissolution from the cathode, which can cause increases in overpotential (see Figs 8 and 9). The higher ohmic resistance of stainless steels compared to pure nickel will also determine their applicability in plant use, and may in any case restrict their use to bipolar cells.

7.4 Effects of Copper and Lead Contamination

Early on in the trials of the electrodes some degree of inconsistency was noted in the poison test results. XPS analysis of used electrodes indicated that trace amounts of lead and copper were associated with enhanced rates of iron deposition, possibly in a form which resulted in high overpotentials for hydrogen evolution. Accordingly, a series of experiments were carried out to examine the effects of

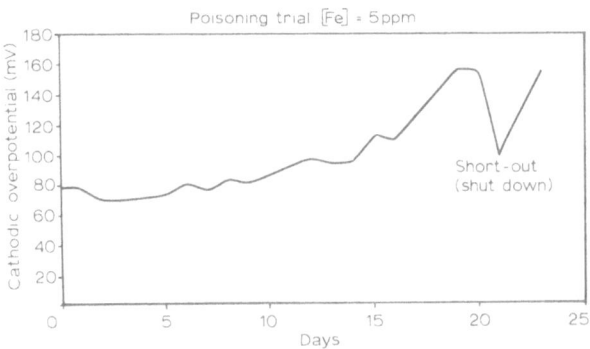

Fig. 8.   Medium-term trials on mesh electrode on stainless steel 310 substrates.

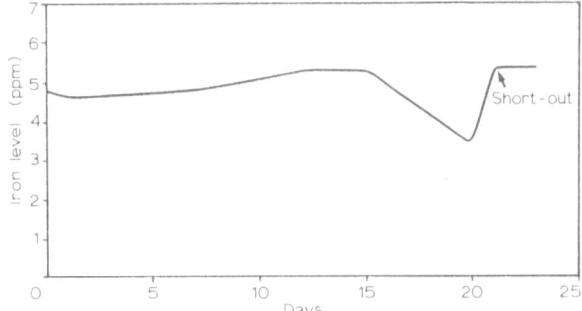

Fig. 9.   Iron levels measured during medium-term trials.

adding lead and copper to electrolytes, also containing iron contamination. The results of these tests are shown in Figs 10 and 11. From these it was deduced that even small amounts of lead are deleterious to poison tolerance to iron, while copper contamination is a much less serious problem (see Figs 11 and 12). From other analytical results it has been concluded that trace amounts of lead result in iron being deposited in a different form (with a high overpotential) in the presence of lead.

8 PLANT OPERATOR TEST RESULTS

Johnson Matthey has collaborated with several major contractors and operators of chlor-alkali plants who have carried out extended laboratory trials at current densities of 2–4 kA/m². These tests have also included deliberate current reversals to evaluate the durability of the coatings.

With the modified coatings, other collaborators have shown low overpotential and stable operation (within a 30 mV voltage range) over test periods of 250 days at

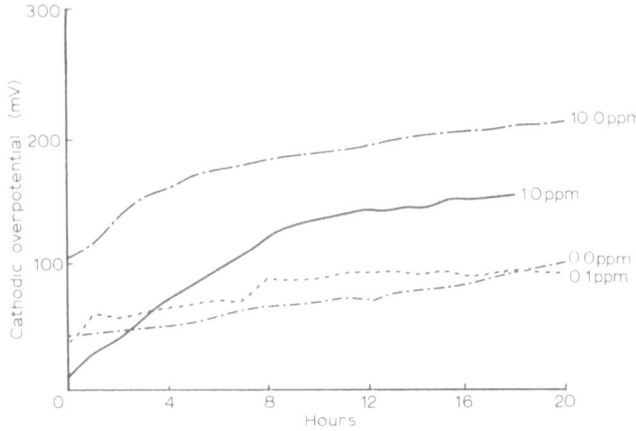

Fig. 10.   Effect of lead additions to iron-containing electrolyte, caustic containing 8 ppm iron.

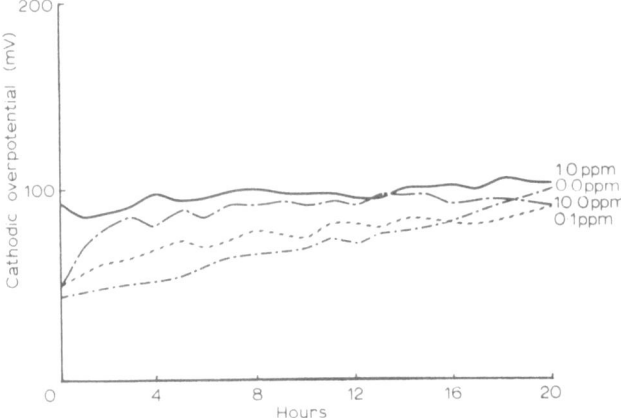

Fig. 11.   Effect of copper additions to iron-containing electrolyte, caustic containing 8 ppm iron.

$3 \, kA/m^2$ cathode current density, with soluble iron levels of 0·5–0·8 ppm in brine electrolysis cells.

The modified platinum/ruthenium catalysed electrodes have been operated with low overvoltage (less than 70 mV at $3 \, kA/m^2$) over extended periods in electrolyte containing 5 ppm of iron. Reversal of cell current, far from having a deleterious effect on performance, merely has the effect of removing iron deposited, with a corresponding improvement in cell voltage.

Fig. 12.   Photograph of used mesh electrodes from 20-day trial—note iron dendrites at base.

9 REASONS FOR POISON TOLERANCE

During the course of these studies it has become evident that the standard Johnson Matthey coatings, utilising platinum and ruthenium as the catalysing medium, already have a degree of resistance to iron poisoning.

They are, however, not immune from iron deposition during operation, and these deposits are manifested as high surface area dendrites which grow from the surface. This causes occlusion of the catalyst surface, but at the same time dramatically increases the surface area of the electrode per unit of geometric area with a spongy growth. This in turn decreases the current density per unit of real area of the electrode. The net result is that the electrode overpotential tends to remain largely unaffected by these spongy iron growths. Some of these growths can be seen in Fig. 12. The tendency to form heavier deposits of iron at the bottom rather than the middle or top of the electrodes can be explained by agitation caused by hydrogen bubbles in the electrolyte. These either prevent iron deposition or break off the long, thin dendrites as they form. During operation there may well be a continuous precipitation, removal and redissolution of iron occurring in the system. As long as a hard, coherent iron coating which has a high overpotential does not form, the electrode is not observed to be 'poisoned' by the iron present.

The catalyst layers modified by the present treatment, however, are more truly iron tolerant in that in iron contaminated electrolyte iron is deposited much more

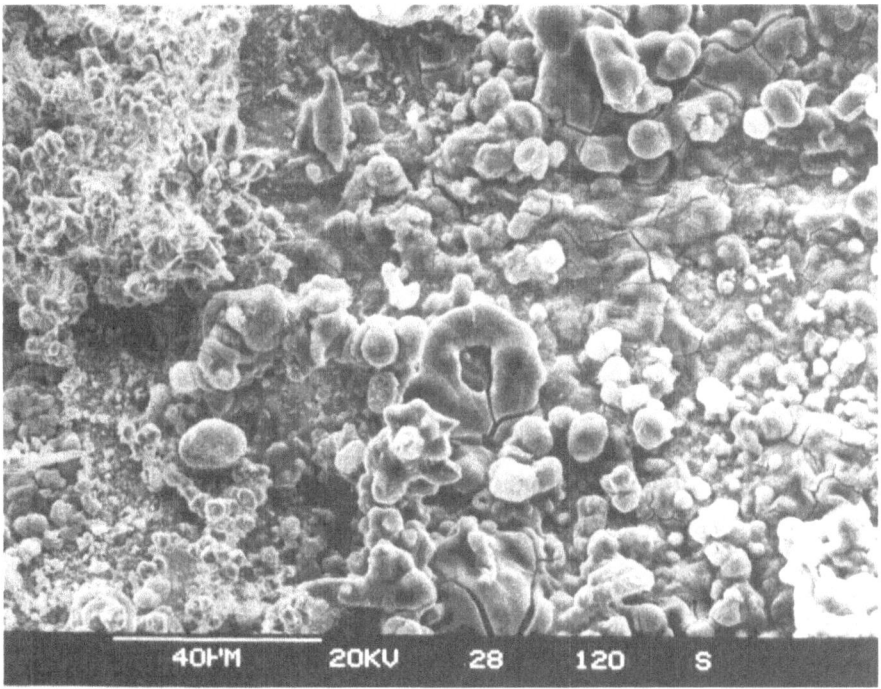

Fig. 13.   Scanning electron micrograph of iron deposition on Pt/Ru catalysed electrodes—low area form.

Fig. 14.   Scanning electron micrograph of iron deposition on Pt/Ru catalysed electrodes—high area
form.

slowly from solution, as evidenced by constant iron analysis figures from the
electrochemical cell results. The reasons for this are not entirely clear. One theory is
that the modification treatment not only occludes part of the platinum/ruthenium
coating but also exposes more of the substrate nickel surface. The discovery of some
platinum/ruthenium deposits on and around the iron dendrites may also indicate
that the iron deposits themselves are catalysed during their formation. However,
this is less likely since there is little sign of the precious metal losses which this would
imply, if a constant separation of the iron dendrites were occurring. It is probable
that the modification to the catalyst coating has the effect of inhibiting iron growth
on the catalyst surface as a coherent layer, and encourages dendritic iron deposits.
The role of lead in dramatically increasing the rate of iron poisoning may be due to
suppressing dendritic iron deposition. Examples of the two different forms, with low
surface area and high area dendrites, are shown in the scanning electron
micrographs of Figs 13 and 14.

10 CONCLUSIONS

Precious metal coated cathodes with low overpotentials for hydrogen evolution
are well established as an effective means of combining high efficiencies with
mechanical stability and long lifetimes. Initial costs are balanced by improved

efficiency in use, excellent service life, and recovery techniques which have been developed to recycle the precious metal employed.

This concept has now been extended to modified catalysed cathodes capable of operation in iron contaminated catholytes. These electrodes have been tested in a range of impurities, including the addition of copper and lead to iron, and it has been demonstrated that lead should be excluded from the system.

In the future there is likely to be a trend towards operation of cells at increasingly high current densities to improve utilisation of cell hardware. One of the side effects is the increased rate of heat evolution in the cells due to ohmic resistance and overpotential losses. Catalysed cathodes not only increase cell efficiency but minimise heat evolved as a by-product of the electrolysis process. Thus cells with catalysed cathodes typically run cooler.

The additional cost of treating the platinum/ruthenium cathodes to render them iron tolerant amounts to no more than 20–25% of the standard unmodified cathode coatings.

It is anticipated that these modified electrodes will be operational at full plant scale shortly. The system is applicable to monopolar and bipolar cells, and requires no activation process before use or special precautions during operation, being robust and resistant to deactivation by reverse polarity operation.

REFERENCES

1. Thomas, V. H. & Rudd, E. J., Energy-saving advances in the chlor-alkali industry. In *Modern Chlor-Alkali Technology*, Vol. 2, Chap. 10, ed. C. Jackson. Ellis Horwood, Chichester, 1983.
2. Grove, D. E., Precious metal activated cathodes for chlor-alkali cells. In *Modern Chlor-Alkali Technology*, Vol. 3, Chap. 19, ed. K. Wall. Ellis Horwood, Chichester, 1986.

11

THE DESIGN AND OPERATING EXPERIENCES OF AZEC ELECTROLYZERS AND RECENT DEVELOPMENT OF FLEMION MEMBRANES

T. Yamashita, Y. Sajima and H. Ukihashi

Asahi Glass Co. Ltd, Japan

1 INTRODUCTION

Asahi Glass is a pioneer in the development and commercialization of the membrane process in the chlor-alkali industry. We supply both the electrolyzer, which is called AZEC, and the membrane named Flemion as a major supplier in each market.

One of our features is that we have been developing the electrolyzer and the membrane side by side. While a deep understanding of membrane characteristics has accelerated improvements in the electrolyzer, knowing the requirements in designing the electrolyzer has in turn brought the effective advancement of the membrane.

This chapter introduces some essential factors of the design of membrane electrolyzers along with the range of the AZEC electrolyzers available.

2 THE RANGE OF AZEC ELECTROLYZERS

In late 1982 AZEC, an energy-saving membrane cell system, was commercialized. Since then many AZEC plants have been supplied. Table 1 shows the chlor-alkali plants using our Flemion membrane process. The total plant capacity has reached about 1 900 000 t of caustic soda production capacity per annum, which includes the ones under construction and contracted ones. This capacity is composed of 26 plants, which includes eight large plants with a capacity of more than 100 000 t/year each.

There are basically four types of AZEC electrolyzers to meet the various requirements of customers, and they are listed in Table 2.

2.1 AZEC-M Electrolyzer

The AZEC-M electrolyzer is the original AZEC system, which is composed of rubber frames and uses relatively small size membranes. The active area of each

TABLE 1
AZEC electrolyzer supply record as of March 1988

Number	Start-up	Clients' name	Plant capacity (t/year)	Location
1	Aug. 1978	Asahi Glass Co. Ltd Kansai Factory	10 000	Osaka, Japan
2	Nov. 1980	Nippon Carbide Co. Ltd	17 000	Toyama, Japan
3	July 1981	Thai Asahi Caustic Soda Co. Ltd	24 400	Bangkok, Thailand
	Aug. 1982	(Phases I, II and III)		
	Mar. 1985			
4	Dec. 1982	Tsurumi Soda Co. Ltd	18 000	Kanagawa, Japan
	Jan. 1987	(Phases I, II and III)		
5	July 1983	Kashima Chlorine & Alkali Co. Ltd	280 000	Ibaragi, Japan
	July 1984	(Phases I, II and III)		
	Dec. 1985			
6	Oct. 1983	Nankai Chemical Industry Co. Ltd	15 100	Kochi, Japan
	July 1985	(Phases I and II)		
7	Apr. 1984	Kansai Chlor-Alkali Co. Ltd	44 000	Osaka, Japan
8	Apr. 1985	Central Chemical Co. Ltd	55 000	Kanagawa, Japan
9	Oct. 1985	Mitsubishi Chemical Industries Ltd	115 000	Okayama, Japan
10	Nov. 1985	Shinetsu Chemical Co. Ltd	42 000	Niigata, Japan
	Apr. 1986	(Phases I and II)		
11	Nov. 1985	Confidential	33 000	Japan
12	Nov. 1985	Hokkaido Soda Co. Ltd	15 200	Hokkaido, Japan
13	Nov. 1985	Yee Fong Chemical & Ind. Co. Ltd	27 200	Taipei, Taiwan
	Aug. 1987	(Phases I and II)		
14	May 1986	Hokkaido Soda Co. Ltd	128 000	Hokkaido, Japan
	Oct. 1987	(Phases I and II)		
15	May 1986	Nankai Chemical Industry Co. Ltd	32 000	Wakayama, Japan
16	July 1986	Shanghai Tian Yuan Chemical Works	10 000	Shanghai, China
17	1986	Asahi Glass Co. Ltd	216 000	Chiba, Japan
18	Apr. 1987	Han Yang Chemical Corporation	60 000	Yeosu, Korea
19	Sept. 1987	Egyptian Petrochemical Company	66 000	Alexandria, Egypt
20	Apr. 1988	Taiwan Chlorine Industry	115 000	Kaohsiung, Taiwan
21	1989	Thai Plastic and Chemical Co. Ltd	26 000	Rayon, Thailand
22	1989	ISK Singapore	8 000	Singapore
23	1989	Asahimas Shentra Chemical Co. Ltd	138 000	Anyer, Indonesia
24	1990	Shanghai Wu Jing	150 000	Shanghai, China
25	1990	Confidential	200 000	
26	1990	Jin-Xi General Chemical Factory	40 000	Liao-Ning, China
		Total	1 884 900	

TABLE 2
Range of AZEC electrolyzers

Type	Frame	Active area of a membrane sheet	Bus bar connection
		(width (m) × height (m))	
AZEC-M	Rubber	$0.20\,m^2$ (0.20×1.0)	Side
AZEC-MD	Rubber	$0.20\,m^2$ (0.20×1.0)	Side
AZEC-F1	Metal	$2.88\,m^2$ (2.40×1.2)	Bottom
AZEC-F2	Metal	$1.71\,m^2$ (1.14×1.5)	Side

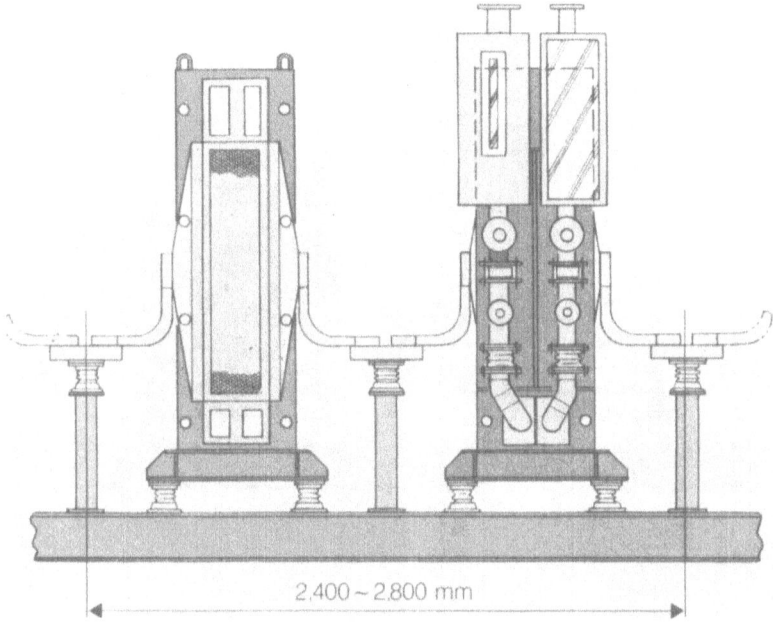

Fig. 1. AZEC-M electrolyzer.

membrane sheet is 0·2 m wide by 1·0 m high or 0·2 m². The bus bars are connected side by side. The pitch of an electrolyzer is from 1·2 to 1·4 m. In order to keep sufficient natural circulations of anolyte and catholyte, there are the gas–liquid separators for the anodic side and the cathodic side, as shown in Fig. 1. An example of a plant which uses AZEC-M electrolyzers is shown in Fig. 2. The Mitsubishi Chemical's plant in Mizushima, Japan, has a caustic soda production capacity of 115 000 t/year, and it has been operated since 1985.

2.2 AZEC-MD Electrolyzer

The AZEC-MD is a modification of the M electrolyzer. It is composed of two M electrolyzers connected together side by side and compressed by one set of fastening

Fig. 2. AZEC-M plant.

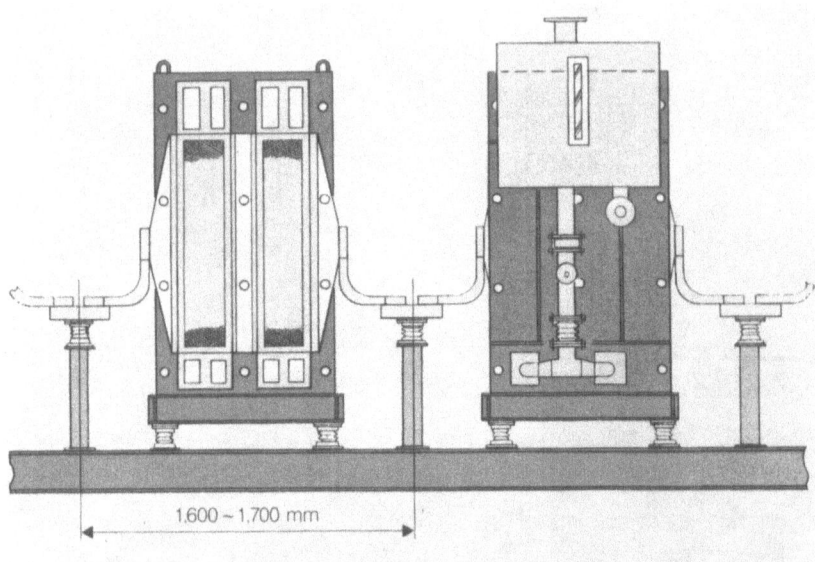

Fig. 3.   AZEC-MD electrolyzer.

plates. A schematic configuration of the electrolyzer is illustrated in Fig. 3. This configuration eliminates one set of intercell bus bars and flexibles, which decreases the voltage drop along with fabrication costs. The pitch of two electrolyzers is decreased to about two-thirds that of two M electrolyzers. The MD electrolyzers, as shown in Fig. 4, have been operating in a circuit at Asahi Glass Chiba Factory since 1986. The factory has a total capacity of 216 000 t/year of caustic soda.

2.3 AZEC-F1 Electrolyzer

The AZEC-F1 electrolyzer is composed of metal frames and uses relatively large size membranes. The active area of each membrane sheet is 2·4 m wide by 1·2 m high or 2·88 m². The bus bars are arranged in a transverse direction against the membrane sheets and are located below the cell frames. In Fig. 5 a two-stack type F1 electrolyzer is shown. Sixteen sheets of membrane make one stack and both stacks are compressed by one set of fastening devices. A current of 150 kA is supplied to each stack in series. The anodic and the cathodic gas separators are placed over the electrolyzer. A picture of the Hokkaido Soda plant is shown in Fig. 6. The plant previously operated Hooker H-4 type diaphragm cells. The cells were replaced with F1 type electrolyzers from 1986 to 1987 in three stages. Now they are operating 37 two-stack type F1 electrolyzers with 150 kA capacity each, and they have a production capacity of 130 000 t/year of caustic soda.

2.4 AZEC-F2 Electrolyzer

Recently the AZEC-F2 electrolyzer was commercialized. This electrolyzer is composed of metal frames and also uses relatively large size membranes. The active area of each sheet is 1·71 m². A schematic configuration of the electrolyzer is

Fig. 4.   AZEC-MD plant.

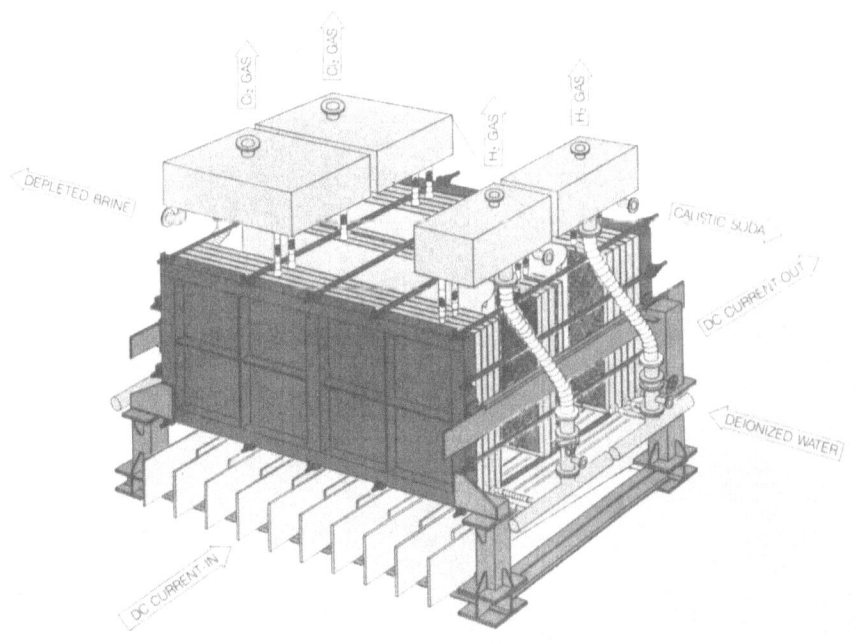

Fig. 5.   AZEC-F1 electrolyzer.

Fig. 6. AZEC-F1 plant.

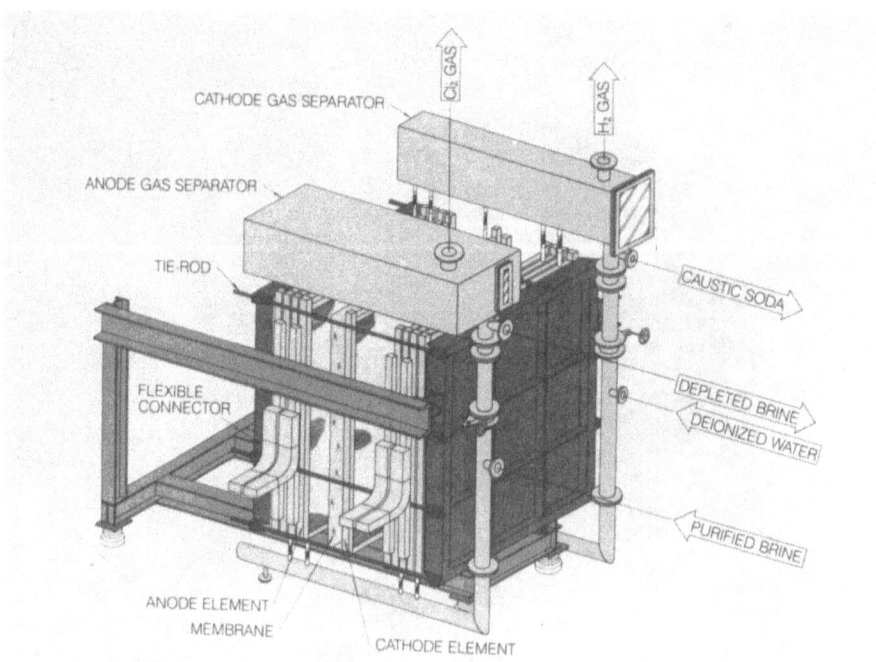

Fig. 7. AZEC-F2 electrolyzer.

illustrated in Fig. 7. In this case the bus bar connection is carried out side by side, the same as the M type design. The cell frames can be moved on the slide arms of the supporting unit of the electrolyzer, so that assembling and disassembling the electrolyzer are very easy. A large circuit with the F2 electrolyzers is now under construction and will be started in 1989 with a capacity of 138 000 t/year of caustic soda.

3 PERFORMANCE OF AZEC PLANTS

Through long-term operation, up to 5 years, with the AZEC electrolyzers the reliability of the system has been verified. The change in operating performance with time in the commercial plants which have been operated more than 2 years is summarized in Fig. 8. The current efficiency is still more than 94% after 2 years, and

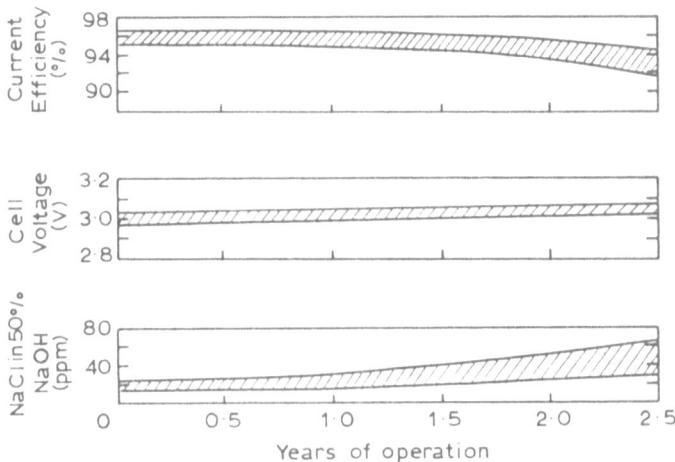

Fig. 8.   Performance of AZEC plants. Membrane: Flemion 790 series. Electrolyzer: AZEC-M3. Conditions: $CD = 3.0 \, kA/m^2$, $NaCl = 210 \, g/liter$, $NaOH = 35 \, wt\%$, temp. $= 87$–$90°C$.

the cell voltage is shown to be very stable. The salt content in caustic soda tends to increase gradually with time, but not very much. The changes of current efficiency and NaCl content in NaOH seem to be affected by the stability of the operation. For example, the plants which adopt operations with daily current cycling are apt to show a little earlier decline of current efficiency and increase of NaCl content in caustic soda than those with constant current operation. However, even with the current cycling, the performance is kept within the ranges shown in the graphs.

Before the planned replacement of membranes, it has been very rare for the membranes to have developed pinholes or splits in the AZEC systems. The stable performance and excellent reliability of the system are largely due to the improvements in strength and quality control of the membrane, along with the established technology for membrane use.

4 IMPORTANT FACTORS IN DESIGNING MEMBRANE ELECTROLYZERS

In the course of developing and improving the electrolyzer and the membrane, much know-how has been accumulated to achieve the stable performance of a membrane in commercial electrolyzers. In this chapter some of them are introduced focusing on the design factors that affect the membrane life.

4.1 Major Factors which Damage the Membrane

There are three major factors relating to design of an electrolyzer which damage a membrane, in addition to mechanical damage. They are listed in Table 3. Figure 9 shows the effect of the chlorine gas phase upon the deterioration of a membrane. A stable chlorine gas phase causes salt crystallization inside the membrane. Chlorine gas from the anode-side and concentrated caustic soda from the cathode-side diffuse into the membrane and form NaCl within. The generated NaCl is almost diffused out from the membrane. However, when the NaCl concentration in the membrane becomes high, in combination with a relatively high caustic concentration, the result is crystallization in the membrane. Severe salt crystallization in a membrane causes its strength to decrease and the formation of pinholes. This kind of problem is observed just where a stationary chlorine gas phase exists. There is no salt crystallization problem in the foam phase, where the membrane faces mainly chlorine gas rather than anolyte, because the membrane is kept wet.

Figure 10 shows the effect of overdepletion of NaCl in anolyte. The membrane used in this accelerated test was Flemion 753, which was developed as a low ohmic drop membrane in the early stage of our AZEC system work. That membrane is apt to be affected more by an overdepleted brine. Continuous operation at 130 g/liter NaCl in anolyte caused the membrane gradually to become overswollen and resulted in changes in performance, as shown in the graphs. Along with those changes, the blisters in the membrane increased with time, and many small blisters were present in almost all of the active area of the membrane after 4·5 months of operation.

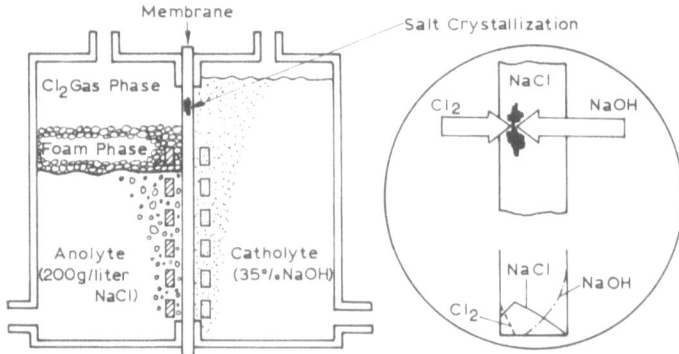

Fig. 9.   Effect of chlorine gas phase. Severe salt crystallization in a membrane causes decrease of strength in pinhole.

TABLE 3
Major factors which damage the membrane
(relating to design of electrolyzer)

1. Chlorine gas phase
2. Overdepletion of NaCl in anolyte
3. Overconcentration of NaOH in catholyte
4. Mechanical damage

In the early stages of our development of the Flemion membrane we evaluated a membrane, which was numbered as 230, with 40% caustic soda for almost 1 year. The current efficiency decreased and the cell voltage increased considerably, as shown in Fig. 11. The reasons for these changes in performance were understood to be as follows: a long-term operation with overconcentrated caustic soda caused a gradual decrease of ion-exchange capacity on the cathodic surface of the membrane. In addition, we could see many blisters in the area of the membrane where it was exposed to a higher chlorine gas void fraction on its anode-side surface. The higher caustic concentration operation made the membrane more sensitive to the chlorine gas phase effect. Although these effects have been reduced for the recent commercial membranes, which have been much improved compared with the earlier ones, it is very important to avoid these phenomena in an electrolyzer in order to realize a stable performance.

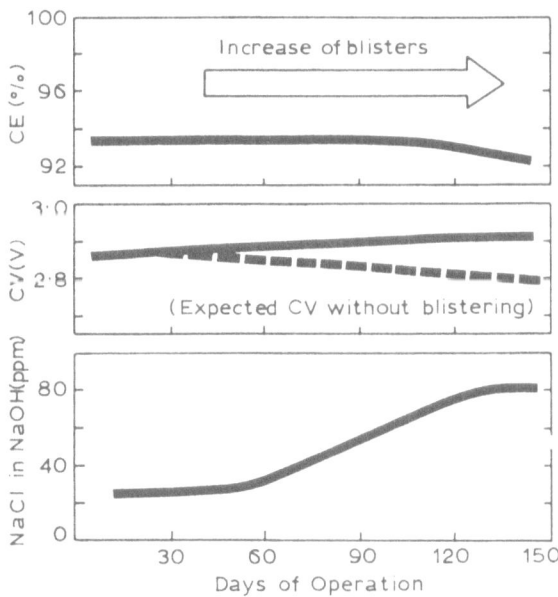

Fig. 10.   Effect of overdepletion of NaCl in anolyte. Membrane: Flemion 753. Cell: 1·5 dm². Laboratory conditions: 3 kA/m², 130 g/liter NaCl, 35% NaOH, 90°C. Long-term operation with overdepleted brine causes overswelling and blistering in a membrane.

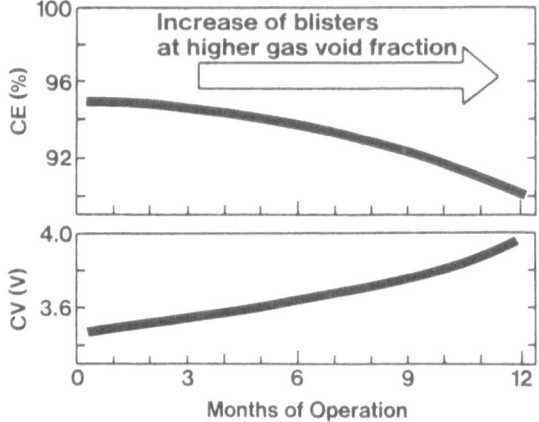

Fig. 11.   Effect of overconcentration of NaOH in anolyte. Membrane: Flemion 230. Cell: 1·5 dm². Laboratory conditions: 3 kA/m², 200 g/liter NaCl, 40% NaOH, 90°C. Long-term operation with overconcentrated caustic causes a gradual decrease of ion-exchange capacity on the surface of a membrane. A higher caustic concentration operation makes a membrane less resistant to the chlorine gas phase effect.

4.2 Undesirable Phenomena in Commercial Electrolyzers

Some examples of undesirable phenomena which can be seen in commercial electrolyzers are shown in Figs 12–17.

The effect of insufficient electrolyte flow in a cell chamber is shown in Fig. 12. If the design of an electrolyzer is poor or the electrolyte flow is insufficient, some dead space for flow is formed in the chamber. That dead space will cause stagnation of the electrolyte and gas pockets to form. The stagnation in the anode-side chamber causes an overdepletion of NaCl, and that in the cathode-side causes an overconcentration of NaOH. It is essential to provide uniform bulk concentrations in designing an electrolyzer.

Even if one can provide those uniform concentrations of bulk electrolytes, it is still required to pay attention to other possible local problems, too.

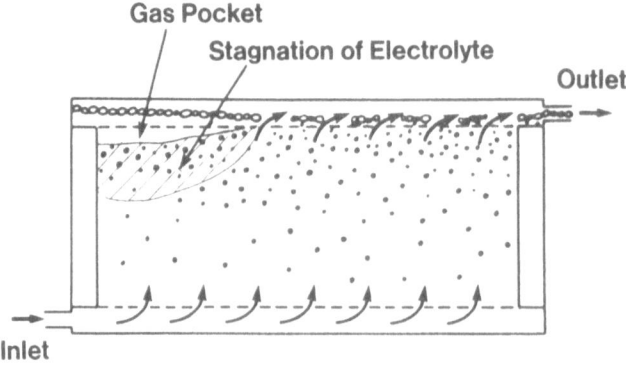

Fig. 12.   Effect of insufficient anolyte flow. An insufficient flow of electrolyte into an electrode chamber causes a gas pocket and/or stagnation of electrolyte in electrolysis area of a membrane.

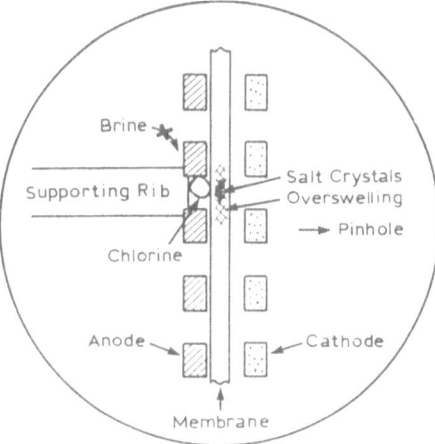

Fig. 13. Effect of dead anode openings. Blinding of opening area of anode mesh with its supporting rib causes membrane overswelling and gas pocket.

As is indicated in Fig. 13, when some openings of anode mesh are blocked by something, for example by an anode supporting rib, a dead space for anolyte is formed and causes a chlorine gas pocket and overdepletion of the anolyte. This condition results in salt crystallization and overswelling in the membrane locally, which then further results in a pinhole when conditions become severe.

A similar effect is observed when the width of the strand of anode is too wide, as illustrated in Fig. 14. The wide strand anode hinders the feeding of brine into the inner surface of the strand, which causes membrane overswelling.

There are several types of mechanical damage to the membrane that can occur and which we can observe in commercial-size electrolyzers. One of these is an abrasion of the membrane by the electrodes. It is normal to have pressure

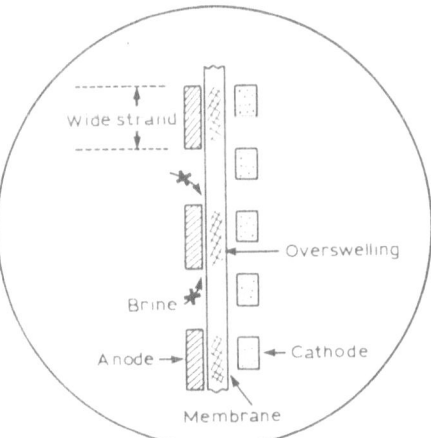

Fig. 14. Effect of wide anode strand. Wide strand anode hinders feeding of brine into the inner position of the anode, which causes membrane overswelling.

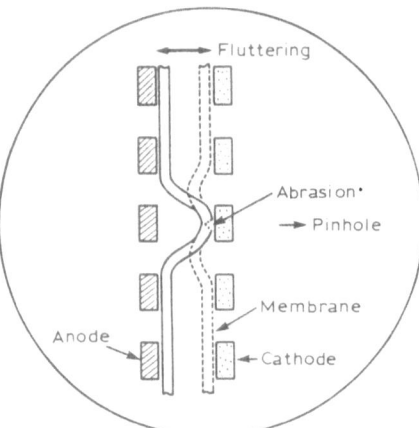

Fig. 15.   Abrasion of membrane with electrodes. Membrane fluttering between electrodes due to
pressure fluctuations causes abrasion of a membrane.

fluctuations caused by gas–liquid two-phase flow in the outlet pipes of the cell frames. When the differential pressure from the cathode-side is insufficient, the membrane flutters between the anode and the cathode. In addition, if there is a wrinkle in the membrane it accelerates the abrasion of the membrane at the wrinkle and results in pinholes, which is shown in Fig. 15.

Figure 16 shows effects of excessive compression and sharp edge. It may be obvious but excessive compression of anode and cathode surfaces, or the presence of sharp edges of electrodes, can generate pinholes in a membrane. It is important to select proper shape, hardness and compression ratio of gaskets for an electrolyzer. A soft gasket with overcompression can cause a crack in the membrane under the gasket area, as shown in Fig. 17. All of the factors mentioned here, along with other factors which are not mentioned in this chapter, have been considered and the solutions have been applied to all types of AZEC electrolyzers. These considerations

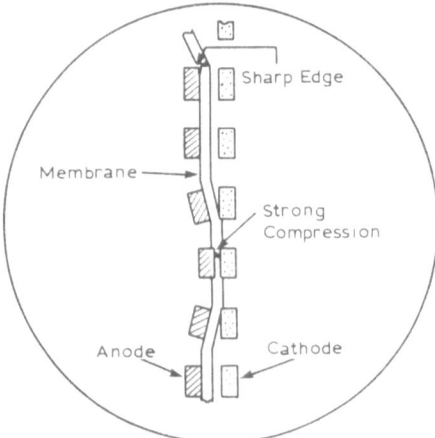

Fig. 16.   Effect of strong compression and sharp edge.

Before Compression

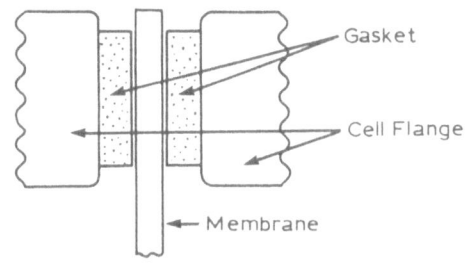

After Compression

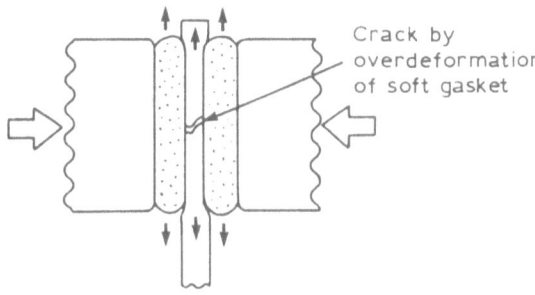

Fig. 17. Overcompression with gasket.

for the design of the electrolyzer achieve preferable conditions, not only for a standard membrane in order to prolong its service life with stable performance but also for future high functional membranes which may be more sensitive than standard ones at present.

5 FLEMION MEMBRANE FAMILY

Before the end of this chapter, our various Flemion membranes are introduced in the following section. Table 4 summarizes the Flemion membrane family. The

TABLE 4
Flemion membrane family

Number	Application
F-793, F-795	AZEC electrolyzers
F-855, F-865	Filter press type electrolyzers
F-811, F-854	Retrofitting diaphragm cells
F-733, F-737	KOH production
Experimentals	Low conc. NaOH ($\sim 23\%$)
	High conc. NaOH ($\sim 50\%$)
	High current density ($\sim 6\,kA/m^2$)
	Etc.

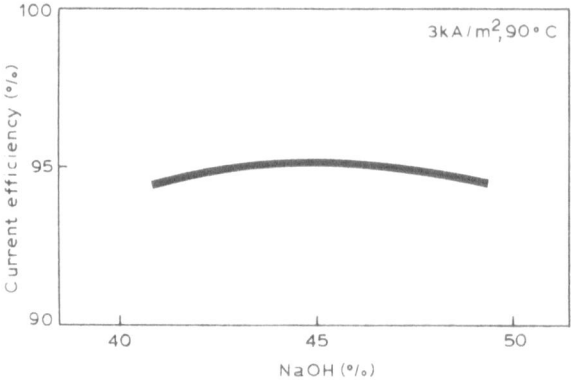

Fig. 18. Characteristics of high concentration caustic membrane (laboratory cell).

commercialized ones are the first four listed. Flemion 793 and 795 are the membranes which have a low ohmic drop and are mainly used in AZEC electrolyzers. Flemion 855 and 865 are mechanically stronger and are less sensitive to operational upsets, so that they are mainly used for general filter press type electrolyzers. Flemion 811 and 854 have high resistance to damage from folding and are mainly used in retrofitting diaphragm cells. Flemion 733 and 737 are designed for KOH production and exhibit high performance and excellent product purity.

In addition, we have some modification techniques on these membranes to meet customer needs. Meanwhile, we have continued to develop and improve our Flemion membranes. Some of these new developments are listed last in Table 4 as experimentals.

Figure 18 shows an example of the characteristics of the membrane developed for production of high concentration caustic soda. Although it was not easy to achieve

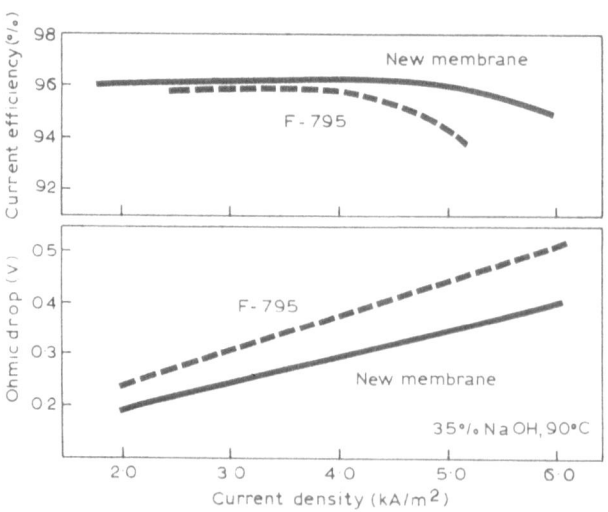

Fig. 19. New low voltage drop membrane (laboratory cell).

an acceptably high current efficiency with about 50% caustic soda, we did find a way to do it. This membrane has been operated in a laboratory cell for nearly 2 years with a stable performance.

Figure 19 shows the characteristics of our new low voltage membrane. In comparison with the lowest ohmic drop membrane at present, which is Flemion 795, the ohmic drop of the new membrane is 50 mV lower at 3 kA/m² current density and 100 mV lower at 5 kA/m². We are now evaluating this membrane in a few commercial-size electrolyzers.

6 CONCLUSION

As a major supplier of both the electrolyzer and the membrane, we believe we are able to fulfill our responsibility to continually supply better membranes along with the electrolyzer to meet the customer's needs.

12

BRINE, IMPURITIES, AND MEMBRANE CHLOR-ALKALI CELL PERFORMANCE

James T. Keating

El Du Pont de Nemours and Company, USA

and

Klaus-Jochen Behling

Du Pont de Nemours (Deutschland) GmbH, FRG

INTRODUCTION

For over 10 years Nafion* research has studied the effects of impurities on the performance of membrane chlor-alkali cells. The aim of this chapter is to summarize what we have learned about impurities: what impurities affect membranes, how impurities degrade performance, and approaches to dealing with impurities.

In the membrane chlor-alkali cell, the membrane is a separator, preventing mixing of the caustic and hydrogen with the brine and chlorine. The membrane also allows passage of sodium ion (or any cation) and water or any other uncharged species of molecular dimensions from the anode to the cathode compartment, but it resists the passage of anions, such as chloride and hydroxide (Fig. 1). The membrane is not porous. Work by Gierke and others [1, 2] shows a polymer with channels 1–5 nm in diameter, not many times larger than the water, uncharged species, or hydrated ions that pass through them (Fig. 2).

In a chlor-alkali plant operated at recommended conditions, membrane cells will perform well for 2 years and more. Membranes we have tested after extended service at 90°C, 32% caustic, 3 kA/m² current density show no deterioration in the physical properties, tensile or tear strengths. The stability of fluorocarbon polymers is so great that one can be confident that chemical degradation will not be a cause of membrane failure, or even of deteriorating performance.

Why should membranes ever suffer reduced performance or failure in a chlor-alkali cell, apart from mechanical damage such as tearing during gasket replacement?

* Du Pont's registered trademark for its perfluorinated membranes.

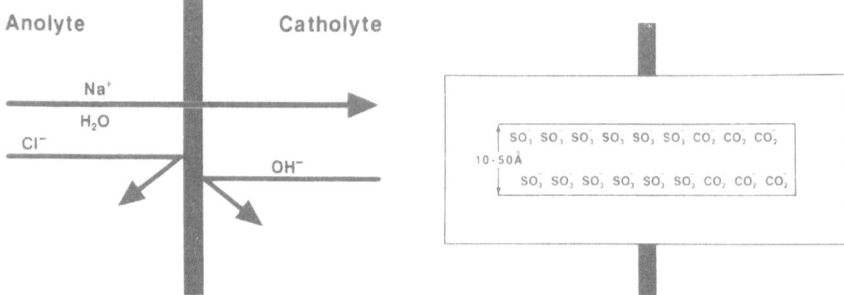

Fig. 1.   The membrane cell.   Fig. 2.   The membrane pore size.

A major source of performance decline is the accumulation of solid material in the membrane. Before going into detail on this subject, let us consider the magnitude of the load the membrane carries during its operating life.

Under typical operating conditions, 200–250 kg/day of sodium ions and water are transported by each square meter of membrane. A square meter of membrane typically weighs 400 g. Therefore, in the course of its normal 2 + year operating life, a membrane will have carried one half million times its weight (Fig. 3). Allowing for density differences, the membrane is carrying close to a million volumes of fluid over this time.

Impurity accumulation of 0·1–1% by weight is usually enough to cause significant loss of performance. Therefore, if the fluid passing through the membrane contains as little as 1:100 000 000 to 1:10 000 000 (10–100 ppb) of impurity that collects in the membrane, performance can suffer.

The rough calculations show why the required levels for many impurities in brine fall into the 10–100 ppb range, especially for cations, which the membrane cannot reject. Specific impurity levels are dependent upon membrane, cell, operating conditions, the impurity itself, and other impurities present.

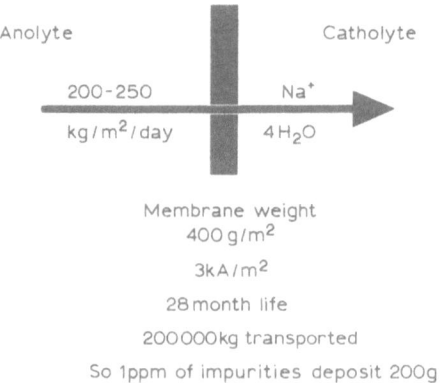

Fig. 3.   Membrane mass transport.

SOURCES OF CONTAMINANTS IN THE CHLOR-ALKALI CELL

Salt

The salt is the principal source of brine impurities. Multivalent cations and sulfate are common, but iodide and silica can occur too. Salts that are mined have impurities characteristic of their sources and are of relatively consistent quality. There are distinct geographical patterns to the distribution of some impurities [3, 4]. Solar salt from the sea is potentially of more uniform composition, but the actual composition depends on how the salt is harvested [3]. If cropping practices change, variations in salt quality (and specific contaminants) can change too.

Water

The water used to make the brine can bring with it impurities, and it must be treated before use to remove cations and excessive concentrations of silica.

Hardware

Materials of construction and cell components can introduce contaminants. To mention a few items here from the many possibilities:

- Gaskets must be chosen to avoid barium and magnesium, common fillers in elastomers.
- Titanium from the anode or anode chamber is rarely seen on the membrane and never, to our knowledge, associated with damage.
- PVC piping can be a source of lead, a common stabilizer in PVC. We have seen traces of it in membranes, but it does not seem to affect performance.

Process

Finally, there are materials in brine that are not so much contaminants as characteristic products of the process. Chlorate is one example. It is formed naturally during cell operation and can accumulate in the brine. It does not affect the membrane or the electrodes, but does affect caustic purity, and at high levels can depress the solubility of sodium chloride in the brine. Barium is sometimes introduced to control sulfate in the brine. Mercury is a third example: in a mercury plant the recycled brine contains mercury. If a membrane plant shares the brine system, it too will be exposed to mercury.

THE MEASURE OF MEMBRANE DAMAGE

The practical indicator of membrane damage is *performance*. Increase in power consumption is a signal that something has changed, though many things besides the membrane can influence power consumption. Usually a change is picked up first in one of the components of power consumption: voltage or current efficiency. This narrows the possible causes of the change, but is still a long way from a diagnosis of impurity as the cause.

The *appearance* of membrane that has been in chlor-alkali service is not a reliable indicator of damage. A membrane can absorb large amounts of certain impurities on the surface and even inside without suffering significant performance declines. Similarly, blistering can occur in certain areas with no influence on performance or membrane integrity.

On the other hand, impurities introduced during temporary loss of brine purity can precipitate in the membrane, causing damage, and then redissolve when brine purity is restored. Membrane analysis will show the effects but it will be too late to identify the cause. Membranes can also sustain physical damage without impurity involvement. There are instances too of poor performance with no apparent physical change.

Some of these effects are outside the scope of this chapter and are mentioned only to emphasize the importance of monitoring performance. In any case, analysis of the membrane can only be done after the cell is dismantled. If problems are to be detected while there is time to act, we must follow cell performance closely.

VOLTAGE INCREASE

Though many impurities are associated with current efficiency decline, voltage can also be affected [5, 6]. However, increases in cell voltage are not easy to interpret because of the number of factors that contribute to voltage.

Thermodynamic, overvoltage, and ohmic components of cell voltage can be analyzed:

- The thermodynamic component of voltage is a function of concentrations (activities), temperatures, and pressures of the reactants and products of the chlor-alkali cell. These include sodium chloride, sodium hydroxide, chlorine, hydrogen, and sodium hypochlorite.
- The overvoltage component depends on the composition of the anode and cathode, any coatings or treatments they have to reduce overvoltage, as well as current density.
- The ohmic component includes the resistance of the anolyte, catholyte, and membrane, and whatever 'hardware' is between the points of voltage measurement and current density.

In principle the contribution of each component can be measured because each responds differently to changes in current density. The thermodynamic component is independent of current, overvoltage is a function of the logarithm of the current (the Tafel relation), and ohmic resistance is a direct function of current (Ohm's law). This approach has been discussed in the literature [7, 8] and will not be described further here.

Anode Voltage Increase

The anode is susceptible to change or damage from brine impurities and the result will be increasing voltage. Materials that simply coat the anode may raise the voltage. Barium sulfate can do this. Organic impurities in the brine also seem to be able to raise the voltage. The variety of organic contaminants in brine has so far prevented precise identification of the offending organics and the tendency in the industry is to set the limit for total organics at a few parts per million.

Fluoride in brine causes no problems for the membrane that we have seen either in laboratory tests or in reports from the field. Fluoride can, however, affect the anode coating.

High pH will also damage anodes, even to the point of dissolving the titanium metal. This is usually due to holes in the membrane through which catholyte can flow, driven by the higher catholyte pressure. Anode damage can also be a secondary effect of poor current efficiency which allows hydroxyl ion migration to the anolyte.

Other problems with organic impurities are:

(1) Foaming—if the organic impurity promotes foaming this will reduce the anolyte concentration, increasing resistance and raising voltage. The Bruggeman equation shows the dependence of voltage on a power of anolyte density [9].

(2) Acidity—if the organic material is rich in hydrogen, oxidation by chlorine will produce hydrochloric acid and can affect the pH of the anolyte.

Cathode Voltage Increase

Mercury

Mercury can be a brine contaminant in chlor-alkali plants that combine mercury technology with membrane technology. If both sides of the plant share a common brine system, mercury can be carried into the anolyte feed to the membrane side. Typical ion exchange resins do not remove mercury well, presumably because it exists as the anionic tetrachloride complex. The membrane is not affected by mercury in the 10–20 ppm range according to our tests, but voltage does increase (Fig. 4).

Some of the mercury passes through the membrane and deposits on the cathode. The high hydrogen overvoltage of mercury is well known and has its effect here, increasing the cell voltage. We have some evidence that the effect is at least partially

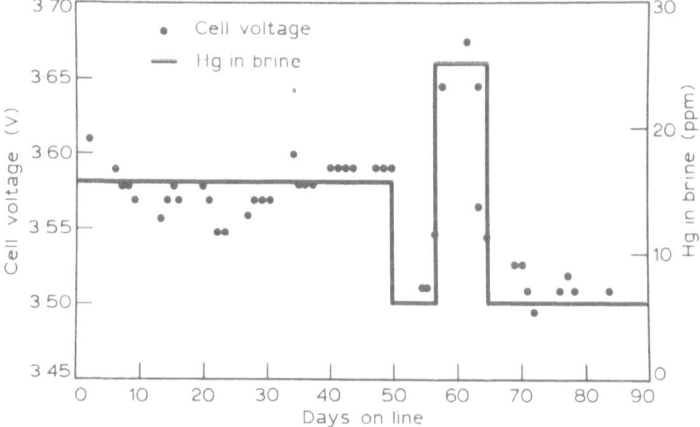

Fig. 4.   Mercury in brine: reversible effect on cell voltage.

reversible. If the mercury to the anolyte is stopped, the cell voltage may come down depending upon the cathode material [10].

Iron

Iron in the caustic (iron in the brine is discussed below) can affect voltage if the cathode is activated with a precious metal coating or finely divided nickel (Raney Nickel). As with mercury, overplating is the mechanism, and the cathode behaves as if it were iron with consequent higher overvoltage. Sources of iron can be cell components, especially where nickel-plated iron is used rather than an all-nickel construction. Iron outside the cell, in the caustic loop, can also cause trouble.

Oxidation

Certain precious metal-coated cathodes are susceptible to damage by hypochlorite during shutdowns, and also to the 'reverse current' that can arise when a cell is 'jumpered out' and becomes galvanic. When ruthenium is used in the previous metal formulation, reverse current can lead to caustic colored orange by the ruthenium ion. As the precious metal coating is destroyed the cathode overvoltage rises to that of the substrate metal, usually nickel [11–14]:

$$Ru^0 + 8OH^- \rightarrow RuO_4^{2-} + 4H_2O + 6e^-$$

Membrane Voltage Increase

Nickel

Nickel is one of the commonest contaminants that can cause an increase in membrane cell voltage. It rarely comes from the brine. Ion exchange resins are very efficient in removing nickel from brine and if the brine impurities such as calcium are under control, nickel will be no problem. Where we have seen nickel in anolyte, the cause has been a nickel fitting installed downstream from the brine treatment.

Though nickel in anolyte is rare, nickel in membrane is not. The principal source is nickel from the cathode. The most likely mechanism is attack on the cathode by

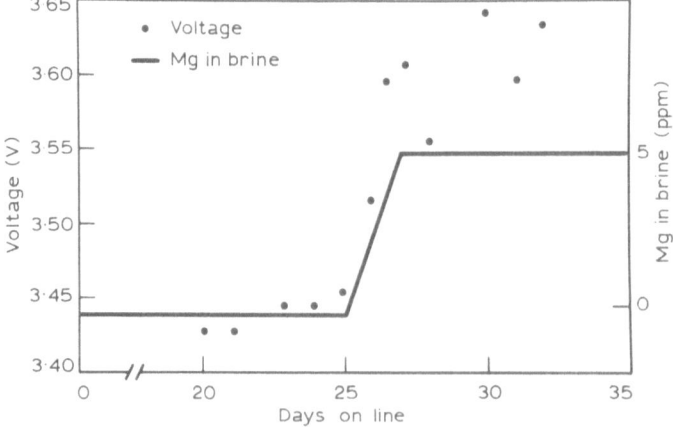

Fig. 5.   Magnesium in brine: effect on cell voltage.

hypochlorite diffusing through the membrane during a cell shutdown. The membrane absorbs the nickel ion. Nickel can sometimes be seen as black spots, or a black or gray cast on the membrane. The black is black nickel oxide, reported to be NiO(OH) [15, 16]. This is not stable and the black color can fade with time [17, 18].

Simple qualitative tests for absorbed nickel include bleaching of the black color with sulfite solution, or darkening of the color with hypochlorite.

Nickel from the cathode requires of course a nickel cathode, and is more prevalent in zero-gap cell configurations, as would be expected from the proposed mechanism.

Magnesium

Magnesium is not normally a problem in anolyte because it is removed efficiently by the ion exchange resin in the secondary brine treatment. However, when its tolerable level in anolyte is exceeded magnesium causes membrane voltage to rise (Fig. 5). It is unlike calcium, which has its effect on the current efficiency. (However, with perfluorosulfonate membranes, magnesium affects only current efficiency, and calcium affects both voltage and current efficiency [19]. At high concentrations both calcium and magnesium will raise the voltages of perfluorocarboxylate membranes [20].)

CURRENT EFFICIENCY DECLINE

Unlike voltage escalation, which can have many causes, most not associated with the membrane, current efficiency declines are almost always membrane-related. Leaks, punctures and tears will of course affect current efficiency, but these are mechanical problems, relatively simple if not easy to correct, and outside the scope of this chapter.

Impurities cause current efficiency decline by reducing the membrane's ability to reject anions, specifically to prevent the hydroxyl ion from moving from the cathode compartment through the membrane to the anode compartment. This decrease in anion rejection is usually the result of physical damage to the membrane. This damage is not a loss of membrane integrity: the membrane still prevents mixing of chlorine and hydrogen—the electrolyzer can still operate safely.

The effect is not subtle. A membrane seriously damaged by impurities can be seen to be changed under low magnification. The mechanism is the precipitation of the impurity within the membrane and the destruction of the anion rejection capacity [19].

The primary cause of the problem is not a mystery. The environment in the membrane changes from an acidic (pH 2–4) 5 mol salt solution to a 10 mol caustic solution over the 150–250 μm thickness of the membrane. Many ions that are soluble in the acidic, more dilute anolyte will become insoluble in the more concentrated caustic.

Since the question is simply whether the impurity will remain soluble during its passage through the membrane, and since the line between solubility and precipitation is sharp, the safe limit for a single impurity can be clearly defined from laboratory tests. For example, in Nafion 50 ppb calcium in brine is the maximum concentration that can be tolerated.

Cations

The periodic table is filled with elements that form cations, but most do not occur in brine. Of those that do, some are so soluble in both acid and base as to pose no problem. Potassium is an example: membranes work as well with potassium chloride as with sodium chloride. At the other extreme are elements whose cations are so insoluble that they cannot penetrate the anode face of the membrane. Iron is the prominent example, precipitating at pH 2–3.

Iron

Iron is a common impurity in plants that receive salt in bulk. Potassium ferrocyanide is added to the salt in part-per-million quantities to prevent caking. The iron-carrying ferrocyanide, being an anion, is not removed by the ion exchange resin and passes into the anolyte where the cyanide ligands are oxidized, releasing the iron to the acidic brine. However, the ferric ion is soluble only below pH 2–3, and is precipitated at the anode face of the membrane and on the surfaces of the anode compartment as ferric oxide. We have seen membrane brown with iron oxide that would wipe off easily from the anode side, and that had had no effect on membrane performance.

The example of iron shows that it is not enough for an impurity to be present in sufficient quantity; it must also be soluble enough to make its way into the membrane and precipitate there to have a chance at affecting performance.

Calcium

Calcium is the best known of the cationic impurities because it is the ion that is usually keyed on in monitoring the condition of the ion exchange beds. Magnesium is held more firmly than calcium by most ion exchange resins. Barium and strontium are held less strongly, but can be tolerated at higher concentrations because of the higher solubility of their hydroxides.

Aluminum

Aluminum is the remaining common cation that can threaten membrane performance. The source of aluminum is usually clay (aluminosilicates) associated with the salt. If the clay is not filtered out, it can dissolve in the alkaline brine, passing through the ion exchange beds as suspended clay or as aluminate anion and into the acidic anolyte chamber where the aluminum cation can be released. Alone, or even in the presence of other cations, aluminum is amphoteric and generally no problem. It enters the membrane as a highly soluble cation and is transformed to an equally soluble anion. It is only when silica is also present that aluminum is dangerous. We will discuss this further when we come to silica impurity.

Anions
Sulfate

Sulfate is the principal anionic impurity in brine, usually present in gram-per-liter quantities. O'Brien has discussed the control of sulfate at the 1985 SCI Symposium

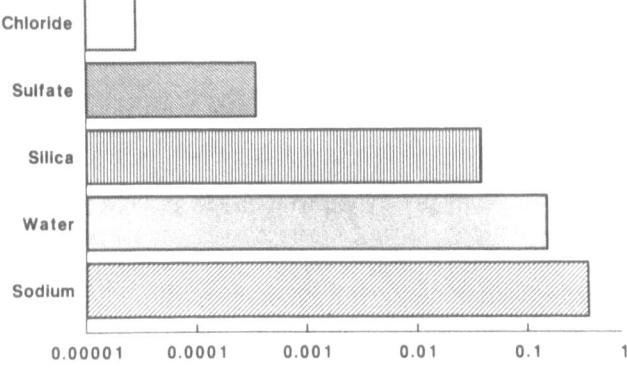

Fraction Passing Through Membrane

Fig. 6. Relative transport of brine species.

and it is enough to say here that the allowable level in brine is a balance of cost of removal and tolerance for sulfate damage [3].

Sulfate can be tolerated in parts-per-thousand concentration as opposed to the parts-per-billion. The reason for this great difference is the rejection of the anion by the membrane and the resistance offered by the electric potential field to anion progress toward the cathode. As an illustration of this effect, half the cations in the brine will go into the membrane during any passage of brine through the anode compartment, but only about one part-per-thousand sulfate goes in [21, 22] (Fig. 6). This rejection, and the substantial solubility of the sulfate salt even at high pH and caustic concentration, result in the high acceptable concentration for sulfate in brine.

There are several unusual aspects of sulfate deposition in a membrane, including evidence for its anionic rather than ion-paired character and its reflection of current distribution, that have been reported elsewhere [23–25]. The phenomenon of co-ion fractionation has been described by Bissot [26] and accounts for the *increased* transport of sulfate as current density and membrane thickness are increased. This behavior is interesting in itself and also has practical consequences for efficient operation of membrane cells in brine that contains sulfate. This will be discussed later.

Halides

The other significant anionic impurities are the halides. Fluoride causes no problems even at the 50 ppm level in our tests, though precipitation of iron fluoride will occur in the anolyte compartment if iron is present and, as mentioned above, fluoride can damage anode coatings.

Bromides do not affect performance either. In our tests 5 ppm bromide had no effect on current efficiency or voltage. Nor have we seen bromide in membrane returned from any of the plants that use Nafion around the world.

Iodide is another story entirely. Iodide can damage the membrane and reduce current efficiency by precipitation. It reacts in the chlor-alkali cell to form a compound that till now has been mainly a curiosity, lurking in obscure paragraphs of inorganic chemistry texts—the insoluble sodium salt: sodium paraperiodate.

$$Na_3H_2IO_6$$
Sodium Paraperiodate

$$I^- + 4Cl_2 + 4H_2O \rightarrow IO_4^- + 8Cl^- + 8H^+$$

Iodide is oxidized to periodate by chlorine in the anolyte or at the anode [27]. Periodate is soluble enough in water, but the least rise in pH is enough to convert it to paraperiodate, which precipitates. In its progress through the membrane toward higher pH the iodate from the anolyte becomes paraperiodate, the solubility of the sodium salt decreasing with increasing basicity [28].

Potassium periodate is different: it becomes more soluble as pH increases (Fig. 7). We can have some confidence from these data that iodide in potassium chloride membrane electrolysis will not cause current efficiency problems.

Iodide in salt varies with the source. There is no well-tested method for removal of iodide in brine. As a result, when a salt source is found to have iodide in quantities above typical warranty limits (400 ppb), thought must be given to its use in membrane cells. One option may be operation at lower caustic strengths, where sodium paraperiodate solubility may be greater and therefore tolerance of iodide higher.

Analysis for iodide has been greatly improved. Until very recently there was no good colorimetric method for analyzing iodide and its higher oxidation states (iodate, periodate) in brine. Polarography was the only satisfactory technique, but polarography is ill-suited to routine plant use. However, a colorimetric method is now available [29].

Nonionic Impurities
Silica

Silica is the prime example of a nonionic impurity. When it occurs in the salt, there is no practical way to remove it. When silica occurs in the water, its concentration can be reduced with ion exchange resins. The use of sand filters for the removal of suspended solids in brine, and storage of salt on concrete pads can also introduce silica into the brine.

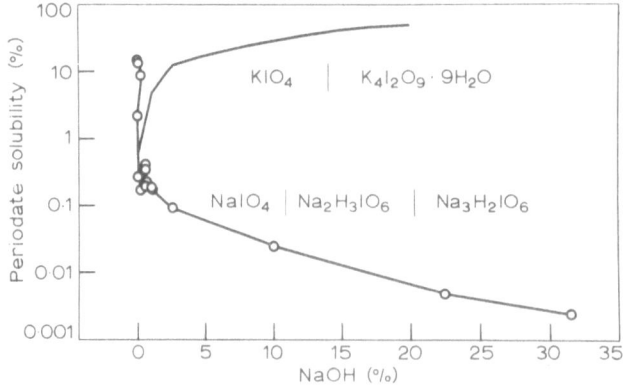

Fig. 7.   Solubility of sodium and potassium periodates in water and caustic at 25°C.

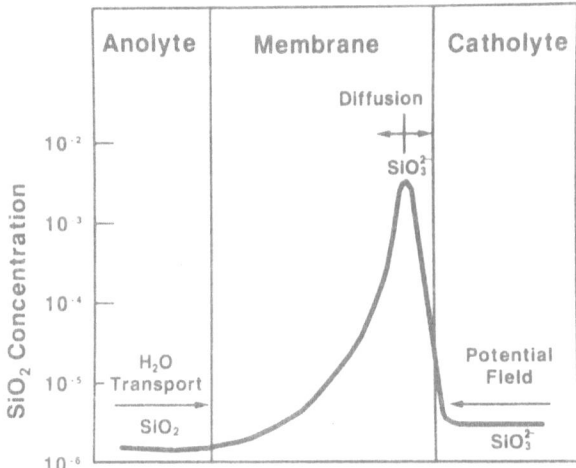

Fig. 8. SiO_2 concentration in the anolyte, membrane and catholyte.

Silica by itself causes no difficulty in the membrane chlor-alkali cell. It is more soluble in catholyte than in anolyte. The mechanism of silica transport through the membrane has been proposed by Bissot [21, 22].

Silicic acid in the brine, un-ionized, is swept into the membrane by convection toward the cathode which is driven by the motion of sodium ions carrying current [30]. As the pH rises the silicic acid ionizes, perhaps in several stages, to silicate. As the negative charge increases the progress of the ion is retarded by the potential field. The end result is the accumulation of silicate within the membrane. The concentration increases until the influx of silica is balanced by the diffusion of silicate out (Fig. 8).

This reservoir of silicate within the membrane can act as a trap for other impurities. Calcium is an example. Typically calcium above a critical concentration will precipitate as calcium hydroxide, since hydroxide is the principal anion in the membrane near the catholyte. However, if silicate is concentrated there also, calcium silicate will precipitate. Silicate sensitizes the membrane to calcium.

Silica also sensitizes the membrane to aluminum. The chemistry is interesting. Like silica, aluminum becomes negative (aluminate) as it passes through the membrane, and is retarded by the potential field, and concentration builds up in the membrane to the point at which influx and diffusion balance. With silicate, aluminate and sodium hydroxide present, this hot aqueous environment becomes a 'zeolite reactor' for it is under similar conditions that synthetic sodium aluminosilicates are produced. Bissot has found sodalite and faujacite crystals in membrane damaged by brine containing silica and alumina [21].

PRACTICAL APPLICATIONS OF IMPURITY RESEARCH

In the early stages of our research into the identity of brine impurities and their action on membranes, the recommendations we could make were largely negative: limits on impurities were established and brine had to be brought within these limits through purification, usually ion exchange and purging of brine. Impurity limits are still necessary, but our understanding of the mechanisms of some forms of impurity damage has given us other methods for dealing with brine impurities. Increased purification is no longer the only remedy. Where a choice of approaches is available, the customer, membrane supplier, and hardware supplier can work together to select the best alternative.

Sulfate

A simple case of this type is sulfate. This can be a troublesome impurity both in its initial reduction to tolerable levels, and the maintenance of those levels. Because only a fraction of a percent of sulfate passes through the membrane and because it is not removed by the ion exchange columns, sulfate accumulates in the brine.

Bissot's work on the transport of sulfate through membrane not only revealed the mechanism of co-ion fractionation, and the dependence of sulfate in the membrane on current density, membrane thickness, and sulfate concentration in brine. He also showed how to increase membrane tolerance for sulfate. By selecting the proper membrane and operating conditions, sulfate tolerance is improved. The selection has been put on a quantitative basis [31]:

$K =$ membrane thickness × sulfate concentration × current density
If $K < 5000$, current efficiency decline $< 2\%$ in 2 years

As the equation shows, there are several controllable factors that can be manipulated to increase sulfate tolerance of the membrane. Thinner membranes are preferred. Current density is also a critical variable, and again the choice is between the cost of further brine purification and the sacrifice in productivity from operation at lower current density.

Silica

A more complex case is the control of silica. As stated above, silica itself is not a purity problem in membranes. Only in the presence of certain cations such as calcium and aluminum can precipitates form and damage be done to the membrane. The concept of the solubility product has been adapted to deal with the problem of silica and to give several methods of control [32].

In a saturated solution of an ionic compound the product of the concentrations of the ions will be a constant at any temperature. For example, in a saturated solution of compound AB, the concentrations of A^+ and B^- ions are constant:

$$K_{sp} = [A^+][B^-]$$

In the membrane precipitation will occur when the concentrations of silicate ion and the multivalent cation exceed the solubility limit. Applying the concept of solubility product constant to this case, we can define a 'membrane' solubility product, K'_{sp}. For silica and aluminum

$$K'_{sp} \sim [SiO_3^{2-}][Al^{3+}]$$

We can estimate the concentration of silicate in the membrane from the silica found in the caustic. Like sulfate, though for different reasons, the transport of silica

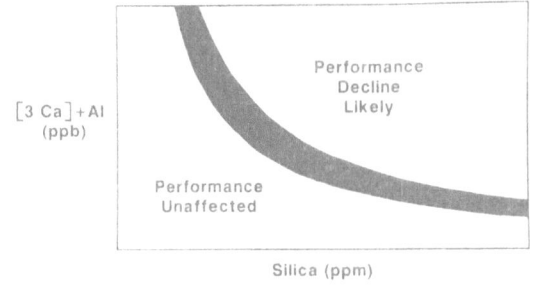

Fig. 9. Membrane operation in brines containing silica, aluminum and calcium.

through the membrane is a function of current density (CD) and membrane thickness (T), and of course silica concentration in the brine (SiO_2):

$$[SiO_2/50\% \ NaOH] = k*[SiO_2]^{0.5}[CD]^{0.75}[T]^{0.5}$$

The current density as used here is *effective* current density and is determined by the membrane reinforcement, which masks a fraction of the membrane: less reinforced membranes are better in this application.

As a practical matter, we have found that the concentrations of aluminum and calcium ion can be combined, with a weighting factor for the greater sensitivity to calcium. The resulting K_{sp}' equation is

$$X = [SiO_2/50\% \ NaOH][3\,Ca + Al]$$

From experiment we can estimate the K_{sp}'. Therefore we know that when the product X is less than a certain value the membrane will not be damaged by silicate precipitation. If X is greater, there will be damage, a decline in performance, and there is a borderline region.

Figure 9 shows the relation of silica and aluminum-plus-calcium [33, 34]. This emphasizes the number of variables that can be adjusted to find optimum operating conditions. For instance, where there is a silica problem, calcium levels can be held to 20 ppb, rather than the normally allowable 50 ppb, to increase silica tolerance. Normal cation exchange resins can easily achieve this level. Similarly, if aluminum is a problem, one can consider reducing calcium or silica concentration. If silica is chosen, the membrane or the current density can be selected to reduce silica concentration, or further brine treatment for silica may be possible.

From the examples given here it is plain that there are many combinations we can consider when facing problems with brine purity. Further purification is only one of these, and in some cases it will not be the preferred method.

SUMMARY

This chapter has reviewed the relation of brine and impurities to membrane damage and chlor-alkali cell performance. The years of research have yielded a picture that is complex but no longer confusing. There are still obstacles to be overcome, but there should be fewer surprises and with our understanding are coming alternatives to the former single prescription: purify the brine.

ACKNOWLEDGEMENTS

Much of the experimental work cited here was done at Du Pont. However, without the close cooperation of our customers in the chlor-alkali industry and the many cell manufacturers with which we work, our understanding could not have progressed so far. Their work with us is confidential and is not of course cited here, but we owe a great deal to them for their careful studies and generous sharing of information. In 1979 Dan Maloney of the NAFION group encouraged our colleagues in the industry to publish data on membrane performance [35]. This is still a sound recommendation.

The Du Pont work was done at the Experimental Station Laboratory (ESL) in Wilmington, Delaware, and at the Customer Service Laboratory (CSL) at Fayetteville, North Carolina. We cannot name all those who have contributed over the years, but a brief list of some of those most involved in our recent research includes D. E. Maloney, S. Keimatsu, T. C. Bissot, D. L. Liczwek, D. P. Crowley, M. K. Gordon, T. N. Norton, J. C. Gaval and R. H. Ruesswick.

REFERENCES

1. Gierke, T. D., Munn, G. E. & Wilson, F. C., in *Perfluorinated Ionomer Membranes*, Chap. 10, ed. A. Eisenberg & H. L. Yeager. ACS Symposium Series #180, American Chemical Society, Washington, DC, 1982.
2. Gierke, T. D. & Hsu, W. Y., in *Perfluorinated Ionomer Membranes*, Chap. 13, ed. A. Eisenberg & H. L. Yeager. ACS Symposium Series #180, American Chemical Society, Washington, DC, 1982.
3. O'Brien, T. F., in *Modern Chlor-alkali Technology*, Vol. 3, Chap. 23, ed. K. Wall. Ellis Horwood, Chichester, 1986.
4. Lefond, S. J., *Handbook of World Salt Resources*. Plenum Press, New York, 1969.
5. Hine, F., Tilak, B. V. & Viswanathan, K., in *Modern Aspects of Electrochemistry*, No. 18, Chap. 5, ed. R. E. White, J. O'M. Bockris & B. E. Conway. Plenum Press, New York, 1986.
6. Hine, F., Ogata, Y., Kojima, T., Uchiyama, S. & Yasuda, M., in *Proceedings of the Symposium on Electrochemical Engineering in the Chlor-Alkali and Chlorate Industries*, Vol. 88-2, ed. F. Hine, W. B. Darlington, R. E. White & R. D. Varjian. The Electrochemical Society, Philadelphia, 1988, p. 298.
7. Bergner, D., *Chem.-Ztg*, **109** (1985) 177. (Also see this volume, pp. 159–70.)
8. Hayes, M., Kuhn, A. T. & Patefield, W., *J. Power Sources*, **2** (1977/8) 121.
9. MacMullin, R. B., in *Chlorine*, Chap. 6, ed. J. S. Sconce. ACS Monograph Series #154, American Chemical Society, Washington, DC, 1962, p. 164.
10. Yamaguchi, K., Ichisaka, T. & Kumagai, I., presented at the Chlorine Institute's 29th Plant Operations Seminar, Tampa, FL, February 5, 1986.
11. Nidola, A. & Schira, R., *Extended Abstracts*, Vol. 85-1, 167th Society Meeting of the Electrochemical Society, Toronto. The Electrochemical Society, Philadelphia, 1985, p. 612.

12. Grove, D. E., in *Modern Chlor-Alkali Technology*, Vol. 3, Chap. 19, ed. K. Wall. Ellis Horwood, Chichester, 1986.
13. Burke, L. D. & Whelan, D. P., *J. Electroanal. Chem.*, **103** (1979) 179.
14. Burke, L. D., in *Electrodes of Conductive Metallic Oxides*, Part A, ed. S. Trasatti. Elsevier, Amsterdam, 1980, pp. 141, 158.
15. Falk, S. V. & Salkind, A. J., *Alkaline Storage Batteries.* John Wiley, New York, 1969, p. 55.
16. Cotton, F. A. & Wilkinson, G., *Advanced Inorganic Chemistry*, 4th edn. John Wiley, New York, 1980, p. 795.
17. Milner, C. P. & Thomas, U. B., The nickel–cadmium cell. In *Advances in Electrochemistry and Electrochemical Engineering*, ed. C. B. Tobias. John Wiley, New York, 1967, p. 1.
18. Briggs, G. W. D., The nickel hydroxide and related electrodes. In *Electrochemistry*, Vol. 4, Chap. 3. The Chemical Society, London, 1974.
19. Molnar, C. J. & Dorio, M. M., 152nd Society Meeting of the Electrochemical Society, Atlanta, October 1977.
20. Ogata, Y., Kojima, T., Uchiyama, S., Yasuda, M. & Hine, F., *Extended Abstracts*, Vol. 87-2, 172nd Society Meeting of the Electrochemical Society, Honolulu, HI. The Electrochemical Society, Philadelphia, 1987, p. 1651.
21. Bissot, T. C., *Extended Abstracts*, Vol. 86-1, 169th Society Meeting of the Electrochemical Society, Boston. The Electrochemical Society, Philadelphia, 1986, p. 643.
22. Bissot, T. C., *Proceedings of the Symposium on Diaphragms, Separators, and Ion-Exchange Membranes*, Vol. 86-13, ed. J. W. Van Zee, R. E. White, K. Kinoshita & H. S. Burney. The Electrochemical Society, Philadelphia, 1986, p. 42.
23. Keating, J. T., in *Proceedings of the Symposium on Electrochemical Engineering in the Chlor-Alkali and Chlorate Industries*, Vol. 88-2, ed. F. Hine, W. B. Darlington, R. E. White & R. D. Varjian. The Electrochemical Society, Philadelphia, 1988, p. 311.
24. Keating, J. T., *Extended Abstracts*, Vol. 87-2, 172nd Society Meeting of the Electrochemical Society, Honolulu, HI. The Electrochemical Society, Philadelphia, 1987, p. 1653.
25. Herrera, A. & Yeager, H. L., *J. Electrochem. Soc.*, **134** (1987) 2445.
26. Bissot, T. C., *Extended Abstracts*, Vol. 85-1, 167th Society Meeting of the Electrochemical Society, Toronto. The Electrochemical Society, Philadelphia, 1985, p. 594.
27. Ibl, N. & Vogt, H., in *Comprehensive Treatise of Electrochemistry*, Vol. 2, ed. J. O'M. Bockris, B. E. Conway, E. Yeager & R. E. White. Plenum Press, New York, 1981, p. 220.
28. Hill, A. E., *J. Am. Chem. Soc.*, **50** (1928) 2678.
29. *Handbook of Brine Analysis*, 2nd edn. Hach, Loveland, Colorado, 1988, p. 29.
30. Helfferich, F., *Ion Exchange.* McGraw-Hill, New York, 1962, p. 328.
31. Bissot, T. C., US Patent 4 722 772, Process for Electrolysis of Sulfate-Containing Brine, February 1988.
32. Bissot, T. C., US Patent 4 648 949, Process for Electrolysis of Silica-Containing Brine, March 1987.
33. Liczwek, D. L., *Proceedings of the Symposium on Electrochemical Engineering in the Chlor-Alkali and Chlorate Industries*, Vol. 88-2, ed. F. Hine, W. B. Darlington, R. E. White & R. D. Varjian. The Electrochemical Society, Philadelphia, 1988, p. 284.
34. Liczwek, D. L., *Extended Abstracts*, Vol. 87-2, 172nd Society Meeting of the Electrochemical Society, Honolulu, HI. The Electrochemical Society, Philadelphia, 1987, p. 1648.
35. Maloney, D. E., in *Modern Chlor-Alkali Technology*, Vol. 1, Chap. 14, ed. M. O. Coulter. Ellis Horwood, Chichester, 1980.

13

DESIGN, INSTALLATION AND OPERATION OF ION EXCHANGE MEMBRANE CHLOR-ALKALI PROCESSES

Yoshiaki Tominaga, Tsutomu Kanke, Kazuo Takagi and Takaya Miyazaki

Asahi Chemical Industry Co. Ltd, Japan

1 INTRODUCTION

The official program for the demercurization of chlor-alkali plants in Japan was mandated in 1973 and completed in 1986. The program was actually executed in two phases. In the first phase, lasting until 1976, conversion progressed as originally scheduled. The second phase was carried out under a revised schedule, to allow time for establishment of the ion exchange membrane process technology which had emerged during the first phase of conversion [1].

With the establishment of the process technology and the recognition of its superior performance and reliability, the ion exchange membrane process was adopted for all of the second phase conversion, and some facilities initially converted to the asbestos diaphragm process earlier in the program were subsequently reconverted to this process [2]. As a result, its share of the total chlor-alkali plant capacity in Japan at present is 80%, including retrofitted process plants, with the remaining 20% held by the asbestos diaphragm process.

The advantages of the ion exchange membrane process for chlor-alkali production are now widely recognized, and the total production capacity of membrane process plants worldwide is now nearly 5 million tonnes, as indicated in Table 1.

Asahi Chemical played a central role in the conversion program, as a pioneer in the development and commercialization of the membrane process. Its Nobeoka plant both proved the technology and served as a model throughout the conversion program. It was the site of the world's first commercial-scale membrane process unit, which came on stream in 1975 with an annual production capacity of 40 000 t.

The Asahi Chemical process technology has since been licensed worldwide. To date it has been adopted for a total of 22 plants with an annual capacity of about

TABLE 1

Regional production capacity of ion exchange membrane process plants (1987)

Region	Annual plant capacity (000 t NaOH)	Share of total capacity (%)
Japan	2900	75
North America	650	5
West Europe	630	5
China	100	4
Korea and Taiwan	150	21
Other Asia	160	8
North and South America	30	2
Oceania	30	15
Africa	90	17
Total	4740	11

1·4 million t of caustic soda. Nineteen plants have been in operation for periods ranging to 10 years or more, and three are currently under construction.

To meet the broad range of technical requirements and restraints represented by these plants, Asahi Chemical has constantly developed modifications and refinements related to the process design, installation, operation of the electrolyzers and their peripherals, while maintaining the integrity of the basic design concepts. The resulting flexibility and adaptability to many different situations is described here, particularly with respect to electrolyzer power feeds and utilization of existing cell buildings.

2 DEVELOPMENT OF ASAHI CHEMICAL'S MEMBRANE PROCESS

Asahi Chemical's ion exchange membrane process employs bipolar electrolyzers and forced circulation of anolyte and catholyte [3]. With the forced circulation, the bipolar electrolyzer developed by Asahi Chemical is particularly characterized by ease of operation and maintenance.

A bipolar electrolyzer design was included from an early stage of the process development at Asahi Chemical, because of the promised ease of operation and maintenance and its inherent capability for high current densities ($4·0$–$5·0 \, kA/m^2$). Relatively low caustic soda concentrations were employed in early operations, but were gradually increased, with the development and installation of later generation electrolyzers in subsequent plant conversions, to a current level of up to 35%. The first operational electrolyzers were composed of $1·2 \times 1·2 \, m$ cells. Standard cell sizes of $1·2 \times 2·4 \, m$ and then $1·5 \times 3·6 \, m$ were subsequently developed and installed. At present each electrolyzer using the $1·5 \times 3·6 \, m$ cells has an annual production capacity of 20 000 t [4].

A development by Asahi Chemical which proved definitive in establishing the superiority of the membrane process was its invention of the carboxylic acid membrane, produced by chemical modification of one side of a sulfonic acid membrane to obtain a region of carboxylic acid groups. This led to the attainment of

TABLE 2
Early and recent performance of membrane process

	1975	*Present*
Cell caustic conc. (%)	17	30–35
Current efficiency (%)	80	95
Current density (kA/m^2)	5·0	4·0
Electrolysis power, d.c. (kWh/t NaOH)	3 500	2 150–2 230

high current efficiency, of more than 95%, with no increase in electrical resistance. Membranes of this type were first used commercially at the Nobeoka plant, where they replaced sulfonic acid membranes, the only type commercially available until then.

Other developments which have greatly enhanced energy efficiency have included heightened membrane performance, anode optimization and activated cathodes. Typical electrolysis performances in commercial operation of the early and the most recent ion exchange membrane process units are indicated in Table 2.

3 CONNECTION OF ELECTROLYZERS AND RECTIFIERS

3.1 Nobeoka Plant

The present configuration of the electrolyzer power supply at the Nobeoka plant is shown in Fig. 1. Early conversion involved operation in parallel with existing mercury cells, with a new transformer–rectifier array installed to power the 1·2 × 1·2 m and 1·2 × 2·4 m electrolyzers. As shown on the left, the array now includes four thyristor-type rectifiers.

The system on the right, consisting of three 1·5 × 3·6 m electrolyzers fed in parallel by two diode-type rectifiers previously used for the mercury cells, was installed in the final stage of the conversion.

Each electrolyzer can be started up and shut down independently through control of the rectifier output itself or the d.c. switch on the rectifier output line. Operation in

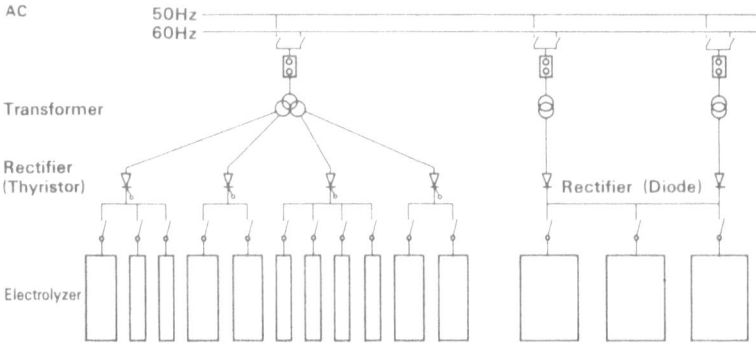

Fig. 1. Nobeoka rectifier connection.

this configuration has continued trouble-free for some years, with frequent current shifting for optimum economy in power consumption [4].

3.2 Other Plants

Among the plants utilizing Asahi Chemical technology under license, six employ rectifiers from existing mercury or diaphragm process systems, essentially in the configurations shown in Figs 2–6.

As shown in Figs 2–4, current for one to four or more electrolyzers may be supplied by one transformer through one, two or more rectifiers.

As shown in Figs 5 and 6, multiple rectifiers may be connected in parallel to as many as seven or eight electrolyzers. In addition, the dual-supply system shown in Fig. 6 may be employed to allow utilization of low voltage rectifiers and minimization of investment in brine circulation and other associated systems. In each case the optimum utilization of existing rectifiers can be effectively achieved through selection of the most appropriate electrolyzer number, cell number and current density.

As evident from these examples, Asahi Chemical's bipolar electrolyzers allow flexibility in the utilization of the existing rectifiers. They also provide a high degree of safety and reliability. Operation in all of these rectifier–electrolyzer configurations has been trouble-free in all cases. The safety control systems uniquely developed for the Asahi Chemical bipolar electrolyzer process have been fully incorporated into all of these plants.

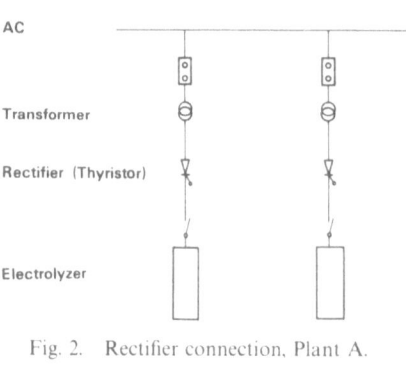

Fig. 2.   Rectifier connection, Plant A.

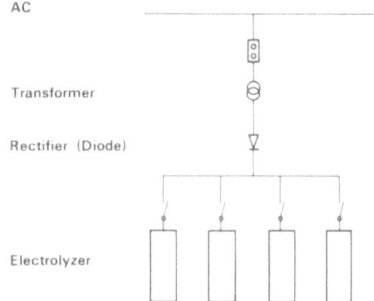

Fig. 3.   Rectifier connection, Plant B.

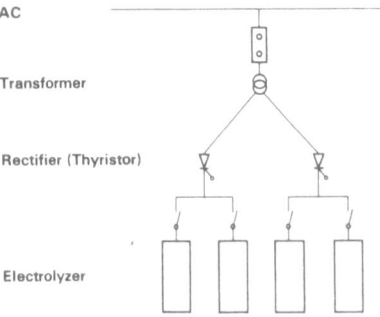

Fig. 4.   Rectifier connection, Plant C.

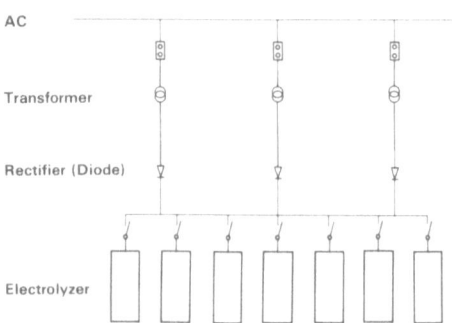

Fig. 5.   Rectifier connection, Plant D.

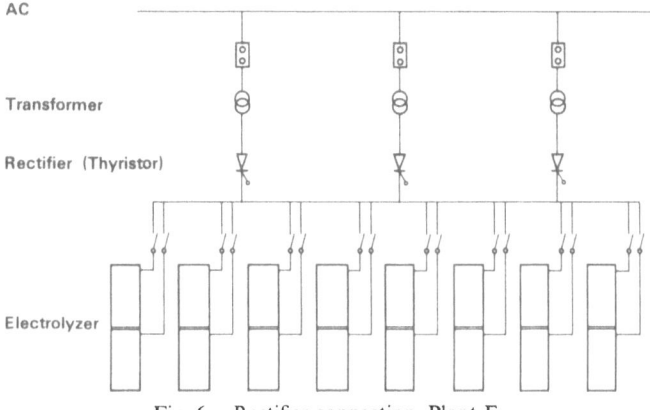

Fig. 6.   Rectifier connection, Plant E.

4 UTILIZATION OF EXISTING CELL ROOMS

With Asahi Chemical's membrane process the required area for the cell room is small. Therefore many conversions utilize the existing cell building. This is the case in ten plant conversions, half from the mercury process and half from the diaphragm process.

The electrolyzer may be installed at ground or elevated levels, in accordance with the structure of the existing building and/or the operational requirements. For conversions, as well as for grass roots plants, the optimum electrolysis building design is effectively developed by comprehensive consideration of operational requirements, construction economics, soil conditions and many other factors.

In a conversion utilizing an existing cell building, it is generally essential to complete the replacement of the electrolyzers and auxiliary systems within a short period. In many cases the chlor-alkali plant is a key part of a plant complex, in which its products are used as basic chemicals or important intermediates. It may therefore be necessary to complete the conversion within the period of a scheduled complex shutdown. This can be readily accomplished with the Asahi Chemical process. Two conversions, each involving more than 80 000 t of annual production capacity, have been achieved during 30-day annual maintenance periods, from shutdown of mercury or diaphragm electrolyzers to start-up of the membrane process.

5 OPERATION

Throughout the 13 years since the start up of the first commercial-scale membrane process plant in 1975, the plants which have now grown in number to 19 (three more are currently under construction) have operated smoothly and trouble-free under various operation conditions. In some of these plants anode replacement has been carried out. It is very important for the confirmation of the durability of the electrolyzers to go through these works. Thus the long life and high durability of the Asahi Chemical's electrolyzer have been confirmed.

This record of many years also demonstrates their operational reliability. In particular, the safety interlocking systems employed in Asahi Chemical's membrane process have been completely effective in preventing any operational difficulties serious enough to require lengthy plant shutdown [4, 5].

All of these plants have shown the membrane process to be characterized by:

(1)   Ease of operation and maintenance [6, 7].
(2)   Flexible production rate operation within a wide current-shift range.
(3)   Safety and reliability.

The bipolar electrolyzer developed by Asahi Chemical has been shown to be particularly characterized by:

(1)   Easy start-up and shutdown of electrolyzer(s), either in combination or individually.
(2)   Freedom from need to remove electrolyzer cell frames for maintenance or membrane replacement.
(3)   Minimization of number of components, instruments, materials, spare parts and inspection points.
(4)   High degree of safety inherent in design and structure, through rigid cell frames, external porting of electrolytes and interlocking.

For current-shift operation, the Asahi Chemical electrolyzers allow simple and effective adjustment to power input in a broad range, and thus facilitate current-shift operation for optimum economy of electric power use.

At the Nobeoka plant, day–night current shifts have been employed since 1980. The shift is performed for all of the electrolyzers at the same time, usually four times and occasionally up to ten times a day, with corresponding changes in the current density between $1 \cdot 5$ and $4 \cdot 0 \, kA/m^2$. To all plants in the Nobeoka complex, power is supplied from external and internal combined sources based on both thermo- and hydro-electric generation. This supply is controlled by centralized computer facilities, one terminal of which is located inside the control room in the caustic soda plant to provide reference data for optimizing current utilization and shift points. Many other plants employing Asahi Chemical's membrane process also utilize current-shift systems in their operations [6, 7].

Finally, Asahi Chemical membrane electrolyzers are operated under elevated pressure, yielding pressurized hydrogen and chlorine as gas products, thus making it easy to combine the gas product pipelines with distant existing gas handling facilities. This also effectively lowers the electric power consumption required for the compression of these product gases.

6 SUMMARY

It is now widely recognized that ion exchange membrane processes for chlor-alkali production have various important advantages over the conventional processes. This recognition has led to their adoption in plants with a total annual production capacity of nearly 5 million t of caustic soda worldwide. This capacity will continue to increase rapidly, with new construction projects and conversion projects throughout the world. Based on its longstanding experience, Asahi Chemical will also continue to achieve further improvements in design, installation and operation of the ion exchange membrane process.

REFERENCES

1. Ministry of International Trade and Industry, Report on the Long-term View on the Chlor-Alkali Industry, Kagaku Kogyo Nippo-sha, July 15, 1980.
2. Miyazaki, T., *Proceedings of the Symposium on Electrochemical Engineering in the Chlor-alkali and Chlorate Industries*, Vol. 88-2. The Electrochemical Society, Inc., Pennington, NJ, USA, 1988, p. 65.
3. Seko, M., *Ind. Engng Chem., Prod. Res. Dev.*, **15**(4) (1976) 286.
4. Ono, H. & Yamamoto, K., *Proceedings of the 11th Symposium on Caustic Soda Technology (Japan)*. The Electrochemical Society of Japan, Tokyo, Japan, November 1987, p. 95.
5. Scarfe, T. & Patel, H. M., Safety aspects of Niachlor membrane cell plant. Presented at 31st Chlorine Plant Operations Seminar, New Orleans, LA, March 1988.
6. Pearson, E. F. & Miyazaki, T., *Proceedings of 2nd World Congress of Chemical Engineering*, Vol. IV, October 1981, p. 346. AIChe, New York, USA.
7. Walkier, J. A. & Zwinkels, J. L. F. M., *Soda and Chlorine (Japan)*, **37** (1986) 16.

14

THE MECHANISM OF THE ACTIVITY LOSS OF RuO$_2$–TiO$_2$ ANODES IN SATURATED NaCl SOLUTION

Shen Manli and Chen Yanxi

Tianjin University, People's Republic of China

1 INTRODUCTION

Although RuO$_2$–TiO$_2$ anodes (DSAs) have been widely used in electrochemical technology since the late 1960s, and the work to find the mechanism of its activity loss has been extensive, the published papers contain very few descriptions relating to the degradation caused in the brine electrolyte. The aim of the present work is to investigate the mechanism of the activity loss of the DSA in a saturated NaCl solution whilst operating at a high current density. It is obvious that the results obtained here will match closely the conditions existing in the chlor-alkali industry, which is the most significant application of DSAs. During anodic polarization a detailed investigation, especially of the variations in composition and structure of the oxide layer, was carried out in terms of the electrochemical and physical surface methods until the end of the service life of the anode. Based on the data obtained, the reasons for the degradation of RuO$_2$–TiO$_2$ anodes in NaCl aqueous solutions are discussed.

2 EXPERIMENTAL

RuO$_2$–TiO$_2$ mixed oxide electrodes were prepared by a thermal decomposition method as outlined elsewhere [1]. The Ti wire ($\phi = 3$ mm), pretreated by mechanical and chemical methods, was painted with a solution of RuCl$_3$ and Ti (OC$_4$H$_9$) in n-butyl alcohol containing a few drops of 20% HCl. The catalyst loading of the Ru component was 0·8 mg/cm^2 and the mole ratio was kept at Ru:Ti = 1:2. To accelerate the service life test, the surface content of Ru and Ti component was decreased to 1/8 of its normal value. After drying below 100°C and heating at 450°C for 10 min, the coated Ti wire was painted again, and the procedure was repeated several times. The final annealing treatment was carried out at 450°C for 1 h. The geometric electrode surface area was 1 cm^2.

All electrochemical experiments, including cyclic voltammetry, a.c. impedance, etc., were performed in saturated NaCl solutions using a standard three-electrode arrangement with a saturated calomel electrode (SCE) as a reference.

The electrolytic experiments were carried out with a current density, $i = 1\, A/cm^2$, at 80°C. To keep the concentration constant, the electrolyte was renewed every 8 h. A Pt gauze with a much larger surface area was used as the cathode, so the cell voltage was determined by the anode. As usual the service life of the RuO_2-TiO_2 anode was defined as the time of operation before the cell voltage increased significantly.

The surface layer was also studied by SEM, X-ray diffraction and SIMS.

3 RESULTS AND DISCUSSION

3.1 The Variation of DSA Catalytic Activity

A typical cell voltage–time curve is shown in Fig. 1. It is seen that the cell voltage decreases slowly at first, caused by the activation of the inner surface in the porous anode; then the voltage remains unchanged for a long period of time, and then rises steeply at the end of the service life, exhibiting the inherent characteristics of DSAs.

The cyclic voltammetry was measured at different time intervals between the potentials $E = -0.36$ and $+1.16$ V (versus SCE) before the evolution of H_2 and O_2. A large maximum was noted at the peak current of anodic chlorine evolution ($E = +1.16$ V) versus time plot (Fig. 2), i.e. the peak current initially increased with time, but after a certain time it decreased, especially at the end of the service life. This means that the catalytic activity of the DSA is not constant with time but changes continuously.

3.2 The Variation of the Pseudo-Capacitance

Typical time dependence of the pseudo-capacitance, C_p, obtained by voltammetry between the potentials $E = -0.20$ and $+0.85$ V (versus SCE) for an RuO_2-

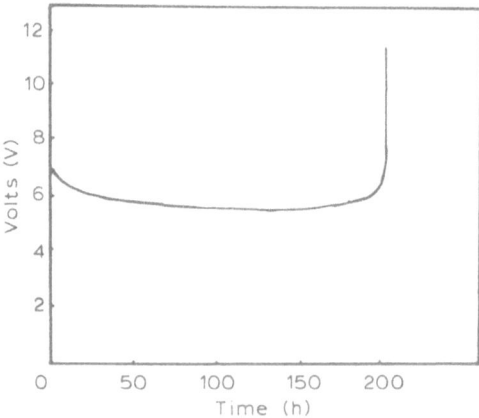

Fig. 1. Typical voltage–time curve for RuO_2-TiO_2/Ti electrode in saturated NaCl solution (80°C, $i = 1\,A/cm^2$).

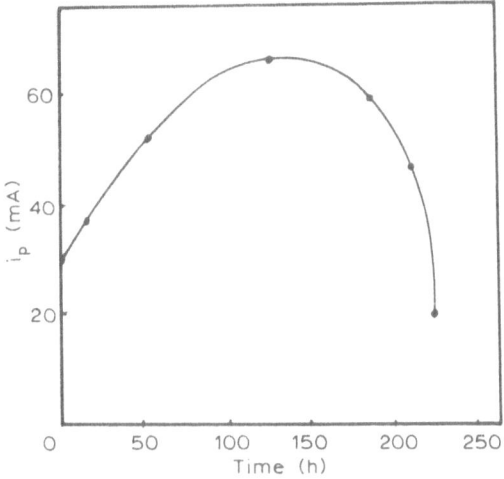

Fig. 2. The time dependence of the anodic peak currents at $E = +1{\cdot}16$ V (versus SCE) in cyclic voltammograms for RuO_2–TiO_2/Ti electrode (80°C, saturated NaCl solution).

TiO₂ anode is shown in Fig. 3. It is noted that the plot has virtually the same shape as in Fig. 2, and shows a maximum value of C_p. It is obvious from these results that there is a close correlation between the pseudo-capacitance and the catalytic activity of the DSA.

It was suggested that the basic relationship between the voltammetric charge (Q), electrode surface area (A) and the RuO₂ content (mol%) can be expressed by the following equation [2]:

$$Q = KA[RuO_2] \tag{1}$$

where K is a proportional constant.

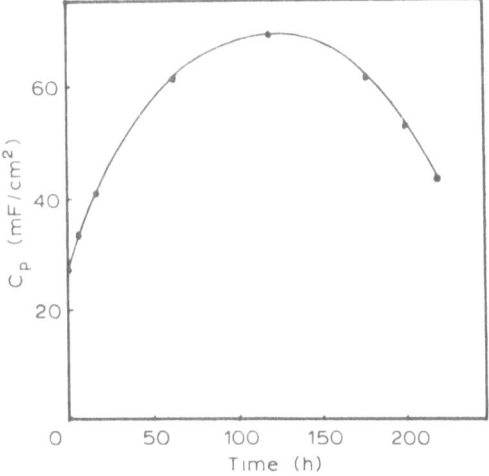

Fig. 3. The time dependence of the pseudo-capacitance, C_p.

Therefore, if the surface area and RuO_2 content in the oxide-coated layer are changed oppositely and with a different rate, the maximum value of the voltammetric charge (or the pseudo-capacitance) will be expected. This is confirmed by our experimental results.

As is well known the film surface of the DSA, prepared by the thermal method, has a cracked, dried mud-like morphology, i.e. there are a lot of crevices in the coated layer. In the depth of the crevice, the concentration of Cl^- will decrease gradually during electrolysis, while the ratio of the oxygen discharge reaction will increase. This results in a decrease of the pH value of the solution, and the acceleration of chemical and electrochemical corrosion in the crevices by the following reactions:

$$RuO_2 + 4H^+ = Ru^{4+} + 2H_2O$$
$$TiO_2 + 2H^+ = TiO^{2+} + H_2O$$
$$RuO_2 + 4H^+ = Ru^{8+} + 2H_2O + 4e$$

The cracking corrosion was confirmed by the following experimental facts.

As shown in Fig. 3, the pseudo-capacitance increases with increasing electrolytic time before the maximum appears. It becomes a linear function when these data are replotted in Fig. 4. The relationship between them can be expressed by [3]

$$\frac{C_{p,t} - C_{p,t=0}}{C_{p,t=0}} = Kt^{1/2} \tag{2}$$

where $C_{p,t=0}$ and $C_{p,t}$ denote the values of C_p before electrolysis and at electrolytic time t respectively; K is a proportional constant equal to 0·145 according to Fig. 4.

The fact that the pseudo-capacitance varies with the square root of the time t suggests that a diffusion process, which may be expected in the cracking corrosion, is rate controlling in this case. In other words, the increase in pseudo-capacitance results mainly from the increase of the effective electrode surface area caused by the cracking corrosion. This effect, as shown in Fig. 3 and eqn (1), plays a predominant

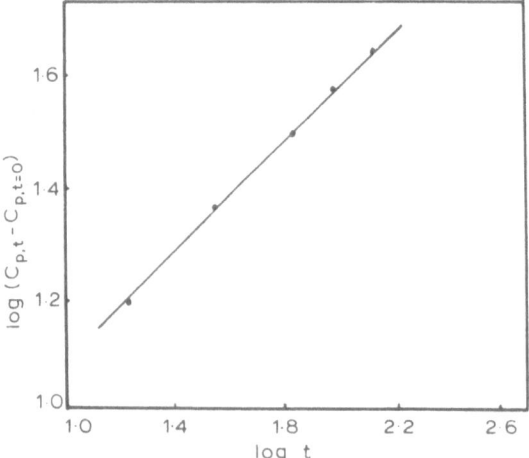

Fig. 4.   The time dependence of the change in pseudo-capacitance.

role during the initial period of electrolysis, and the catalytic activity of the RuO_2–TiO_2 anode increases simultaneously, as shown in Fig. 2.

3.3 The Corrosion Behavior of Ru and Ti Components

The results of an in-depth analysis of the coated layer for Ru and Ti components, conducted by means of SIMS, are shown in Fig. 5. A comparison of the plots in Fig. 5(a), (b), (c) indicates that the Ru species, which centered before electrolysis by the surface of the coated layer, was dissolved off continuously with time, and the signal of the Ru species decreased significantly by the end of the service life (it should be noted that the value of the ordinate and the amplification of the Ru plot is different in Fig. 5(c) from that in Fig. 5(a), (b)).

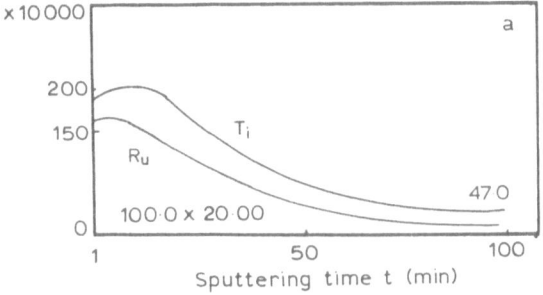

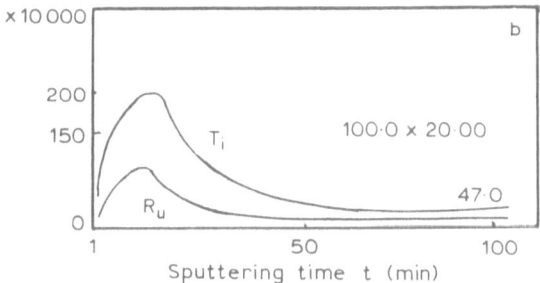

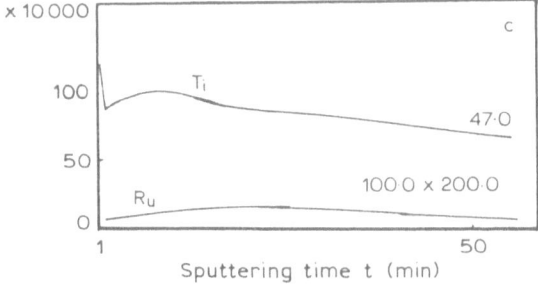

Fig. 5.   Typical in-depth distribution profiles of Ru and Ti in RuO_2–TiO_2/Ti electrode. (a) Before electrolysis; (b) after anodization for 130 h; (c) after anodization for 280 h.

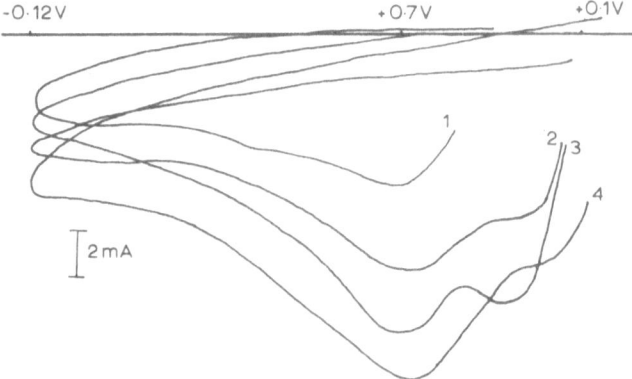

Fig. 6.   Effect of the upper scan limit on the change of the reductive peak currents at 0·70 V versus SCE.
The potentials at which the positive scan was reversed are: 1, +1·28 V; 2, +1·80 V; 3, +2·15 V; 4, +2·22 V
(25°C, $V = 184$ mV/s).

The voltammetric plot shown in Fig. 6 indicates the potential dependence of the corrosion rate of RuO_2. It is seen that RuO_4, formed by the corrosion of RuO_2, will be reduced in the negative voltammetric scan at +0·70 V (versus SCE), and the reductive peak current will be larger when the upper scan limit is towards more positive potentials. These results are consistent with earlier work [4]. Evidently the polarization of the DSA will be increased during electrolysis. As a result, the rate for Ru dissolution will be increased and will accelerate significantly, especially by the end of the service life. Under this case the dissolution of the Ru component alternatively becomes the predominant factor in eqn (1), and the decrease in the pseudo-capacitance will occur (Fig. 3).

Therefore it seems likely that two effects—the cracking corrosion and the dissolution of the Ru component—predominate at different times of electrolysis

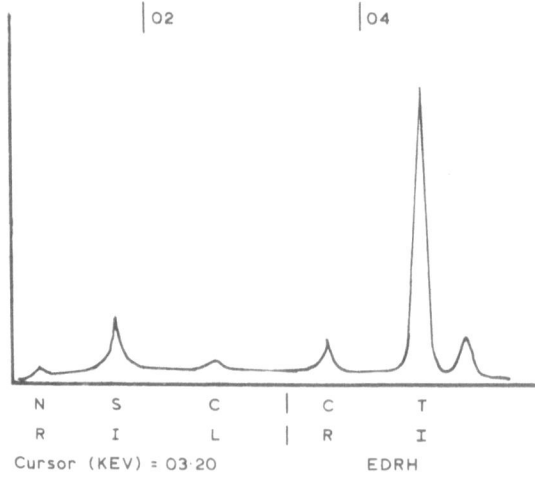

Fig. 7.   X-ray energy dispersive spectrum of the white deposition.

and result in the appearance of a maximum in C_p versus t or i_p versus t plots (Figs 2 and 3) of the DSA.

The inactive Ti component in the surface layer of the DSA also dissolves during the anodic polarization (Fig. 5), but the rate of dissolution is less than those of the Ru species. At the end of the service life, however, the Ti content decreases, obviously.

Unfortunately, the titanium hydrous product as a white deposition (Fig. 7) will cover the electrode surface and the activity of the anode will be lost immediately. The fact that the white deposition only appears at the end of the service life, even though the electrolyte was renewed every 8 h, proves definitely that the dissolution of the Ti component increases at the end.

3.4 The Measurement Results of a.c. Impedance

Figure 8 shows the time dependence of the reactive resistance and the ohm resistance of the DSA respectively, obtained by a.c. impedance measurements.

The time dependence of the reactive resistance (Fig. 8(a)) is surprisingly similar to that of the cell voltage (Fig. 1). On the other hand, a sharp increase with time does not appear in the ohmic resistance, as it will result if the passivation of the electrode

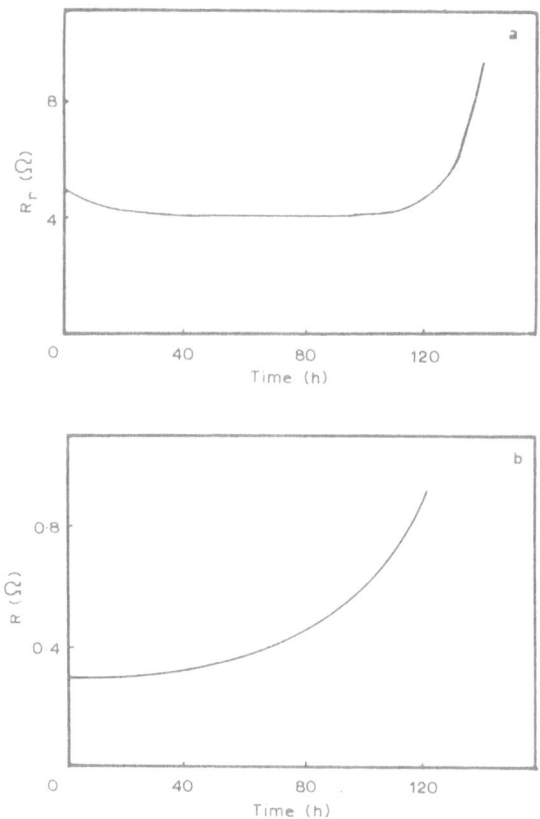

Fig. 8. The time dependence of the resistance of the RuO₂–TiO₂/Ti electrode. (a) Reactive resistance; (b) ohmic resistance.

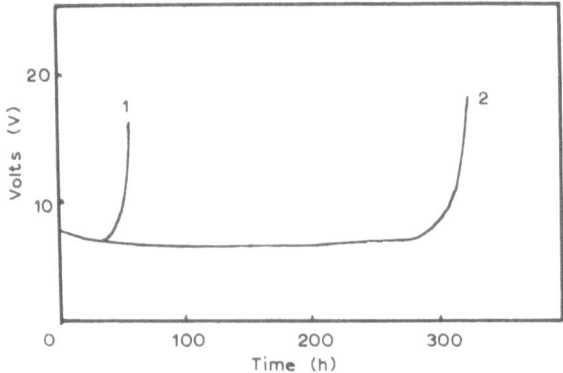

Fig. 9.   Effect of Ir component in the coated layer on the service life of the electrode. 1, $RuO_2 + TiO_2$; 2, $RuO_2 + TiO_2 + IrO_2$.

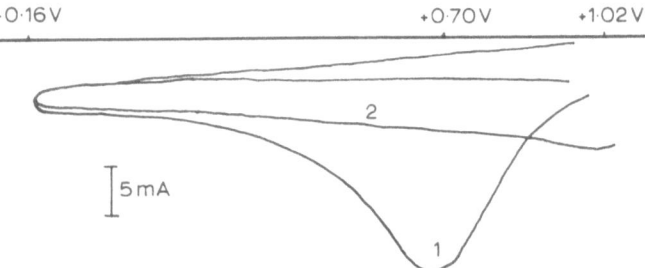

Fig. 10.   Effect of the Ir component in the coated layer on the voltammetric reductive peak current at $+0.70$ V (versus SCE) with the same positive limiting scan potential. 1, $RuO_2 + TiO_2$; 2, $RuO_2 + TiO_2 + IrO_2$.

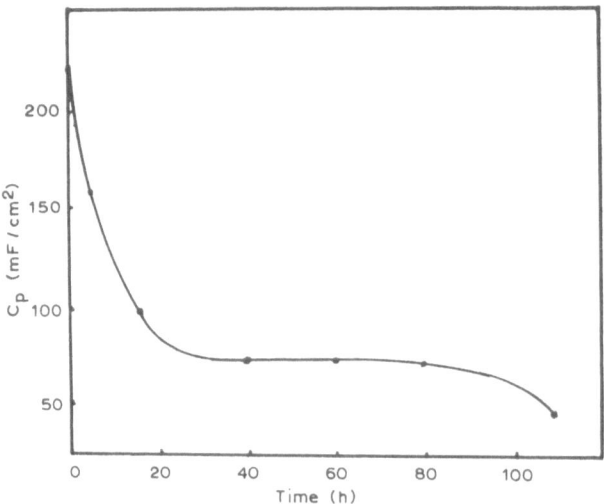

Fig. 11.   The time dependence of the pseudo-capacitance, C_p.

occurs. Moreover, the value of the ohmic resistance is about one order less than that for the reactive resistance. All these results support the view that the dissolution of the Ru component is the main reason for the loss of activity of RuO_2–TiO_2 anodes in a saturated NaCl solution.

3.5 The Influence of the Ir Component

Experimental results indicate that the service life of the DSA becomes significantly longer when Ir is added to the coated layer (Fig. 9). The reductive peak current at $+0.70$ V (versus SCE) (Fig. 10) is also decreased. Figure 11 shows the time dependence of the pseudo-capacitance of a DSA with a RuO_2–TiO_2–IrO_2 coated layer. It is noted that the pseudo-capacitance C_p decreased monotonically with anodic polarization, and the maximum as shown in Fig. 3 disappeared. So it is suggested that the addition of IrO_2, forming a solid solution with RuO_2–TiO_2 and with a higher oxygen overpotential, will suppress the cracking corrosion and the dissolution of the Ru and Ti components will be effectively decreased [5].

4 CONCLUSION

The results obtained in this investigation have demonstrated the following.

(1) The variation of the composition and the structure in surface layer of DSA occurs from the beginning of electrolysis, although it is not clearly observed in the cell voltage versus time plot, because the reactive resistance of anodic chlorine evolution is relatively low.

(2) The degradation of the RuO_2–TiO_2 anode in saturated NaCl solution is a process which takes place gradually. The main cause for it is the dissolution of the Ru component in the active layer, although some other reasons are also possible, such as mechanical damage by gas products, etc.

It is also found that a white titanium hydrous compound is precipitated on the anode surface at the end of the electrode service life. This decreases the effective electrode surface area and accelerates the degradation of the anodes significantly.

(3) Cracking corrosion is also one of the important reasons for the loss of activity of DSA in a NaCl solution. It occurs in the coated layer prepared by the thermal method, resulting in the acceleration of the corrosion rate of Ru and Ti components, and the decrease of the coated layer adhesion.

(4) There is no direct evidence for the passivation of the Ti substrate.

REFERENCES

1. Novak, D. M., Tilak, B. V. & Conway, B. E., *Modern Aspects of Electrochemistry*, Vol. 14. Plenum Press, New York, USA, 1981, p. 200.
2. Burke, A. D. & Murphy, O. J., *Electroanal. Chem.*, **112** (1980) 39–50.
3. Tomiya Kishi, Yoshiharu Sugimoto & Takashi Nagai, *Surf. Technol.*, **26** (1985) 245–51.
4. Kötz, R., Stucki, S., Scherson, D. & Kolb, D. M., *J. Electroanal. Chem.*, **172** (1984) 211–19.
5. Shen Manli, Chen Yanxi & Ke Li, to be submitted.

15

VOLTAGE–CURRENT CURVES: APPLICATION TO MEMBRANE CELLS

D. Bergner, M. Hartmann and H. Kirsch

Hoechst AG, FRG

1 INTRODUCTION

If a voltage–current curve is recorded for a membrane cell, the graph shown in Fig. 1 is obtained. Above 1 kA/m^2 the curve can be interpreted as a straight line with the linear equation

$$U = C + ki \tag{1}$$

At current densities below 1 kA/m^2 the curve is bent downwards, ending at the point of the decomposition voltage U_0. Under the conditions of alkali chloride electrolysis in membrane cells, this voltage has the numerical value of 2·2 V. However, the intercept C in eqn (1) is always larger than 2·2 V, normally between 2·3 and 2·7 V [1].

The cell voltage U of a membrane cell comprises the following quantities:

$$U = E_A + |\eta_A| - E_C + |\eta_C| + \Delta E_A + \Delta E_C + \Delta E_M \tag{2}$$

where E stands for the reversible electrode potentials, η for the overvoltages of the electrodes, and ΔE for the voltage drop across the electrolytes and the membrane. The subscripts A, C and M stand for anode or anolyte, for cathode or catholyte, and

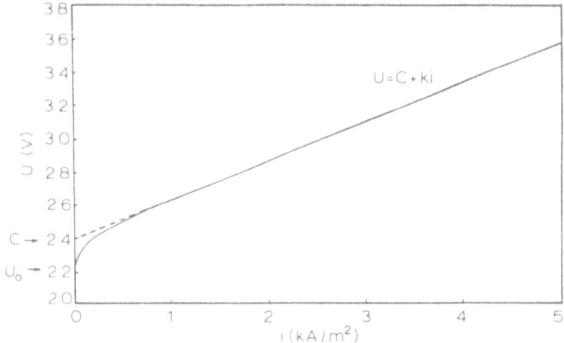

Fig. 1.   Voltage–current curve of a membrane cell.

for membrane. If electrode potential and overvoltage are combined to form E' and the Tafel equation

$$|\eta| = a + b \log i \tag{3}$$

where a and b are constants, is substituted for the overvoltage, eqn (4) is obtained:

$$E'_A = E_A + a_A + b_A \log i \tag{4}$$

Recently it was shown that the current density dependence of the overvoltage above 1 kA/m^2 can be considered linear (cf. Fig. 2 in [1] for the hydrogen overvoltage). This means that the expression $b_A \log i$ in eqn (4) can be set $b_A \log i \approx b'_A i$. This is justified in as much as the expression

$$\frac{\log i}{i} = \frac{b'_A}{b_A}$$

is nearly constant (about 0·15) in the range 1·5–5 kA/m². Hence eqn (4) can be converted to the following linear equation:

$$E'_A = (E_A + a_A) + b'_A i \tag{5}$$

Analogously, the same is true for the cathode with E'_C. On the other hand, the electrode potentials E_A and E_C can be combined to form the decomposition voltage U_0:

$$U_0 = E_A - E_C \tag{6}$$

The voltage drops across the electrolytes can be represented with the electrolyte resistance R or the area resistance R' as being a function of the current strength I or the current density i:

$$\Delta E_A = R_A I = R'_A i \tag{7}$$

The same is analogously true for the catholyte with ΔE_C.

By our own measurements [2] and references from the literature [3–5] it has been proven that the voltage drop across the membrane ΔE_M is not caused only by an ohmic resistance. Rather, in the current density range above 1 kA/m², ΔE_M can be described as a linear function of the current density with an intercept on the voltage axis:

$$\Delta E_M = a_M + b_M i \tag{8}$$

Chandran & Chin [3] measured, at different caustic concentrations, the voltage drop across a NAFION® 901 membrane as a function of current density (2–4 kA/m²). The graph in their Fig. 9(b) shows for each of the caustic concentrations 7·5 and 10 mol/litre four measurement points, through which straight lines with intercepts from 0·1 to 0·2 V can be drawn. The authors have, nevertheless, drawn straight lines that pass through the origin and are therefore based on an equation similar to our eqn (10). Thus Chandran & Chin [3] have assumed ohmic behaviour for the membranes and explain the deviations of the measured values from their straight lines as errors of measurement.

Except for the measurements of Chandran & Chin [3], the non-ohmic behaviour of the membranes is confirmed by the current strength dependence cited by Heitz & Kreysa [4] and by the measurements of Rondinini & Ferrari [5], who found that the specific resistance of the membranes decreases with increasing current density.

By combining eqns (2) and (5)–(8), and by combining the current density independent and current density dependent quantities, one obtains

$$U = (U_0 + a_A + a_c + a_M) + (b'_A + b'_C + b_M + R'_A + R'_C)i \tag{9}$$
$$U = \qquad C \qquad + \qquad ki \tag{1}$$

A comparison of eqns (1) and (9) reveals that the intercept C of the current–voltage straight line becomes larger than U_0 by the overvoltages or by the exchange current densities of the electrodes and by a term a_M dependent on the membrane.

On the other hand, the slope k of the current–voltage curves is essentially determined by the area resistance R' of anolyte and catholyte, and by a term b_M dependent on the membrane.

2 EXPERIMENTAL

A common procedure is to break down the cell voltage of an electrolysis cell into its components and plot a graph of the dependence of these components on the current density [6, 7].

The objective of the work reported here was to:

1. measure the current density dependence of the cell voltage components of membrane cells;
2. compute the linear equations of these dependencies; and
3. add up the linear equations and compare the sum U_Σ with the measured current–voltage curve $U = f(i)$.

This chapter is intended to enable prediction of the current–voltage behaviour of any cell configuration by combining individual data of the cell components.

The measurements were performed on a circular laboratory cell with an electrode area of 36 cm^2. This cell is depicted in Fig. 2 and described in detail elsewhere [2]. By way of two intermediate pieces ('i' in Fig. 2) rotatable Luggin capillaries were introduced into the cell, by means of which all required potentials on the electrodes and membrane could be measured. The components and the operating conditions of this electrolysis cell are compiled in Table 1.

A total of four series of measurements were run under these operating conditions. The following values were determined as a function of current density in the range 0–6 kA/m^2:

E'_A—anode potential versus SHE. According to eqn (4), E'_A is the sum of the reversible anode potential and the overvoltage of the anode.

$E'_A + \Delta E_A$—sum of the anode potential and the voltage drop across the anolyte.

ΔE_M—voltage drop across the membrane.

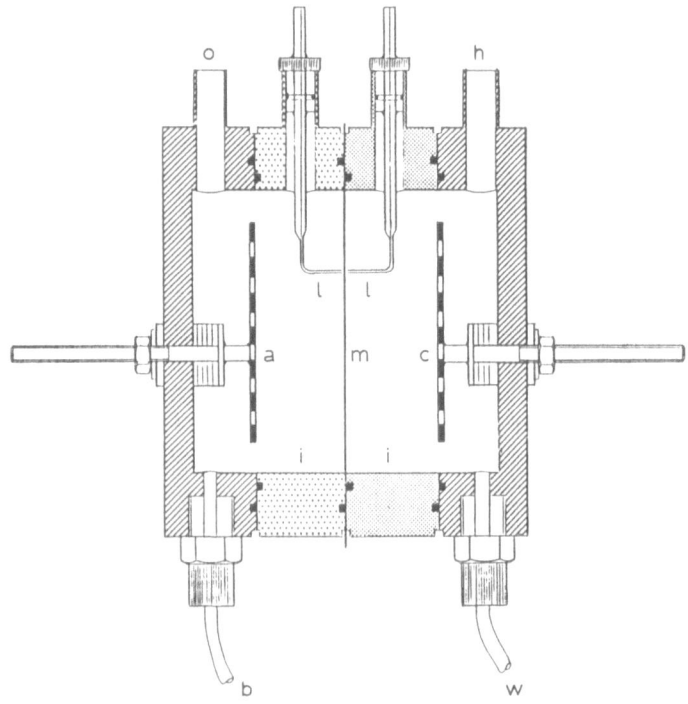

Fig. 2.   Circular laboratory cell: (a) anode, (b) brine feed, (c) cathode, (h) hydrogen/catholyte outlet, (i) intermediate piece, (l) Luggin capillary, (m) membrane, (o) chlorine/anolyte outlet, (w) water feed.

E_C'—cathode potential versus SHE. E_C' is the sum of the reversible cathode potential and the overvoltage of the cathode.

$E_C' + \Delta E_C$—sum of the cathode potential and the voltage drop across the catholyte.

U—cell voltage measured at the electrodes.

The linear equations for these quantities were calculated for each of the four series of measurements in the current density range 1–5 kA/m². Within the individual series the addition of the linear equations only very inaccurately yielded the dependence of the cell voltage on the current density. This is attributed to errors of measurement and to the not always accurately adjustable cell temperature. The desired results were obtained only after all measurements of the four series were combined. The linear equations are compiled in Table 2. Here the function $\Delta E_A = f(i)$ was obtained from the linear equations for $E_A' + \Delta E_A$ and E_A' by subtraction. The same applies analogously to $\Delta E_C = f(i)$.

The equation for the cell voltage, obtained by adding up the components $U_\Sigma = f(i)$, agrees quite well with the directly measured $U = f(i)$. At 3 kA/m² the cell voltage values differ by only 40 mV.

The following should be noted concerning the equations in Table 2:

1. The voltage drop at the membrane does not follow Ohm's law, as discussed above.

TABLE 1
36 cm^2 laboratory cell (cf. Fig. 2), components and operating conditions

Anode	Titanium expanded metal, 36 cm^2; RuO$_2$–TiO$_2$ coating
Cathode	Nickel expanded metal, 36 cm^2
Membrane	NAFION® NX-954 (Du Pont)
Temperature: anode compartment	87°C
cathode compartment	85°C
Concentration: anolyte	18% NaCl
catholyte	33% NaOH
Gap: membrane–anode	30 mm
membrane–cathode	30 mm
Brine feed (26% NaCl)	130 ml/h
Water feed	10 ml/h
Reference electrode	Ag/AgCl in sat. KCl solution (+0·1976 V at 25°C [8] for conversion to SHE)

This is a consequence of a potential difference arising at the membrane in the absence of current. According to Schmid [9], this 'dialysis potential' consists of a smaller part of the diffusion potential and of a larger part of both Donnan potentials at the electrolyte/membrane interfaces. In the absence of current, values of approximately 100 mV for this dialysis potential were measured in the laboratory cell [2]. The deviation of this value from the intercept of Table 2 indicates that the curves are actually slightly bent. The curvature could result from concentration shifts at higher current densities, which then in turn affect the dialysis potential. In the application range from 1 to 5 kA/m^2 the deviations from a straight line are negligible.

2. The voltage drops in the electrolytes do not differ from the values of the gas bubble-free solutions.

If the specific resistance of the bubble-free electrolytes ρ_0 is measured [2], and if the calculations are performed with the layer thickness d used in the experiment (3·0 cm) according to

$$\Delta E = IR = \rho_0 di \qquad (10)$$

the following linear equations are obtained:

$$\Delta E_{A,0} = 0·64i \qquad (11)$$

$$\Delta E_{C,0} = 0·34i \qquad (12)$$

which practically agree with the values for ΔE_A and ΔE_C given in Table 2. A correction using the Bruggeman equation

$$\rho/\rho_0 = (1 - \varepsilon)^{-3/2} \qquad (13)$$

where ε is the gas void fraction, with the bubble content observed in industrial electrolysis cells (at 3 kA/m^2: 18% Cl$_2$ in the anolyte; 8% H$_2$ in the catholyte) is apparently not necessary here. In the 36 cm^2 experimental cell used the electrode gap is large. Thus the mean void fraction is small and practically of no importance.

TABLE 2

Components of cell voltage of membrane cell as a function of
current density (for operating conditions see Table 1)

		$3\,kA/m^2$
	$E'_A = 1{\cdot}29 + 0{\cdot}014i$	1·33 V
30 mm	$\Delta E_A = \qquad\ 0{\cdot}606i$	1·82 V
	$\Delta E_M = 0{\cdot}17 + 0{\cdot}062i$	0·36 V
30 mm	$\Delta E_C = \qquad\ 0{\cdot}293i$	0·88 V
	$E'_C = 1{\cdot}18 + 0{\cdot}034i$	1·28 V
	$U_\Sigma = 2{\cdot}64 + 1{\cdot}009i$	5·67 V
	$U = 2{\cdot}64 + 0{\cdot}966i$	5·63 V

Similar experimental studies to break down the cell voltage components were conducted by Nidola [10]. From his diagrams for a zero-gap cell, that is a membrane cell with an electrode gap of <1 mm, we have calculated linear equations in order to enable a comparison with our results. When comparing Tables 2 and 3 it is apparent that the anode potential and both electrolyte voltage drops (after reduction of the values in Table 2 from 30 to 0·5 mm) show good agreement. On the other hand, the intercept for the voltage drop of the membrane given by Nidola [10], at about 50 mV, is considerably smaller than our value of 170 mV. Presumably the smaller value was measured in the absence of current. The difference between the equations $E'_C = f(i)$ in Tables 2 and 3 is due to the difference in cathode activity. The current–voltage curve $U = f(i)$ of the cell according to Nidola intersects as a straight line the voltage axis at 2·23 V, which is very close to the decomposition voltage. According to our experience this is not possible.

Of interest here is the 'membrane cathodic bubble effect' reported by Nidola [10]. From the diagrams shown there one can calculate for an uncoated (hydrophobic) bimembrane

$$\Delta E_{B,C} = 0{\cdot}1i \qquad (14)$$

TABLE 3

Components of cell voltage of zero-gap cell as a function of
current density (for operating conditions see Figs 7–14 in
Ref. 10)

	$3\,kA/m^2$
$E'_A = 1{\cdot}28 + 0{\cdot}02i$	1·34 V
$\Delta E_A = \qquad\ 0{\cdot}011i$	0·03 V
$\Delta E_M = 0{\cdot}05 + 0{\cdot}078i$	0·28 V
$\Delta E_C = \qquad\ 0{\cdot}005i$	0·01 V
$\Delta E_{B,C} = \qquad\ 0{\cdot}01i$	0·03 V
$E'_C = 0{\cdot}90 + 0{\cdot}042i$	1·03 V
$U_\Sigma = 2{\cdot}23 + 0{\cdot}166i$	2·73 V
$U = 2{\cdot}23 + 0{\cdot}194i$	2·81 V

TABLE 4

Components of cell voltage of membrane cells as a function of current density and their numerical values at $3 \, kA/m^2$

(a) NAFION NX-90209, uncoated nickel cathode, membrane–cathode gap: 2·5 mm

$E'_A = 1·33 + 0·011i$	1·36 V
$\Delta E_A = 0·077i$	0·23 V
$\Delta E_M = 0·18 + 0·054i$	0·34 V
$\Delta E_C = 0·066i$	0·20 V
$E'_C = 1·18 + 0·008i$	1·20 V
$U_\Sigma = 2·69 + 0·216i$	3·34 V

(b) NAFION NX-961, coated nickel cathode, membrane–cathode gap: 2·5 mm

$E'_A = 1·33 + 0·011i$	1·36 V
$\Delta E_A = 0·102i$	0·31 V
$\Delta E_M = 0·14 + 0·037i$	0·25 V
$\Delta E_C = 0·061i$	0·18 V
$E'_C = 0·93 + 0·008i$	0·95 V
$U_\Sigma = 2·40 + 0·219i$	3·06 V

(c) NAFION NX-961, coated nickel cathode, membrane–cathode gap: 0 mm

$E'_A = 1·33 + 0·011i$	1·36 V
$\Delta E_A = 0·038i$	0·11 V
$\Delta E_M = 0·12 + 0·105i$	0·44 V
$\Delta E_C = 0·038i$	0·11 V
$E'_C = 0·93 + 0·008i$	0·95 V
$U_\Sigma = 2·38 + 0·200i$	2·98 V

and for a coated (hydrophilic) bimembrane

$$\Delta E_{B,C} = 0·01i \tag{15}$$

The value from eqn (15) was used in Table 3.

The laboratory cell shown in Fig. 2 was again used to check Nidola's results [10] and for comparison with the data presented in Table 2. In order to obtain realistic electrode compartment depths and electrode gaps the intermediate pieces supporting the capillaries, marked 'i' in Fig. 2, were removed. Two capillaries were mounted in each of the two electrode compartments through holes in the back wall of the cell. One capillary in each compartment was inserted into the expanded metal sheet and ended at the electrode surface facing the membrane. The other two capillaries were also inserted through the openings in the expanded metal, but were placed in contact with the membrane. The operating conditions corresponded to those given in Table 1, except that the electrode gaps were as follows: anode–membrane, 0 mm; cathode–membrane, 2·5 or 0 mm. When using low overvoltage coated cathodes a catholyte temperature of only 68°C was achieved. When using

non-coated nickel cathodes the catholyte temperature was 85°C. NAFION®
membranes NX-90209 (uncoated) and NX-961 (coated) were employed.

The linear equations obtained are compiled in Table 4. As expected, the anode
potentials E'_A were approximately equal regardless of the cell configuration. For this
reason an average for $E'_A = f(i)$ was used in Table 4(a)–(c). The same applies
analogously for the cathode potentials. The potential of the coated cathode is
250 mV lower than that of the uncoated cathode. The difference is determined by the
intercept of the function $E'_C = f(i)$, which contains the exchange current densities.
The voltage drop in the catholyte also behaved as expected. It is of the same size for
both membranes and is reduced by about 40% by the decrease in the membrane–
cathode gap from 2·5 to 0 mm. The behaviour of the remaining quantities ΔE_A and
ΔE_M is largely inexplicable. With NAFION NX-961, ΔE_A declines considerably
upon gap reduction and ΔE_M increases considerably. It is possible that the observed
effects are due to mistakes in the experimental arrangement, because in the
arrangement chosen and with 0 mm electrode gap (actually about 0·5–1·0 mm) the
position of the Luggin capillary cannot be adjusted with sufficient accuracy. It is
interesting, however, that the gap reduction (in Table 4(b) and(c)) and the effect of
membrane properties, cathode coating and gap reduction (in Table 4(a) and (c))
become clear in the functions $U_\Sigma = f(i)$ obtained by addition.

Figure 3 shows graphs of the data of Table 4(a) and (c). Comparing the two ΔE_M
lines it must be assumed that the unexpectedly large voltage drop, ΔE_M, at the

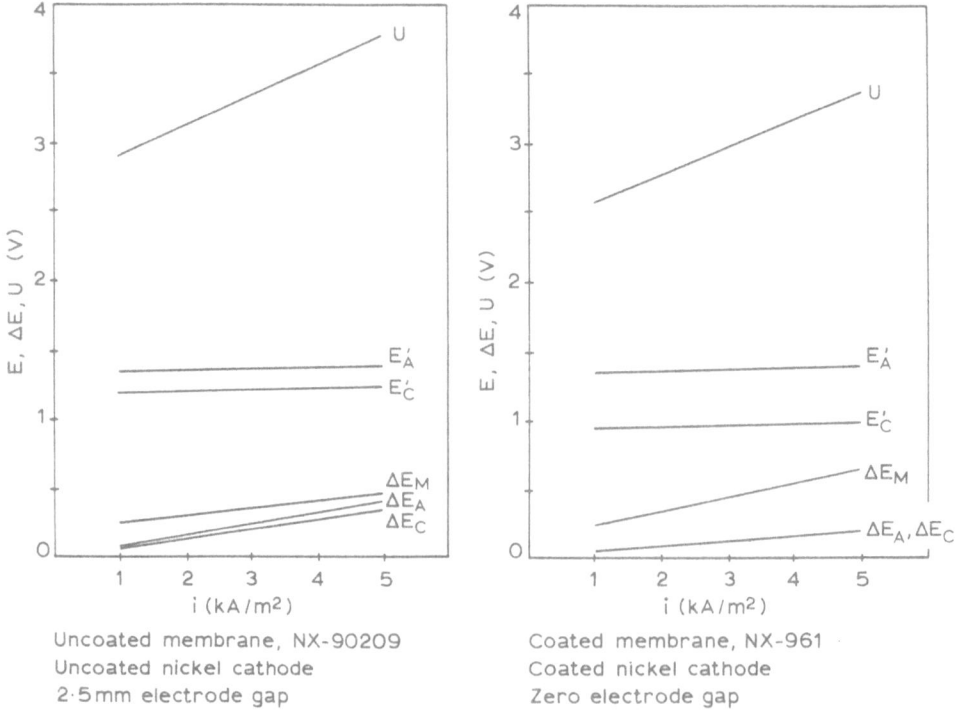

Uncoated membrane, NX-90209   Coated membrane, NX-961
Uncoated nickel cathode   Coated nickel cathode
2·5 mm electrode gap   Zero electrode gap

Fig. 3.   Components of the cell voltage versus current density.

hydrophilic membrane is caused by a bubble effect, i.e. the adhesion of gas bubbles to the membrane surface. Such a 'membrane cathodic bubble effect' is described by Nidola [10], but according to eqn (15) it is very much smaller than the value we observed. As discussed above, however, it is not always possible to determine the effects reliably because of the difficult measurement geometry.

3 APPLICATION

If the cell voltage of a membrane cell changes during continuous operation of, for example, a laboratory cell or production unit under otherwise constant operating conditions, this may be caused by the electrodes as well as by the membrane. The overvoltages of the anode and cathode can change, as can the voltage drop across the membrane. Any significant effects of anolyte or catholyte on cell voltage are unlikely. In order to distinguish the effects of the electrodes and the membrane, eqn (9) has to be considered.

The first expression in brackets contains two variables determined by the electrode overvoltages and one determined by the membrane. The second expression in brackets is determined practically only by the cell resistances; here the membrane resistance in particular can be regarded as a variable quantity. With the following examples of empirical voltage–current curves it is attempted to distinguish between these effects. Since the chlorine overvoltage, owing to its small size and long-term constancy, is of less interest for practical operation than the hydrogen overvoltage, only the latter will be included in our considerations. The effects described for the cathodes, however, also apply analogously for the anodes.

In Tables 5–8 the cell voltage U for $3\,kA/m^2$ is always given in addition to the C and k values. It will reveal the effects of the different changes whose influence on the C and k values is to be explained.

The influence of the cathode material and of the cathode coating on the C value can be seen in Table 5. The cathode has practically no effect on the k value. This is understandable through consideration of eqn (9), since the constant a of the Tafel

TABLE 5
Influence of cathode (36 cm^2 cell; 3 mm electrode gap; NAFION
NX-90209; 85°C)

Cathode material/coating	$U = C + ki$		$3\,kA/m^2$ $U\,(V)$
	$C\,(V)$	$k\,(Vm^2/kA)$	
Nickel	2·68	0·203	3·29
Stainless steel	2·68	0·202	3·29
Stainless steel/coating I	2·61	0·202	3·22
Nickel/coating II	2·49	0·204	3·10
Stainless steel/coating III	2·46	0·204	3·07
Stainless steel/coating IV	2·45	0·203	3·06

TABLE 6
Influence of electrode gap (membrane–cathode gap) ($0.9 \, m^2$
cell; NAFION NX-90209; coated nickel cathode; 90°C)

Gap d (mm)	$U = C + ki$		$3 \, kA/m^2$ $U \, (V)$
	$C \, (V)$	$k \, (Vm^2/kA)$	
3	2·42	0·203	3·03
7	2·41	0·255	3·18

equation is dependent on the nature of the electrode material and the constant b is practically the same for all metals.

Table 6 shows the influence of the cell resistance when the electrode gap is increased from 3 to 7 mm. The C value remains constant and the k value increases because of the increase in the electrolyte resistance R'_C.

The influence of the cell temperature is shown by Table 7. As the temperature increases the resistance between the electrodes, and hence the k value, decreases. The C value, which also falls with rising temperature, is influenced by the decomposition voltage and the overvoltages, which all decrease with rising temperature.

Measurements on a production cell are shown in Table 8. The 32-week-long rise in the cell voltage resulting from the increase in the C value can be interpreted as an ageing of the cathode coating. Direct measurements on the cathodes indicated an increase in potential with about the same magnitude as that of the cell voltages and C value.

According to our experience with membrane cells of all sizes voltage–current curves should always be recorded in the upper current density range and always with the same procedure. Without these precautionary measures a comparison of the linear equations will be rendered more difficult. At low current densities parts of the voltage–current curve that deviate downwards could be included in the measurements and hysteresis phenomena could arise.

An obstacle to a wider application of this procedure could be that it does not always appear possible to record the voltage–current curves in industrial

TABLE 7
Influence of cell temperature ($36 \, cm^2$ cell; 3 mm electrode gap;
NAFION NX-90209; coated stainless steel cathode

Temperature t (°C)	$U = C + ki$		$3 \, kA/m^2$ $U \, (V)$
	$C \, (V)$	$k \, (Vm^2/kA)$	
70	2·66	0·201	3·26
75	2·63	0·194	3·21
80	2·60	0·189	3·17
85	2·55	0·187	3·11
90	2·52	0·185	3·08

TABLE 8

Influence of time of operation (2·7 m² cell; 3 mm electrode gap;
NAFION NX-90209; coated nickel cathode; 90°C)

Weeks since start-up	$U = C + ki$		$3 kA/m^2$ $U (V)$
	C (V)	k (Vm²/kA)	
3	2·36	0·248	3·10
14	2·40	0·244	3·13
32	2·47	0·248	3·21

electrolysers. However, the following possibilities exist for taking the measurements. The measurements could be conducted at periodic load changes (day/night) or in a very narrow current density range (e.g. 2·5–3·5 kA/m² in 0·1 kA/m² steps) or as a rapid procedure with about 3 min per measurement. The more experience obtained, the more familiar will be the behaviour of the cell in question. One should therefore record voltage–current curves as frequently as possible, if for no other reason than to determine the normal range of error of the C and k values. This will avoid false conclusions on the basis of minor random changes.

4 CONCLUSIONS

The components of the cell voltage of membrane cells and their dependence on current density have been discussed. It was determined that within the technically important current density range of 1–5 kA/m² linear functions exist and the adding-up of the individual linear equations show the dependence of the cell voltage on current density.

These conclusions were confirmed by measurements on laboratory membrane cells. Moreover, data taken from the literature were considered under this aspect. The data in the literature and further measurements of our own on three different cell configurations yield a good picture of the current density dependence of the components of cell voltage.

When breaking down the cell voltage into its components it was also determined that the intercept and the slope of the voltage–current curve $U = C + ki$ are influenced by various quantities. The intercept C is essentially determined by the overvoltages of the electrodes. The slope k, on the other hand, is influenced very largely by the resistances of the electrolytes and the membrane.

These relationships were used to interpret the increase in the voltage of a membrane cell using the voltage–current curves. A change in the cell voltage and hence in the C and k values with time frequently permits conclusions concerning the cause of these changes. The influence of the cathode, the cell temperature and the electrode gap was shown by means of examples.

Although the recording of voltage–current curves in industrial electrolysers is not always a simple matter, the method gives an early indication as to the necessary renewal of electrode coating or of membranes.

REFERENCES

1. Bergner, D., *Chem.-Ztg*, **109**(5) (1985) 177–83.
2. Kirsch, H., Diplom-Arbeit, Fachhochschule Darmstadt, 1986.
3. Chandran, R. R. & Chin, D.-T., in Proceedings of a Symposium of the Electrochemical Society, *Advances in Chlor-Alkali and Chlorate Industries*, Vol. 84-11, 1984, pp. 294–324.
4. Heitz, E. & Kreysa, G., *Principles of Electrochemical Engineering*. VCH-Verlagsgesell-schaft, Weinheim, 1986, p. 197.
5. Rondinini, S. & Ferrari, M., Proceedings of the Electrochemical Society, *Diaphragm, Sep., Ion-Exch. Membr.*, Vol. 86-13, 1986, pp. 120–32.
6. Kircher, M. S., in *Chlorine*, ed. J. S. Sconce. Reinhold Publishing Corp., London, 1963, p. 115.
7. Dotson, R. L. & Yeager, H. L., in *Modern Chlor-Alkali Technology*, ed. M. O. Coulter. Ellis Horwood, Chichester, 1980, pp. 153–71.
8. Hamann, C. H. & Vielstich, W., *Elektrochemie I*. Verlag Chemie—Physik Verlag, Weinheim, 1975, p. 114.
9. Schmid, G., *Ber. Bunsenges. Physik. Chem.*, **71**(8) (1967) 778–89.
10. Nidola, A., in *Membr. Membr. Processes (Proc. Eur.–Jpn Congr. Membr. Membr. Processes)*, ed. E. Drioli & M. Nakagaki. Plenum Press, New York, 1986, pp. 281–9.

16

DEVELOPMENT OF LARGE MONOPOLAR TYPE MEMBRANE ELECTROLYZERS

Kenzo Yamaguchi and Isao Kumagai

Chlorine Engineers Corp. Ltd, Japan

1 INTRODUCTION

The ion exchange membrane (IEM) process is now accepted as superior to both mercury and diaphragm processes in energy efficiency as well as environmental impact. The number of membrane cell plants has been steadily increasing in recent years.

Figure 1 shows the process conversion trend in Japan since 1975. The membrane process increased its share to 75–80% by the end of 1987.

Chlorine Engineers Corp. Ltd (CEC) has supplied many mercury and diaphragm cells, and has been engaged in developing various types of membrane electrolyzers such as a retrofit-diaphragm cell (MBC® cell), filter-press type monopolar cell and bipolar cell. CEC maintains a unique position in its contribution to the chlor-alkali industry as a major supplier of advanced electrochemical technologies.

As shown in Fig. 2, caustic soda produced by membrane cells with which CEC is associated either in development or construction amounts to 1 400 000 t/year of caustic soda as of April 1988. It shows CEC's market share in Japan to be 44%. In total, 32 IEM plants ranging from 8 to 370 t/day of NaOH have been operating in the world. Furthermore, CEC made a contract at the end of 1988, with Formosa Plastics Corp., USA, for a 1900 t/day NaOH plant, which is the largest single membrane plant.

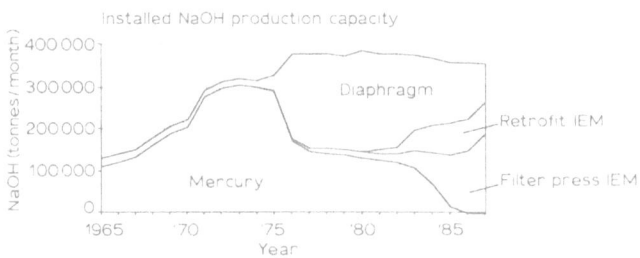

Fig. 1. Trend of process conversion in Japan.

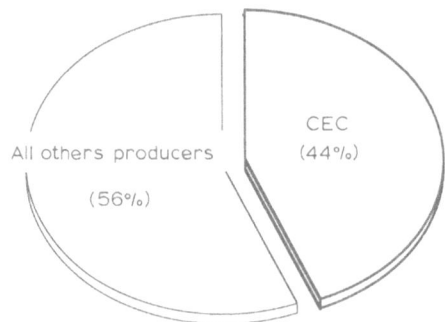

Fig. 2.   Japanese membrane process market share.

2 CEC's MEMBRANE CELL DEVELOPMENT HISTORY

Table 1 shows the history of CEC's membrane cell development.

CEC's first program on the development of the membrane process for commercial use was an installation of a bipolar type pilot cell in 1975 in the era of the first phase of mercury process conversion. However, since membrane cell performance in those days was unable to exceed that of the diaphragm process for industrial use, most chlorine producers adopted diaphragm cell technology.

Although we started our R&D with a bipolar cell for easy evaluation of membrane and cell design concepts in an independent test on each unit cell, as a result of careful study of the forthcoming conversion projects, we moved back to monopolar cell development.

Figure 3 shows a recent market share of various membrane processes in Japan. The share of monopolar cells is 1·8 times larger than that of bipolar cells, reflecting clients' needs for maximum utilization of existing systems and minimum production loss during conversion.

It was in 1977 that two monopolar type pilot cells were installed by us at a design current of 100 kA and, 3 years later, joint development of the membrane process

TABLE 1
Development history of CEC's IEM electrolyzer

Year	Development	Comment
1975	Test cell	Bipolar type
1977	Semi-commercial test plant	Monopolar type (100 kA)
1980	First commercial plant	Monopolar type
	Large size cell with NAFION® N-901 membrane	Monopolar type (60 kA)
1981	Development of retrofit cell	MBC® cell (13 plants)
1984~	CME application for commercial conversion projects	

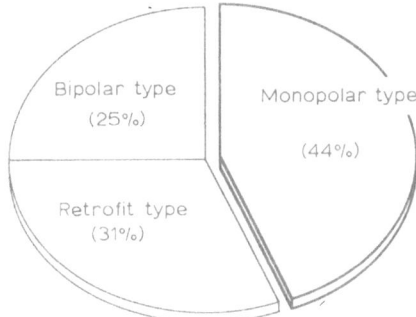

Fig. 3. Membrane process market share in Japan classified by cell type.

started with a leading membrane supplier based on an agreement through which we obtained much useful data for commercialization.

In 1980 the first CME commercial plant of 8 t/day of caustic soda was constructed in Okinawa, Japan. It was a conversion from diaphragm technology. As this client needed only 25 wt % caustic soda, conversion to the membrane process, producing 25–28 wt % caustic soda directly, was very attractive to them.

In the same year another 60 kA monopolar type cell was added to the pilot cell circuit. In this cell Du Pont's high-performance carboxylic membrane, NAFION® N-901, was fitted.

As shown in Fig. 4, since the appearance of NAFION® N-901 and other carboxylic membranes, and improvements in activated cathode coating technology and cell hardware, cell performance has improved dramatically over a short period of time to reach today's high standards.

3 KEY FACTORS FOR THE BEST SELECTION OF IEM ELECTROLYZERS

Recently technology suppliers have been offering a variety of membrane electrolyzer designs: monopolar and bipolar, finite gap and zero gap, metal frame

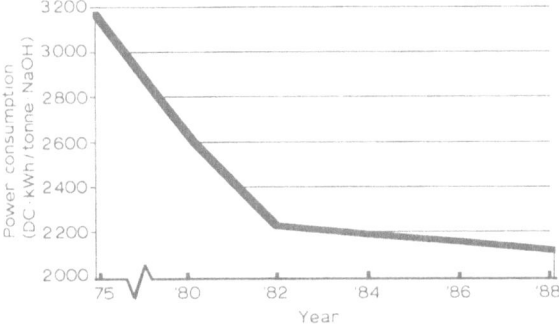

Fig. 4. CME performance improvement.

and rubber frame. However, regardless of cell type, there are certain key factors in selecting any commercial membrane cell for a chlor-alkali plant:

(1) long life with high cell performance;
(2) easy and safe operation;
(3) low maintenance cost; and
(4) flexibility of cell operation.

3.1 Long Life with High Cell Performance

This can be achieved only by the right combination of correct cell design, proper operation, reliable membranes, and excellent activated anodes and cathodes.

A primary factor in cell design is how membranes are positioned for optimum cell operating conditions. In order to do this, local effects due to uneven electrolyte concentration and electrode gap, pressure fluctuation and gas void area should be avoided as much as possible.

Another point is the membrane itself. As already mentioned, today's high performance membranes meet industrial requirements. However, as each commercialized membrane has different properties, users must select carefully the most suitable one on the basis of plant conditions and experience in membrane cell operation, as well as the cell supplier's recommendation. For the user without prior experience in membrane cell operation, it may be more practical to start with a membrane which is less sensitive to operating fluctuations and brine impurities.

3.2 Easy and Safe Operation

In general, the membrane process is safer and easier to operate than mercury or diaphragm processes. This has become more commonly accepted since the improvement in membrane mechanical stability. In addition, with today's various sophisticated process controls and monitoring technologies, it is possible to avoid most operating problems.

3.3 Low Maintenance Cost

Membrane and frame gaskets are major items to be replaced at the end of a cell's life.

From the financial point of view of the membrane cell plant, selection of the correct cell type that offers maximum utilization of membranes and gaskets is therefore of primary importance.

For this reason, we think, in the long term, large size unit cells are more economical than small ones and metal frames result in less investment for frame gaskets than rubber ones.

3.4 Flexibility of Cell Operation

Unlike most pilot test cells, which are carefully monitored, commercial plant operation involves a large number of cells managed by a minimum number of personnel. Therefore a high resistance to operating fluctuations in brine quality and anolyte and catholyte concentrations is very important in commercial cell design.

Our experience has taught us that a finite gap cell has greater flexibility and is safer than a zero gap cell under abnormal conditions.

4 FEATURES OF THE CME ELECTROLYZER

Now we would like to explain the features of the CME electrolyzer, referring to those key factors in membrane cell selection described above. An overview of the CME electrolyzer is shown in Fig. 5.

4.1 Large Size and Effective IEM Utilization

The electrochemically effective area of the CME for the DCM 400 series is as large as $3 \cdot 03 \, \text{m}^2$ per element, resulting in a lower IEM inventory, less assembly and disassembly man-hours and a smaller installation area. Moreover, $2 \cdot 98 \, \text{m}^2$ ($1 \cdot 22 \times 2 \cdot 44 \, \text{m}$) on a dry basis of membrane can fit each element by taking advantage of the expansion properties of the IEM. Membrane utilization is as much as 102%, as shown in Fig. 6. Therefore the user can enjoy a large IEM inventory saving. In order to utilize IEM even more efficiently, we are able to offer the ECM 400 series with $3 \cdot 79 \, \text{m}^2$ per element.

4.2 Use of Durable Material

As membrane cell operating conditions become more severe and the life of the IEM becomes longer, durability of cell component materials becomes more important.

Although there are many economic factors involved in the choice of material construction, CEC has selected high grade materials from its many years of experience in the electrochemical field to ensure a long life for the CME.

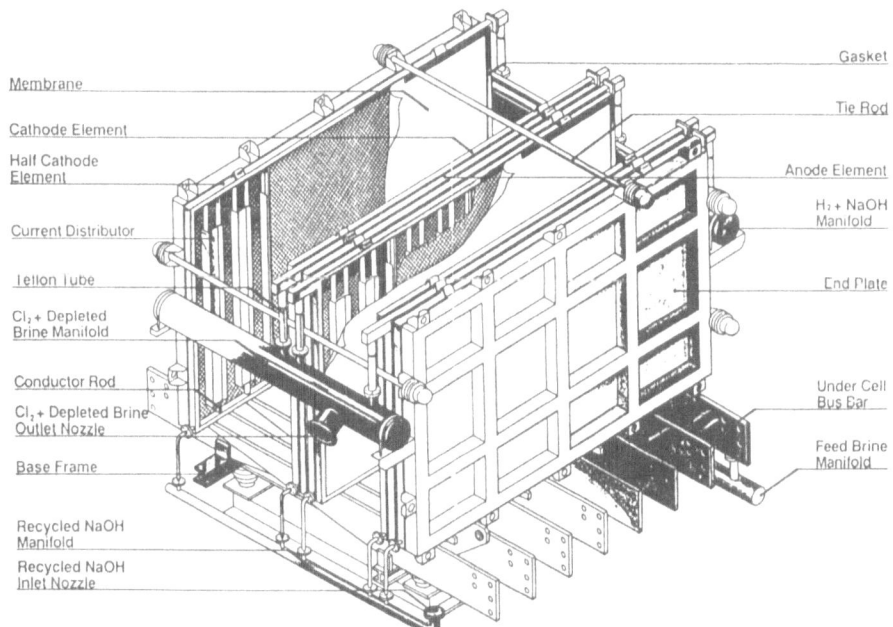

Fig. 5. Structure of CME.

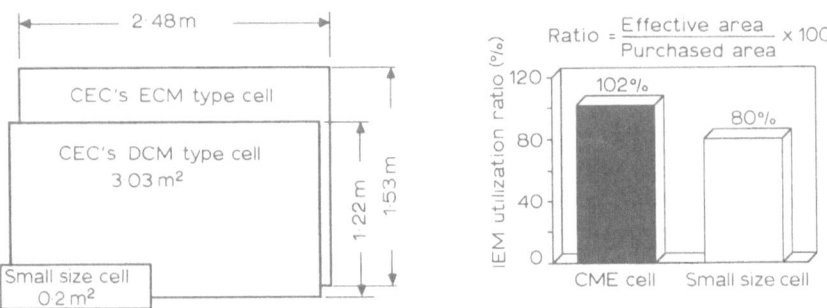

Fig. 6.   Large size and effective utilization.

The following materials have been selected:

	Cell component	Material
(1)	Anode frame	Ti
(2)	Cathode frame	Special stainless steel or Ni
(3)	Anode conductor rod	Cu/Ti
(4)	Cathode conductor rod	Cu/special stainless or Ni

4.3 Uniform Current Distribution

It is said that the bipolar cell is superior to the monopolar cell in terms of uniform current distribution. However, we have overcome this by incorporating the following features into the cell design and have successfully developed high performance large monopolar cells.

(1) As shown in Fig. 7, there are eight conductor rods having high electric conductivity connected with current distributors on a 310-mm pitch located in each element so that electric current can pass through the conductor rods and current distributors to the mesh by the shortest route. In order to achieve this, eight undercell bus bars are arranged at right angles to the elements and current distribution between the elements is then uniform.

(2) Also essential for a uniform current distribution is the preparation of a uniform electrolyte concentration inside the anode and cathode chambers. In response to this requirement, the CME electrolyzer has a unique feature, as shown in Fig. 8. A box type current distributor in the respective anode and cathode elements allows self-circulation of anolyte and catholyte by specific gravity differential between the inside (solution) and the outside (solution and generated gas).

By this means, generated gases (Cl_2-anode chamber, H_2-cathode chamber) leave swiftly from the electrode surface, reducing electric resistance otherwise caused by gases and equalizing the concentration of electrolytes in the anode and cathode chambers.

The volume of self-circulating electrolyte is more than 100 times that of

(a)

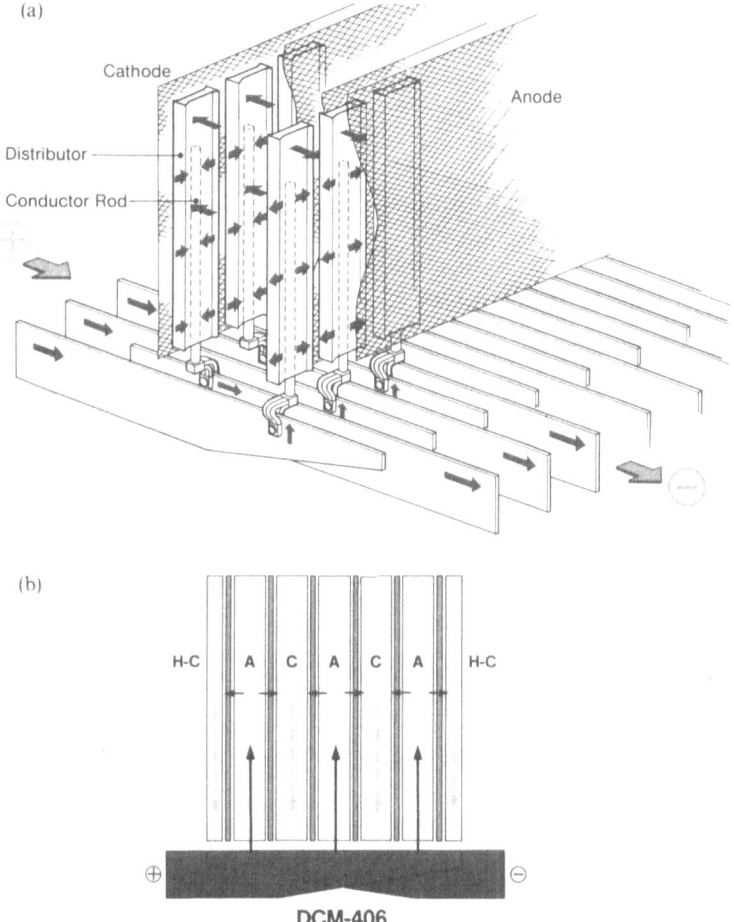

(b)

DCM-406

Fig. 7.   Uniform current distribution: (a) current flow; (b) cross-section.

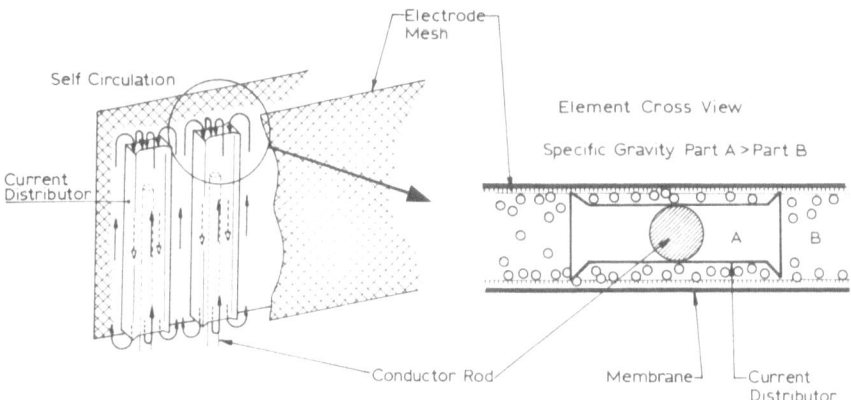

Fig. 8.   Self-circulation.

feed brine, which eliminates the extra power consumed by external circulation methods.

4.4 Overflow Mode

As shown in Fig. 9, gas and electrolyte, separated in the upper hollow frame of each element, leave the cell compartment in overflow mode with little pressure fluctuations and therefore the differential pressure across the IEM can be minimized ($\Delta P = P_{H_2} - P_{Cl_2} = 60\,mm\,H_2O$).

Thus maintaining anolyte and catholyte levels in the upper hollow frame of each element offers the following advantages:

(a) Electrochemical and physical degradation of the IEM, caused by direct contact with chlorine gas, is eliminated as the membrane is completely immersed in electrolyte.

(b) If pinholes do develop in the IEM, H_2 and caustic soda leakage into the anolyte chamber is so limited that there is no risk of an excessive rise of hydrogen concentration in the chlorine gas or of severe damage to the anodes. This compares very favourably with high pressure operation.

4.5 Durable Activated Cathode

Cathode activation technology is one of the key factors in cell voltage reduction. CEC's activated cathodes can provide a constant 200 mV saving for over 3 years and make possible a 5–6 year operating cycle before recoating. Figure 10 shows typical performance data for CEC's activated cathodes.

4.6 Wide Selection of Membrane

Membrane selection is often difficult for users. In order to get successful results, many factors such as economic factors (membrane cost, operating cost), product specifications (caustic strength, quality), operating skills and experience with membrane technology should be taken into consideration. However, CME's superior design concepts provide the user with flexibility and successful results using

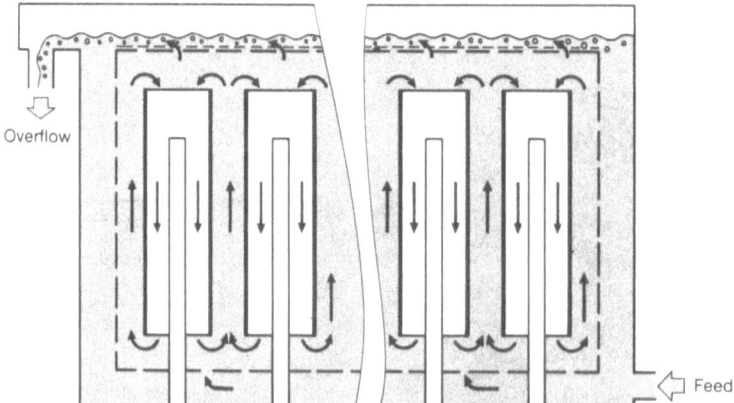

Fig. 9.   Overflow mode.

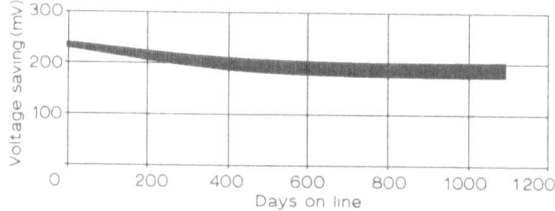

Fig. 10. Durable activated cathode. Operating conditions: 90°C, 32 wt% NaOH, 3 kA/m².

a large variety of membranes such as NAFION® and FLEMION® depending on clients' needs. Figure 11 shows cell voltage curves with different types of NAFION® membranes.

4.7 Safety Considerations
The CME electrolyzer incorporates various safety considerations as follows.

(a) Leakage and Corrosion Free
The CME combination of monopolar cell design, high grade cell fabrication technology and durable cell materials promises the elimination of leakage and corrosion.

(b) Visible Flow
As shown in Fig. 12, unlike an internal duct system, the CME applies an external liquor feed and gas/liquor overflow system of transparent Teflon tubes through which the flow at each element can be visually checked.

If any membrane damage occurs, the tint of chlorine gas in the Teflon tube, which normally is yellow, turns pale. By observing such colour change abnormal cell conditions can be identified.

In addition, abnormal brine flow and catholyte flow also can be detected through these tubes.

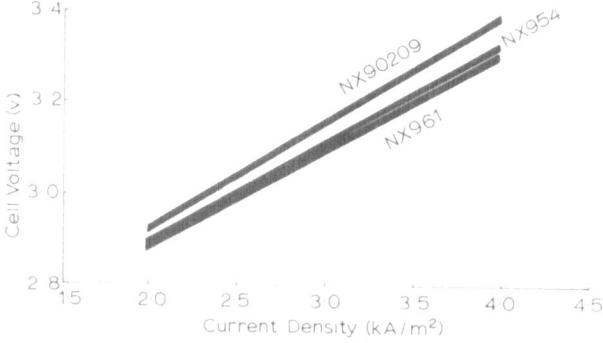

Fig. 11. Wide selection of NAFION® membranes. Operating conditions: 90°C, 32 wt% NaOH, 210 g/litre NaCl, DSA® anode, activated cathode.

Visual Check of

1) Brine feed failure

2) Caustic feed failure

3) Membrane damage

Safe Operation

Fig. 12.   Visible flow.

(c) DDP System

CME cell plants employ the scanning and alarm system of DDP (Data Display & Processing) which detects any abnormality by monitoring cell voltage and cell temperature.

5 APPLICATION EXAMPLES OF CME ELECTROLYZERS

Now we would like to briefly describe two examples of the application of CME technology; firstly, a conversion from a mercury process, and secondly, a grass roots plant will be discussed.

5.1 Process Conversion from Mercury Process

This plant was converted during the period from December 1984 to March 1986. Before the conversion 22 mercury cells had been operated at 200 kA with a production capacity of 150 t/day NaOH.

Thirty membrane cells (DCM 408 × 2) were installed on 15 of the 22 mercury cell bottom plates, thus the total installation area could be reduced by 30% while slightly increasing the capacity to 155 t/day NaOH.

In order to minimize investment, the major existing facilities such as cell foundations, bottom plates of the mercury cells, rectifier, short circuit switches, main piping headers, crane, etc., continued to be used in the converted plant.

During the conversion both mercury cells and membrane cells were operated in the same circuit and 50 wt% caustic soda was produced without evaporation by feeding 32 wt% caustic soda from the membrane cells to the decomposers of mercury cells.

A part of the purified brine from the primary brine system for the mercury process was fed to the membrane cells through a secondary brine purification system. This combined operation had no adverse effect on either CME or mercury cells. The plant has operated smoothly since its initial start-up and a membrane life of 3–4 years is expected.

5.2 Grass Roots Plant

This is a grass roots plant constructed for Chemfab Alkalis Limited, Pondicherry, India, in June 1984.

This is the first membrane process plant in India and has been operating smoothly for about 3 years. All of the initially installed membranes are still in use despite

TABLE 2
CME electrolyzer applications

Plant name	*Location*	*Start-up date*	*Capacity (t/month NaOH)*
Showa Enso Ind.	Okinawa	Nov. 1980	220
Osaka Soda	Amagasaki	June 1981 and Aug. 1982	330
Mitsui Toatsu Chem.	Ohmuta	Dec. 1983 and Feb. 1984	490
Mitsui Toatsu Chem.	Nagoya	Nov. 1984–June 1985	490
Mitsui Toatsu Chem.	Osaka	Dec. 1984	1 860
Showa Enso Ind.	Okinawa	June 1985	20
Chemfab Alkali	India	June 1985	750
Toagosei Chem. Ind.	Nagoya	Oct. 1985 and Dec. 1985	5 080
Osaka Soda	Kokura	Dec. 1985–Feb. 1986	2 150
Mitsui Toatsu Chem.	Osaka	Dec. 1985–Mar. 1986	2 790
Nippon Soda	Takaoka	Mar. 1986	4 180
Kanto Denka	Mizushima	Apr. 1986 and July 1986	1 258
Mitsui Toatsu Chem.	Ohmuta	June 1987 and Sept. 1987	120
Plant A	Korea	July 1987	1 026
Toagosei Chem. Ind.	Tokushima	Sept. 1987 and Oct. 1987	6 125
Plant B	West Germany	Oct. 1987	67
Nippon Soda	Takaoka	Oct. 1987	660
Osaka Soda	Kokura	Dec. 1987	80
Plant A	Korea	Feb. 1988	513
Toagosei Chem. Ind.	Tokushima	Mar. 1988	3 530
Mitsui Toatsu Chem.	Nagoya	Apr. 1988	240
Mitsui Toatsu Chem.	Ohmuta	May 1988	120
Toagosei Chem. Ind.	Nagoya	May 1988	320
Hodogaya Chem.	Koriyama	Aug. 1988	1 155
Plant C	Spain	Aug. 1988	144
Chemfab Alkali	India	Sept. 1988	375
Mitsui Toatsu Chem.	Nagoya	Sept. 1988	120
Plant D	Japan	Oct. 1988	639
Plant A	Korea	Apr. 1989[a]	1 231
Plant E	Japan	Aug. 1989–Mar. 1990[a]	11 840
Plant F	USA	May 1989[a]	838
FPC, USA	USA	End of 1990[a]	57 000
			105 761

[a] Scheduled start-up dates.

many shut-downs due to power failure. This September the plant capacity will be increased from 25 to 38 t/day NaOH.

Table 2 shows the CME electrolyzer applications list.

6 FUTURE TECHNICAL DEVELOPMENT

The present membrane process technology has been established and demonstrated in many commercial plants around the world. However, there is still potential for further improvement, and progress in reducing energy consumption and investment cost is anticipated.

The following are possible targets that will be commercially available within 2–3 years:

(1) high current density cell, 4–5 kA/m^2;
(2) high strength caustic soda operation, 40–50 wt% NaOH; and
(3) lower power consumption, 2000–2050 d.c. kWh/t NaOH at 3 kA/m^2, 90°C, 32 wt% NaOH.

CEC constantly meets the challenge of membrane process technology, contributing to the expanding future of the chlor-alkali industry.

7 CONCLUSION

CEC has succeeded in developing large monopolar type membrane electrolyzers offering stable, safe, high performance operation with minimum maintenance. CEC's superiority has been proven in many commercial plants and is highly appreciated by our clients. With our considerable experience and expertise, we are confident that we can meet the needs of clients anywhere in the world.

REFERENCES

1. Yamaguchi, K., Ichisaka, T. & Kumagai, I., The control of brine impurities in the membrane process in the chlor-alkali industry. In *Proceedings of the 29th Plant Operations Seminar*, Tampa, Florida. The Chlorine Institute, Inc., Washington, DC, USA, 1986.
2. Yamaguchi, K., The progress of process conversion in Japanese chlor-alkali industry. In *Proceedings of the Symposium on Electrochemical Engineering in the Chlor-Alkali and Chlorate Industries*, Honolulu, Hawaii. The Electrochemical Society, Inc., Pennington, NJ, USA, 1987.

17

FM21 ELECTROLYSERS—THEIR APPLICATION TO A RANGE OF OPERATIONS IN CHLOR-ALKALI PLANTS

I. J. M. Girvan, C. Brereton and A. L. Crawford

ICI Chemicals and Polymers Ltd, UK

1 INTRODUCTION

The past 7 years have seen the introduction of ICI FM21 cells into a wider range of chlor-alkali plants than any other membrane cell. The size range of the plant has varied from 100 000 tpa to less than 5000 tpa. The types of operation can be classified into:

Greenfield sites—completely new plants.
Mercury cell conversions.
Diaphragm cell conversions.
Replacement of membrane cells.
Demonstration plants.

The geographical spread of the plants is wide (Table 1) and covers organisations experienced in chlorine manufacture to those starting chlor-alkali units for the first time. FM21 technology has been adapted to this range of operations and technical support varied according to the widely differing needs of our customers.

Turning to the greenfield projects, we can see from Fig. 2 the wide range of capacity covered. It is interesting to note how the number of electrolysers varies from the 52 large double-pack systems used in Finland to the smaller packs of electrolysers used to meet the needs of the smaller producer. By varying the number of electrodes in each pack a sensible electrical arrangement consistent with economical plant layout can be achieved. With some of our remote greenfield projects the emphasis must be on straightforward reliable systems which relatively inexperienced personnel can operate safely. With the larger merchanting applications a greater emphasis on performance is needed.

Table 2(a) highlights the geographical spread, capacity range and numbers of electrolysers.

TABLE 1

Chlor-alkali plants where ICI FM21 technology has been selected

Commercial plants	Capacity (TeNaOH/year)	
Skoghall-Kemi, Skoghall, Sweden	45 000	Conversion of mercury cell capacity. Commissioned December 1983
	5 000	Conversion of bipolar membrane cells. Commissioned 1985
Nokia, Joutseno, Finland	100 000	Greenfield plant. Commissioned April 1984
Elkem, Svelgen, Norway	11 000	Replacement of membrane cell capacity. Commissioned June 1984
CCM, Padang Jawa, Malaysia	10 000	Conversion of mercury cell capacity. Commissioned October 1984
Mondi, Richards Bay, South Africa	18 000	Greenfield plant. Commissioned February 1985
Procter & Gamble, Green Bay, USA	5 250	Greenfield plant. Commissioned March 1985
Pennwalt, Tacoma, USA	90 000	Conversion of diaphragm cell capacity. Commissioned September 1985
Confidential, USA	5 000	Conversion of membrane cell capacity. Start up November 1985
Sabah Forest Industries, Malaysia	11 000	Greenfield plant. Start up 1987
CSBP, Australia	5 000	Greenfield plant. Start up 1987
APPM, Australia	7 000	Conversion of mercury cell. Start up January 1988
Prodesal, Colombia	12 000	Greenfield plant. Start up 1988
ICI, United Kingdom	28 000 (KOH)	Conversion of mercury cell plant. Start up 1988
Chemiabone, Iran	1 000	Greenfield plant start up
China General Petroleum Co.	35 000	Conversion with expansion of mercury cell plant. Start up 1989
PPG, Canada	84 000	Conversion of mercury cell plant. Start up 1990

The mercury cell plant conversions in Table 2(b) have been total or partial conversions and have mostly involved the use of existing rectification equipment. With excellent utilisation of floor area which the FM21 cells give and the capability to vary electrolyser pack size customers have been able to uprate their capacity within existing plant buildings and using existing rectifiers.

Indeed, ICI is in the process of converting its KOH mercury cells to membrane cells.

One large diaphragm cell plant has been converted to FM21 cells as well as part of ICI's own diaphragm cellroom at Lostock here in the UK, where the initial demonstration plant is being gradually enlarged as more diaphragm cells are converted to membrane cells. Bipolar diaphragm cell technology was displaced in one case while monopolar technology was displaced in the other (Table 3(a)).

Two of our plants have replaced bipolar membrane cell types which other suppliers had provided early in the development of the technology (Table 3(b)).

Three demonstration plants are currently operated by customers keen to evaluate the technology thoroughly before committing to major conversions at a time when the need inevitably will arise (Table 3(c)).

TABLE 2(a)
Greenfield sites

Company	Location	Capacity (*TeNaOH/year*)	Start-up date	Number of electrolysers
Nokia	Finland	100 000	April 1984	52 (56)
Mondi	South Africa	18 000	February 1985	18
Procter & Gamble	USA	5 250	March 1985	12
Sabah Forest Industries	Malaysia	11 000	June 1987	12
CSBP	Australia	5 000	November 1987	12
Prodesal	Colombia	12 000	July 1988	14
Chemiabone	Iran	1 000	Late 1990	2

TABLE 2(b)
Conversion of mercury cell plant

Company	Location	Capacity (*TeNaOH/year*)	Start-up date	Number of electrolysers
Skoghall-Kemi	Sweden	45 000	December 1983	50
CCM	Malaysia	10 000	October 1984	18
APPM	Australia	7 000	January 1988	24
ICI	UK	28 000 (KOH)	October 1988	36
China General	Taiwan	35 000	Early 1989	60
PPG	Canada	84 000	December 1990	46

TABLE 3(a)
Conversion of diaphragm cell plant

Company	Location	Capacity (*TeNaOH/year*)	Start-up date	Number of electrolysers
Pennwalt	USA	90 000	September 1985	76

TABLE 3(b)
Replacement of membrane plant

Company	Location	Capacity (*TeNaOH/year*)	Start-up date	Number of electrolysers
Elkem	Norway	11 000	June 1984	22
Confidential	USA	5 000	November 1985	12

TABLE 3(c)
Demonstration facilities

Company	Location	Capacity ($TeNaOH/year$)	Start-up date
ICI, Lostock	UK	6 000	Since 1978
Bayer	West Germany	700	1982
Confidential	Japan	1 000	1984
Confidential	Belgium	200 (KOH)	1985

2 PRACTICAL LESSONS

During the course of designing and commissioning these plants a wide range of experience has accumulated in the FM21 team. Three areas will now be discussed in more detail:

1. Brine systems.
2. Analytical needs.
3. Cellroom layout.

2.1 Brine Systems

Starting with brine purification systems, in most cases, our technology has been matched to the existing primary purification plants to give comprehensive systems capable of providing brine to the rigorous specification demanded by the technology. A few problems have been encountered in getting the combined systems to work efficiently and two case studies will now be described.

In the first case history, a large plant had a unique salt dissolving system using imported salt. The symptoms experienced were a steadily deteriorating energy performance and erratic voltage.

The problem was traced to organics contaminating the brine and the fact that initial salt washings were returned to the brine feed. The reason for the problem was foaming in the cell caused by organics.

The problem ceased immediately with the elimination of organics from the brine and the contaminants from the salt washings, and excellent steady energy performance has been achieved for over 2 years on the plant.

The main lesson here is to ensure that the primary brine system operation is thoroughly checked out before secondary brine design goes ahead.

The second case also relates to the performance of the primary brine purification system.

In this case a small FM21 mercury cell conversion using solar salt experienced slowly deteriorating current efficiency. The symptoms in plant operation were:

1. Very short primary brine filter cycles.
2. Scaling-up in filters.
3. Evidence of Ca/Mg in membrane.
4. Carry over of solids into the brine system.

After studying the problem, the difficulties were solved by improving desuper-saturation and improving washing efficiency in the filters. The result is a well-run plant which has operated at over 95% current efficiency for 2 years.

These two examples show that primary brine systems, which are perfectly satisfactory for mercury and diaphragm cell operation, can give problems when combined with a secondary system for membrane cells, and close attention to operation and design is required prior to secondary brine design and plant start-up.

2.2 Analytical Needs

At first sight the analytical needs of membrane cell plants may seem substantial, particularly to the smaller producer, and even to the larger operator they still represent a large investment in equipment and expertise. You can see from Table 4 that the feed brine and depleted brine need careful regular monitoring since, in order to maintain good and *durable* energy performance, the brine quality has to be kept as high as practicable. It is a condition of warranty on the membrane that the quality of brine is monitored on a regular basis.

Further important analyses are:

—Strength and chloride ion content of caustic streams to measure production rates and product quality.

TABLE 4
Membrane chlor-alkali plant analyses

Matrix	Analyte	Frequency
Feed brine	NaCl	Shift
	Calcium + magnesium	Shift
	Alkalinity/acidity	Shift
	Sulphate	Daily
	Aluminium	Weekly
	Silica	Weekly
	Strontium	Weekly
	Total iodine	Weekly
	Barium	Weekly
	Heavy metals	Weekly
	Mercury	Weekly
	Iron	Weekly
Depleted brine	NaCl	Shift
	Acidity	Shift
	$NaClO_3$	Daily
	NaOCl	Daily
Feed caustic soda	NaOH	Shift
	Iron	Weekly
Product caustic soda	NaOH	Shift
	Chloride	Daily
Water	Conductivity	Daily
Chlorine gas	$H_2/O_2/CO_2$	Shift

TABLE 5
FM21 users and analytical packages

User type	Typical analytical facility	Package content	Package cost (K£ UK)
Major chlor-alkali producer	Analytical department	State-of-the-art high technology equipment	150–200
Minor chlor-alkali producer	Analytical laboratory	Instrumentated laboratory with proprietary dedicated C/A analysis instruments	50–70
Remote plant	Plant laboratory	Wet chemistry laboratory plus toll analyses	15–25

—Gas analyses which, combined with anolyte analyses, yield current efficiency results which are a good indication of the health of the cells (Table 4).

Aware that producer's resources and plants' individual needs vary greatly there is a need to provide a range of graded options which both protect the user from process fluctuations, protect his warranty and yet do not involve him in unreasonable expenditure on analytical equipment.

We see above (Table 5) the order of cost that a customer may expect to encounter depending on the size and facilities available to him. With our experience in a wide range of plant situations it has been possible to design packages of the type shown in this table tailored to specific customer needs.

A major chlor-alkali producer is likely to have an analytical department, and state-of-the-art high tech equipment can readily be used and serviced. The order of cost here is £150–200k. The smaller chlor-alkali plant with an analytical laboratory is more likely to want a package containing more reasonably priced proprietary equipment. This will cost in the region of £60k. In more remote locations the

TABLE 6
Analytical options—calcium + magnesium in feed brine, maximum limit 20 ppb

Method	Instrument cost (£ UK)	Benefits/disadvantages
Titrimetry	20	Quick but very inaccurate (± 25 ppb); high contamination risk
Colourimetry	1 500	Tedious and inaccurate (± 10 ppb); high contamination risk
Graphite furnace AA	22 000	Fast and accurate (± 2 ppb); high degree of manipulative skill—special laboratory environment
Inductively coupled plasma	100 000	Fast and accurate (± 5 ppb); automated and versatile

package would contain a wet chemistry laboratory with other important elements on toll analysis, and subsequently lower initial package costs of around £20k.

If we look more specifically at the options available for calcium and magnesium analysis in brine, we see a wide range of cost and expertise which offer different degrees of protection (Table 6).

The options vary from a rudimentary method using titrimetry for £20 to a fast, accurate automatic and versatile method for £100k.

We have worked with customers and analytical equipment suppliers to find the right package for each customer and, as a user and supplier of membrane cell technology, we are continuously working with suppliers of membranes and analytical equipment to reduce the analytical element in the overall cost consistent with sound plant operation.

Our own analytical facilities in Runcorn are substantial and a great asset in pursuit of our customers' best interests in these matters.

In respect of analytical needs for our own and our customers' operations the ICI service is summarised in this table. Basically the services offered cover:

1. Up-to-the-minute awareness of sensitivities of all commercially available membranes.
2. Ongoing evaluating of state-of-the-art analytical equipment in our own laboratories.
3. Development and evaluating of new analytical procedures.
4. Full analysis of customer/proposed feedstocks.
5. Recommendations for brine treatment plant modification based on a data base of worldwide experience.
6. Trouble-shooting analytical tech service for prompt response to contamination-induced problems.
7. Merchant analytical services for arbitration, reference and start-up purposes.

2.3 Plant Layout

We now turn to plant layouts. The following three existing plant layouts illustrate the adaptability of the FM21 concept.

In Finland we have a cellroom designed from first principles for a greenfield plant. This plant has a capacity of approximately 100 000 tpa using 52 electrolysers. These electrolysers have 116 anodes and 116 cathodes in an FM21 double pack arrangement. The cellroom layout is attractive and ergonomically efficient. Electrolyser switching is designed to be effected from underneath the electrolysers (Fig. 1).

Also in Scandinavia we have a different arrangement at Billerud, where a partial mercury cell conversion also has been effected in a manner to allow gradual conversion from mercury cells and some older bipolar cells to FM21s. The plant was designed by Stearns Catalytic, and in this case electrolyser switching was arranged from switches mounted above the electrolysers (Figs 2 and 3).

For a cost-effective provision of capacity for consumer installations, a computer detailed standard design is available with 12 electrolysers covering 3–12 ktpa Cl_2.

Across the world in Western Australia a relatively small plant (10 000 tpa) was

Fig. 1

commissioned in 1987 for CSBP, as shown in Fig. 4. The plant, designed by one of our preferred contracting organisations, Chemetics of Canada, used this compact computer design for a customer who had limited experience of chlorine production. In this case no switch is used. When necessary, which is very infrequently, all of the cells are switched out. The arrangement as shown in Fig. 5 has proved attractive to a number of our smaller customers with five installations built to date.

There are a number of considerations, of which space utilisation is one, that need to be addressed in retrofitting cells to, say, a mercury cellroom. There is a very significant saving in space which can be achieved with membrane cells and FM21 cells in particular. In Fig. 6 the equivalent of 500 tpd mercury cell capacity can be converted using less than 30% of the existing cell floor space. Even a bipolar cell

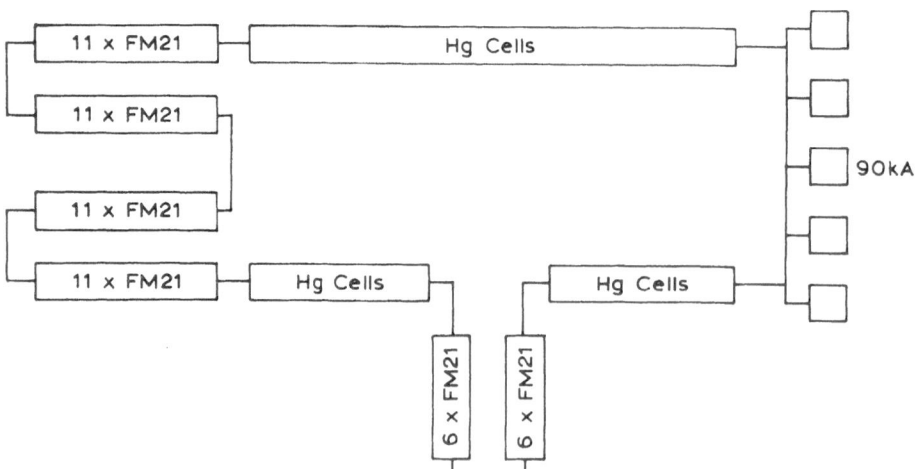

Fig. 2.   Hg cell—partial conversion.

Fig. 3

Fig. 4

Fig. 5

191

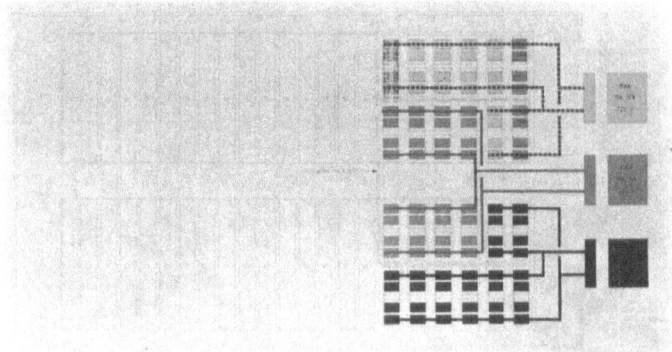

Fig. 6.   Plan of cellroom with bipolar membrane electrolysers.

arrangement might occupy as little as one-third of the equivalent mercury cell space. However, other aspects need to be considered also, viz.:

Use of existing rectifiers.
Operability.
Safety.
Maintenance, etc.

Figure 7 shows a bipolar arrangement recently offered which illustrates arrangements that should be avoided. In this case we have three rectifiers coupled with two piping circuits. This raises the aspect of electrical safety; in particular, when the central electrical circuit is not operating there will be electrolysers with large potential differences less than 1 m apart.

In addition, with liquor feeds common to two different electrical circuits there is a real possibility of liquor feed being stopped to cells when current is still passing. Such an occurrence could happen during maintenance or partial shutdown. In a design such as this we would recommend (1) the use of three electrolyser rows to

Fig. 7.   Plan of cellroom with monopolar membrane electrolysers.

Fig. 8

avoid having mixed electrolysers on the same row; (2) a review of the hazard associated with close proximity of high differential voltages; and (3) repositioning of feeds to identify them more readily with their associated rectifiers.

As an operator as well as a licensor of technology ICI is in a good position to design cellroom layouts to achieve a good balance between performance, capital cost, operability and safety as demonstrated by the FM21 cellroom at the Elkem plant in Norway, as shown in Fig. 8.

3 FM21-SP CELLS

The experience gained in operating the HP cells has been used to develop and improve the cell while maintaining the excellent initial basic design concepts. The

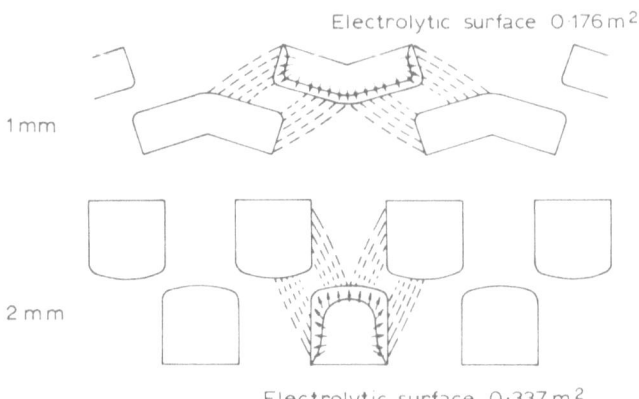

Fig. 9.   Increased area of FM21-SP electrode.

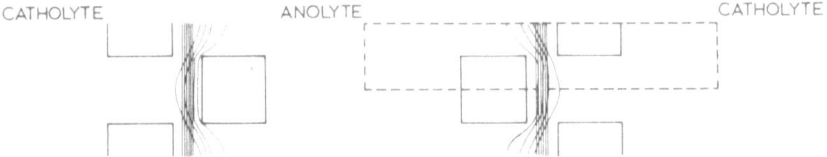

Fig. 10.   Potential contour map of a membrane cell with 2 mm square lantern blades.

new cell, known as the SP cell (superior performance), incorporates the following improvements:

Thicker electrode panels.
Increased surface area in the electrodes.
Wider gaskets to improve sealing for operation at elevated pressures.
Balanced headers for distribution of cell liquors.
Modified panel supports to enable easier assembly and maintenance.

The result is the SP electrolyser.

Improvement in metal pressing technology has enabled heavier pressings to be used, thus reducing the electrical resistance in the surface area of the electrode profile and giving improved current distribution, as shown in Fig. 9.

Mathematical modelling and experimentation have been used to investigate current distribution across the cell for different electrode shapes. An example of this work is shown in Fig. 10. Taking a horizontal section across the cell, Laplace's equation was solved over a complex region defining the boundaries of both the anode and cathode as well as the membrane which was assigned a finite thickness. This made it possible to study primary current distributions in various designs of cell, giving quick screening of potential electrode designs. This work clearly shows that reducing the pitch of the electrode blades improves the distribution of current on the membrane.

The performance of the SP cell, proven in the trials conducted in full-scale demonstration cells, is shown in Fig. 11. The durability of performance of membranes has now improved to the extent that 3-year membrane change-outs may be anticipated. Because the engineering principles behind the cell are straight-

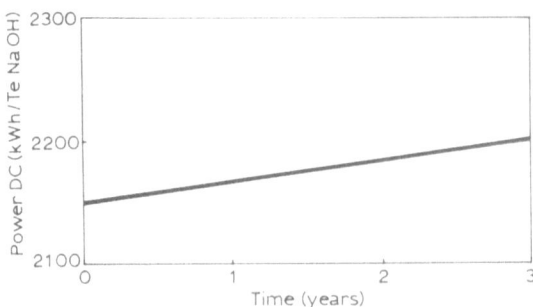

Fig. 11.   FM21-SP performance with NAFION membrane, coated cathodes at 3 kA/m².

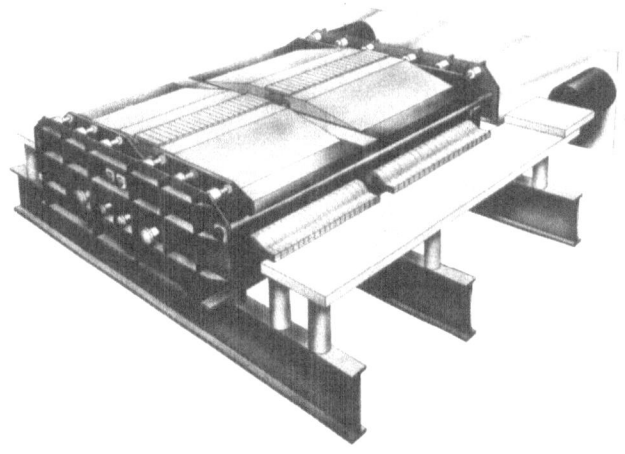

Fig. 12

forward it has been possible to rapidly design, develop, produce and test those new components in the relatively short period of $2\frac{1}{2}$ years.

Our latest development in the FM21 range is the SP100 cell, as shown in Fig. 12. This is in response to customer demand for an FM21 operating unit with a capacity of up to 4000 tpa. It consists of four standard electrolysers grouped together as a single operating unit and fitted with single brine and caustic feeds, and single liquor and gas exit lines.

In conclusion, we believe that ICI's FM21 team with our wide range of experience in meeting customer needs, with well-proven and adaptable products and world-class technical excellence, can offer the best package in all chlor-alkali membrane cell projects.

.

18

IMPROVING MEMBRANE CELL OPERATION

Gregory J. Morris

Oronzio De Nora Technologies SpA, Italy

INTRODUCTION

The purpose of this chapter is twofold: to review the status of commercial membrane cells as they exist to date, and to discuss the advancements in the De Nora Technologies DD-type zero-gap and finite-gap membrane cells.

Over the past 10 years or so there has been a considerable resurgence of interest in the electrolysis processes which use ion exchange membranes. This has been brought about by concerted efforts in the area of cell design, electrode coatings and membrane performance improvements. Because of these efforts numerous cell designs for both bipolar and monopolar configurations have been developed. These developments in cell design will be addressed in this chapter, which will be presented in two sections. The first will be on monopolar cell designs and the second on bipolar cell designs. It is important to keep in mind the similarities in design and operation of the De Nora Technologies monopolar cell and bipolar cell. Such similarities will be clarified in this chapter.

STATUS REVIEW OF COMMERCIAL MEMBRANE CELL TECHNOLOGY

A significant factor in determining electrolytic cell performance is the membrane. Energy consumption of early membrane cells exceeded 4800 kWh/t NaOH. Today, however, membranes are allowing cells to approach 2100 kWh/t NaOH and are showing signs of providing further decreases. Figure 1 shows the trend in energy reduction brought about by improvements in membrane technology.

Even though present-day membranes have provided great improvements in power consumption and durability, it is the electrolyzer and its design that allow membranes to attain their full potential over an extended lifetime. The major problems associated with today's commercial membrane electrolyzers, that affect membrane efficiency, longevity and overall cell performance, are listed in Table 1. These problems still exist in many of the commercial electrolyzer designs being used

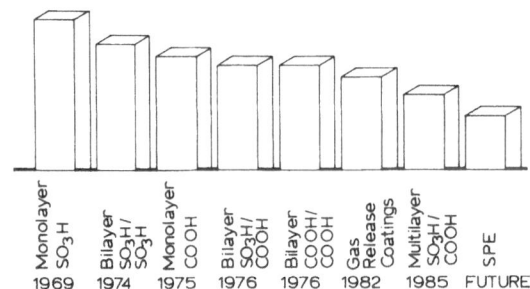

Fig. 1.   Trends in energy reduction with membrane improvements.

in the industry today. Some of these cell designs will now be addressed and it will be shown how the DD-type cell provides solutions to the above problems.

MONOPOLAR MEMBRANE ELECTROLYZERS

The basic design criteria which were initially set by early cell developers were to produce a membrane cell which combined the need for a low cost of manufacture (since membrane efficiency was low the cost of the cell had to be low to justify projects) and easy replacement (since membranes had a relatively short life and coatings had not been perfected for membrane use). From these early criteria two basic types of monopolar membrane cells were developed:

—plate cell
—frame cell

The Plate Monopolar Cell
The monopolar plate cell employs plates of metals that are resistant to the electrolytes. Slits are formed in the interior of the plate to allow electrolyte to flow to

TABLE 1
Problems associated with membrane performance and longevity

Problem	Area of concern	Consequences
Cl_2 stagnation	Top, sides and electrode connection points	Blisters
High structural ohmic losses	Cell body	High power consumption
Nonuniform current distribution	Surface of electrode	Blisters
High electrolyte ohmic losses	Electrode spacing	Power consumption
Membrane fluttering	Top and gas outlet	Membrane abrasion Membrane fatigue
Circulation restrictions	Electrode chamber	Blisters
Masking of membrane	Electrode surface and electrode connection points	Blisters
High gasket to element ratio	Peripheral flange	Waste of membrane
Complexity in fabrication	Cell	Increases cost Decreases reliability

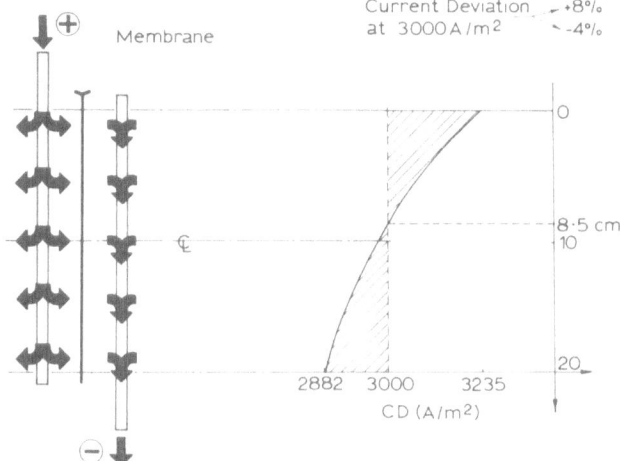

Fig. 2. Schematic of current flow and current deviation in a plate cell.

the membrane. The periphery of the plate is used for sealing and for electrical connection. Because the material used to construct the plates was chosen based on its chemical resistance and not on its current transmission ability, the cell is limited in height and requires a current distribution bus along its entire length. Even with these limitations the cell has a very large current deviation. For a cell whose cross-sectional area is only $0.2\,m^2$ and an electrode plate thickness of 2 mm a deviation of $+8\%$ to -4% at a current density of 3000 A/m^2 can be experienced. Figure 2 shows a pictorial representation of how the current is distributed in this type of cell.

Another problem associated with this cell is that the peripheral structure of the cell is normally constructed of rubber or plastic. During thermal excursions from ambient temperature to 90°C the cell can develop leaks.

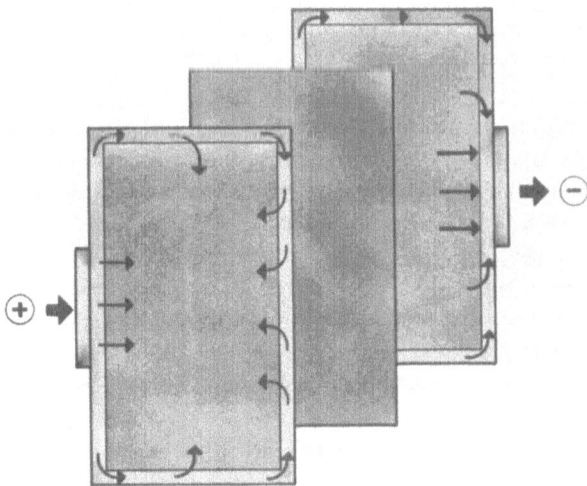

Fig. 3. Schematic of current flow in a frame cell.

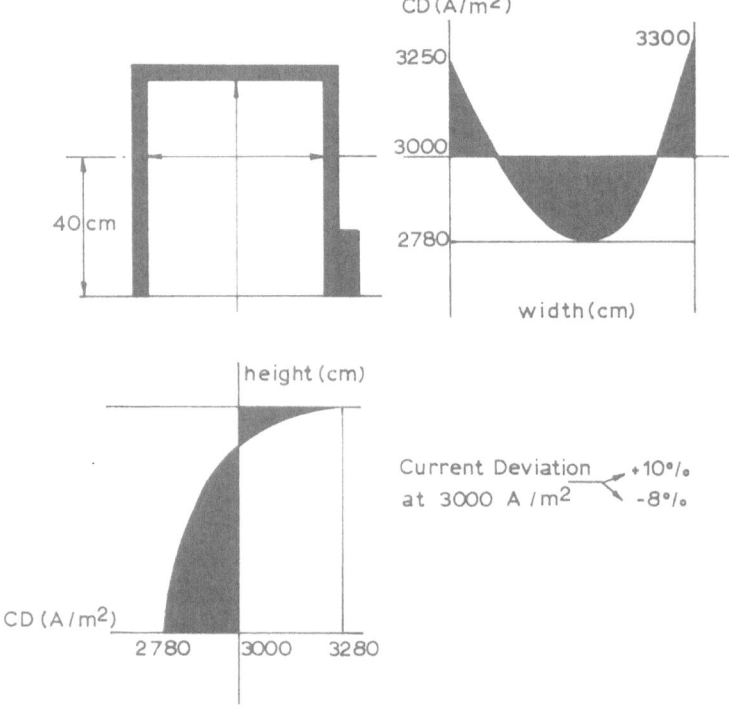

Fig. 4.   Current deviation in a frame cell.

The Frame Monopolar Cell

In the monopolar frame type cell design, tubular material resistant to the electrolytes is used for the frame construction. Foraminous electrodes also resistant to the electrolytes are welded to this tubular frame. Current is carried into the peripheral tubular flange of the cell and distributed within the cell through the corrosion-resistant electrodes (see Fig. 3). This type of cell has size limitations similar to the plate cell, in that the electrode material has a relatively low electrical conductivity and a limit on the cross-section available to carry current.

Upon closer examination of the power distribution in a frame type cell, we find a current deviation at 3000 A/m² for a cell, whose cross-sectional area is 0·8 m² (note: four times larger than a plate cell) and an electrode thickness of 2 mm, to be between +10% and −8% (see Fig. 4).

It should be noted that no matter what the cross-sectional area is a cell that has over 1% current deviation is placing additional workload on the membrane at those areas of higher current density. For the plate cell this will be at the top of the cell, where also we encounter the added problem of gas pockets. For the frame cell this is at the periphery, again in an area where gas pockets can form.

To improve current distribution in a frame cell some designers have introduced current distribution systems in the form of copper-filled electrolyte-resistant tubes (see Fig. 5). These tubes are inserted into the cell and welded to the peripheral flange,

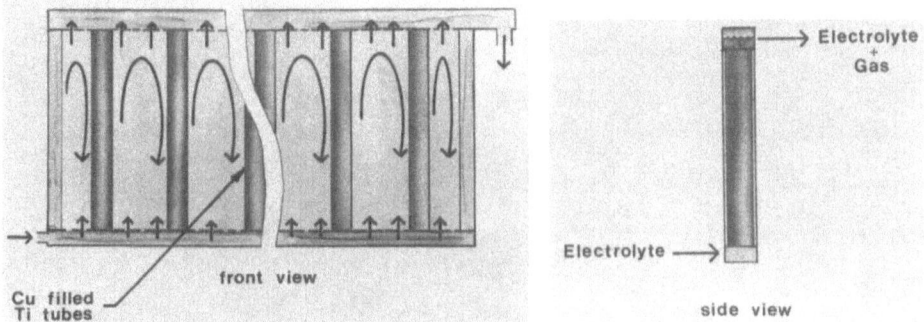

Fig. 5. Pictorial view of electrolyte flow pattern in a frame cell with current distributors.

and to the electrodes. This method improves the current distribution but creates the following problems:

1. Restricts natural electrolyte circulation.
2. Increases likelihood of a cell operating with zones of depleted anolyte, and with caustic concentrations in excess of optimum membrane operating parameters (consequence of problem 1).
3. Increases likelihood of localized hot spots (consequence of problems 1 and 2).

De Nora Technologies DD-Type Monopolar Cell

Under a research agreement in the late 1970s, researchers of the Dow Chemical Company and the Oronzio De Nora Group developed a novel technique for improving the distribution and collection of current in a monopolar cell. This cell design employs a current carrier (see Fig. 6, cathodic/anodic element) inside the cell to supply and distribute current. The current carrier is made of a single casting of metal to simplify construction and reduce cell cost. The casting is equal in surface dimensions to the active area of the electrode, with the plate's periphery extending beyond the active area for use as a sealing surface and a means for connecting power to the cell.

To protect the casting from chemical attack by electrolytes, press-formed corrosion-resistant metal liners are used. A typical cell design will use titanium for the anode lining of the cell and nickel for the cathode lining. These liners are mechanically and electrically connected by spot welding to the ends of the bosses which extend from the current carrier, and also support the electrode structures, which are spot welded to the liners at the bossed extensions. The current is conducted from the cast metal current carrier to the electrode liner and electrode structure via the spot-welded connections.

Due to the excellent current conduction of the metal current carrier, current distribution across the active electrode is very uniform. If we examine the current profile at one boss (see Fig. 7), we find for a $0.8 \, m^2$ cell with a 25 mm thick current carrier a deviation of only $+0.8\%$ to -0.3% at a current density of $3000 \, A/m^2$. For higher current density the thickness of the center board can be adjusted to maintain this even current transmission. Figure 8 shows a comparison of current deviations in each type of monopolar cell described.

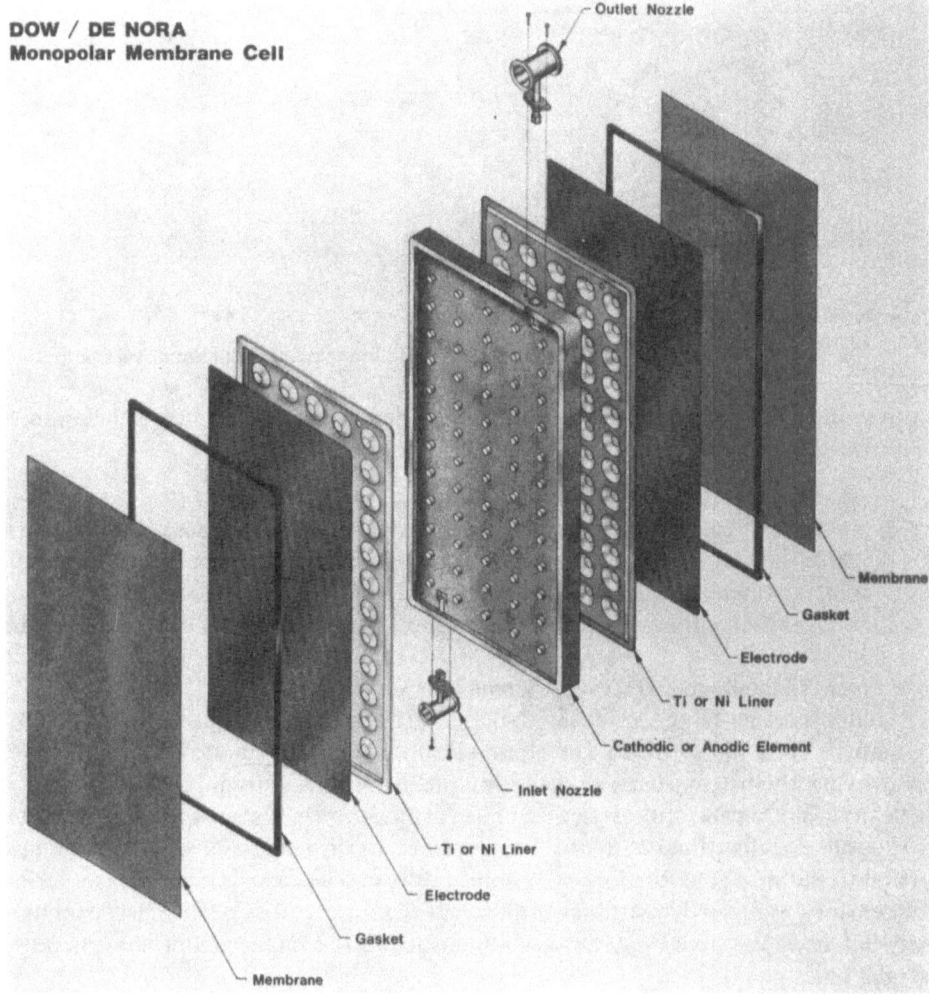

DOW / DE NORA
Monopolar Membrane Cell

Outlet Nozzle

Membrane
Gasket
Electrode
Ti or Ni Liner
Cathodic or Anodic Element
Inlet Nozzle
Ti or Ni Liner
Electrode
Gasket
Membrane

Fig. 6.   Exploded view of DD-type monopolar membrane cell.

Another advantage of the shape of the DD-type cell is that it promotes uniform circulation of the electrolytes. Restriction of electrolyte flow around the extended bosses and the membrane surface is minimized, and only the open electrode is allowed to come into contact with the membrane. Figure 9 shows an exploded isometric view of the major components of a DD-type monopolar electrolyzer.

BIPOLAR MEMBRANE ELECTROLYZERS

There are two basic types of commercial bipolar cells used in the industry to date:

—all-welded
—combination of welded and bolted

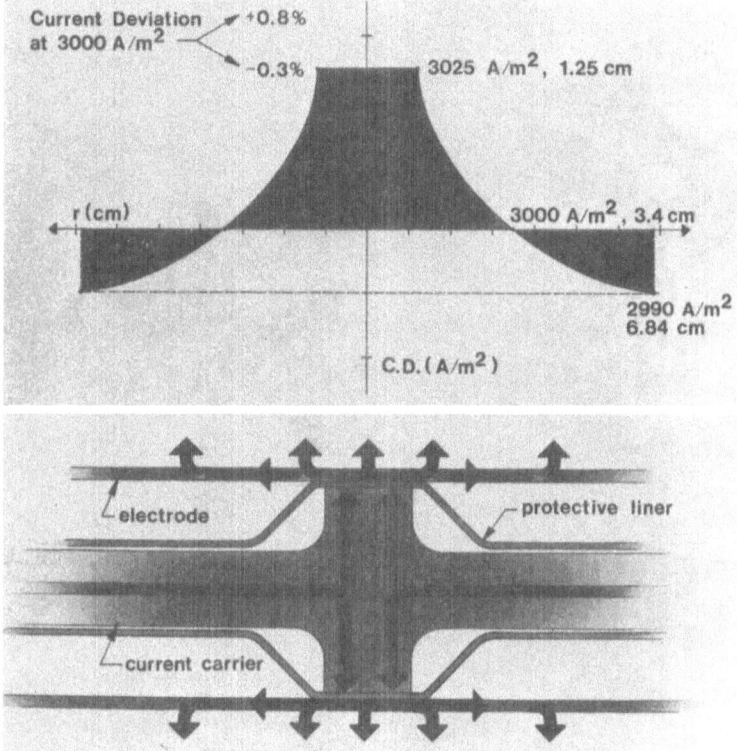

Fig. 7. Schematic of current flow and current deviation in a DD-type monopolar cell.

All-Welded Bipolar Cell

There are some variations in how this structure is assembled, but the most common is to use a double layer, and in some cases a triple layer, of explosion bonded metals. Materials normally used are titanium and steel for low caustic strength and titanium and stainless steel or nickel for higher caustic strengths.

Flanges with protective metal coverings are welded to the periphery of the plate to complete the electrolyte chambers. Ribs of titanium are welded vertically to the anodic chamber plate to carry the current from the center board to the electrode. The anode electrode is welded to these ribs to complete the anode section of the cell. For the cathodic chamber, steel, stainless steel or nickel ribs are welded to the center board. To these ribs the cathode electrode is welded (see Fig. 10) to complete the cathodic section of the cell.

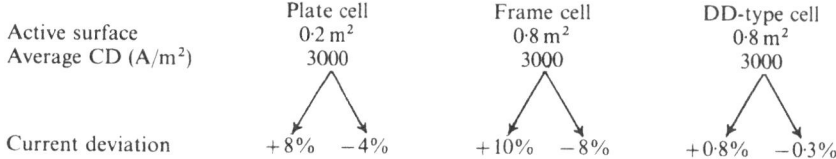

	Plate cell	Frame cell	DD-type cell
Active surface	0·2 m²	0·8 m²	0·8 m²
Average CD (A/m²)	3000	3000	3000
Current deviation	+8% −4%	+10% −8%	+0·8% −0·3%

Fig. 8. Summary of current deviation of the three types of cells.

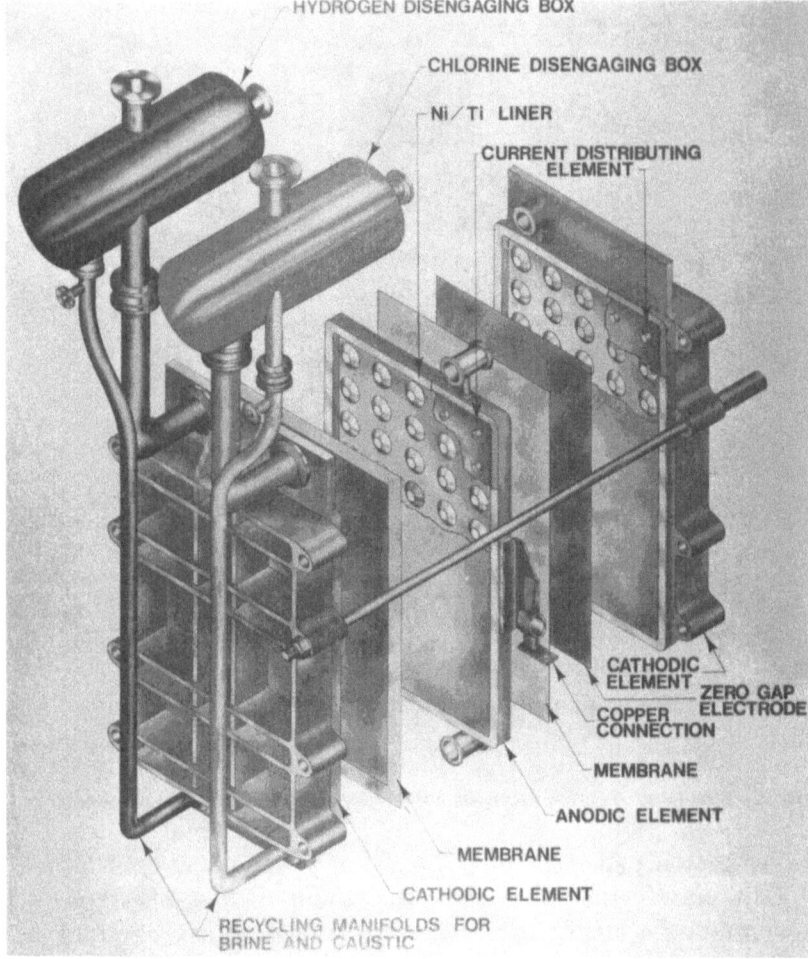

Fig. 9.   Exploded isometric view of DD-type monopolar membrane electrolyzer.

The operation of this cell requires electrolyte to be fed into the bottom of the cell and gas and electrolyte in two-phase flow to be taken out at the top. In this type of cell design we have the same problems with circulation restrictions as we had in the monopolar cell with current distributors (see Fig. 11):

1. Restricted natural electrolyte circulation.
2. Increased likelihood of a cell operating with zones of depleted anolyte, and with caustic concentrations in excess of optimum membrane operating parameters (consequence of problem 1).
3. Increased likelihood of localized hot spots (consequence of problems 1 and 2).

Another under-publicized problem that has not received much attention is the problem associated with titanium being attacked by atomic hydrogen. In the bonding of titanium to steel using the explosion bonding process, stress points that

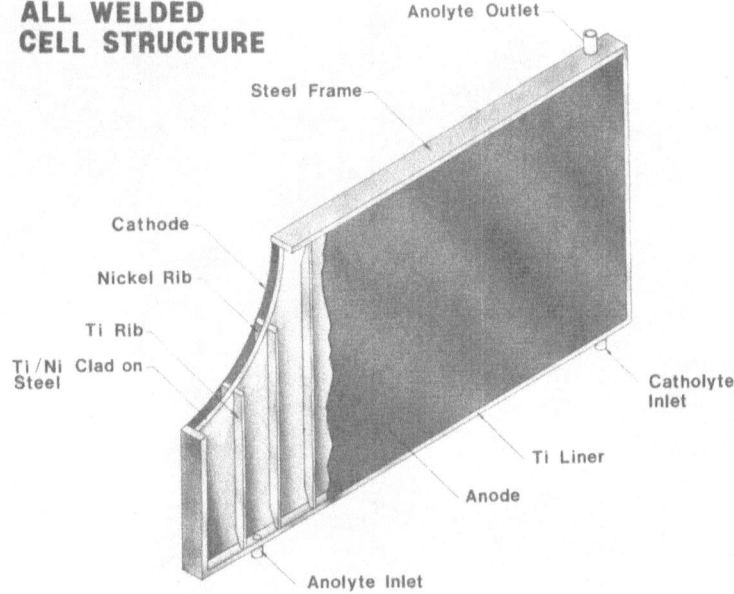

ALL WELDED CELL STRUCTURE

Anolyte Outlet

Steel Frame

Cathode

Nickel Rib

Ti Rib

Ti/Ni Clad on Steel

Catholyte Inlet

Ti Liner

Anode

Anolyte Inlet

Fig. 10. A partially broken away front view of an all-welded ribbed type bipolar cell.

are very susceptible to attack by the atomic hydrogen produced during electrolysis (see Fig. 12) are formed in the titanium liner. This attack by atomic hydrogen can form significant concentrations of hydrides of titanium at temperatures greater than 80°C, which is within the normal range of operating temperatures of standard membrane cells. These hydrides are structurally unsound (very brittle) and resistant to the passage of electricity (high resistance).

Bolted and Welded Cell

This cell design is similar in construction to that of the all-welded ribbed cell except that instead of ribs it employs a screwed-in boss that is protected with

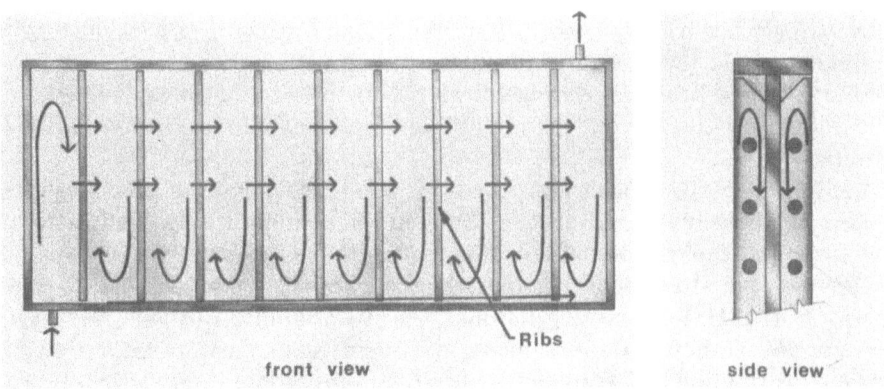

Ribs

front view   side view

Fig. 11. Pictorial view of electrolyte flow pattern in an all-welded cell with ribs.

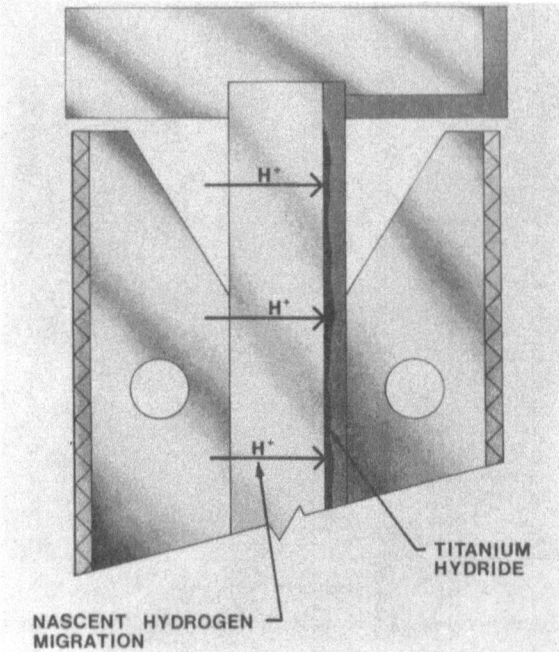

Fig. 12.   Side view of all-welded ribbed bipolar membrane cell with Ti hydride formation.

corrosion-resistant materials. The corrosion-resistant covering material is welded to the liner to complete the seal.

Figure 13 shows a partially broken away view of this type of cell design. While this cell design has substantially minimized the problem of circulation by replacing the ribs with an extended bolted boss, it has increased the linear length of seam weld required to seal the electrolyte from the cell center board and still has the problem with atomic hydrogen attack due to the explosion bonded titanium liner.

Warpage is another undesired side effect incurred during the construction of both the all-welded and the combination bolted and welded type of cells. The excessive length of seam welds needed causes warpage in the cell body. Warpage problems may initially begin before fabrication. When working with large weldments, the individual parts themselves may not be straight, flat, smooth, etc., which will ultimately cause problems during and after fabrication. For proper alignment and positioning of parts, jigs and fixtures often are not adequate to compensate for such problems.

Methods to correct such warpage may include heating/cooling, pressing, heating/pressing and machining. All such methods may in turn induce additional stresses in the structure, causing secondary warpage in the cell element during operation, because of the stress relieving that can be obtained at normal cell operating temperatures. This stress relieving may lead to nonuniform planes between the anolyte and catholyte compartments which can cause a nonuniform electrical current distribution across the active surface of the cathodic and anodic electrodes. Since the distribution of electric current is uneven, the electrochemical reactions are

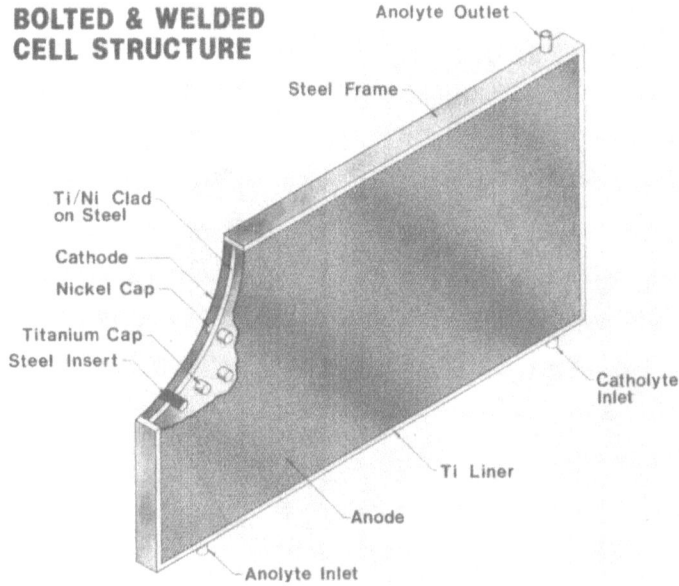

Fig. 13.   Partially broken away view of a bolted and welded bipolar membrane cell.

also nonuniform. The close relationship that may exist between the electrodes and the membrane due to warpage can cause localized areas to have a more vigorous reaction, which in turn can cause hot spots in those areas and a corresponding reduction in current efficiency.

Still another problem associated with nonuniform planes is that the anode and cathode chambers cannot be brought sufficiently close to each other without the risk of puncturing the membrane. Thus a large voltage loss is incurred because of this increased electrode spacing.

In addition to warpage, other concerns which are common to welded structures include: (1) undesirable weld stress within the part, (2) defective welds, (3) corrective welds which are defective, and (4) examination of the weldment for flaws.

De Nora Technologies DD-Type Bipolar Membrane Cell

Again, in this cell design, to minimize the problems listed in Table 1, the Dow and De Nora research team has used the same novel technique of employing a unitized cast metal structure complete with hangers, flanges, current distributors and center board (see Fig. 14) as previously described for a DD-type monopolar cell.

This approach substantially minimizes the problem of complexity since there is only a single part for the central element. It also minimizes the problem with many welded parts and, therefore, the problem of warpage and low and unequal electrical transmission due to the presence of many welded parts and related weld quality. The casting is machined flat, therefore eliminating the need for sophisticated machinery to flatten the cell.

To protect the cell casting from corrosion, press-formed corrosion-resistant titanium and nickel liners are spot welded to the casting at every boss. This assembly

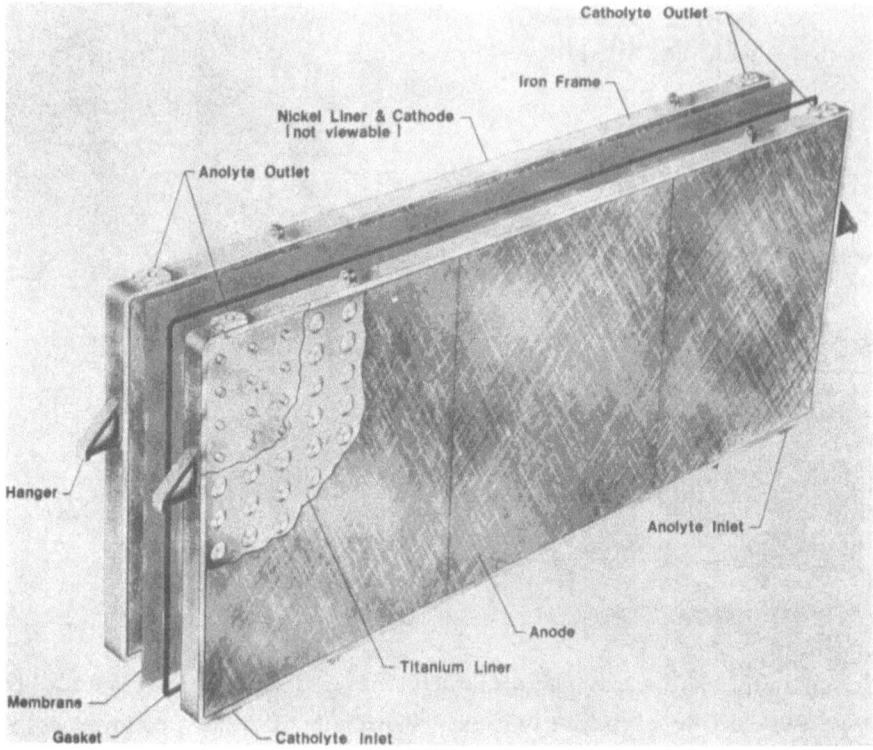

Fig. 14.   Partially broken away view of DD-type bipolar membrane cell.

approach greatly reduces the risk of titanium hydride formation by creating a structure which has a titanium liner with minimal stress points, and by locating these stress points at the farthest distance from the atomic hydrogen source. The liner attachment also allows for an air gap between the casting and the liner so that the atomic hydrogen which might diffuse through the center board can be converted to diatomic hydrogen gas and allowed to escape without harming the liner or the welds. Foraminous electrodes are spot welded to the bosses to complete the main structure of the cell.

COMMON MONOPOLAR AND BIPOLAR PROBLEMS

In addition to the specific problems listed in Table 1, there are some problems that are common to both monopolar and bipolar cells:

—replacement of worn-out electrodes
—unsupported membrane
—IR loss due to gap between electrodes
—binding of electrolyte to membrane
—design for future high current density, high strength caustic membranes

Zero-Gap Electrode Package

De Nora Technologies' approach to minimizing the above problems was the development of a zero-gap electrode package. This approach can be better understood with a review of the various components in Fig. 15.

In the assembly of the zero-gap electrode, an activated fine anode screen is spot welded to the primary anodic electrode. When deactivated this fine anodic electrode can easily be removed and replaced at the plant site using a low amp welder. The cathode package, consisting of activated fine nickel screen and nickel mattress, is installed on the primary cathodic electrode using small plastic clips. To replace the deactivated fine nickel cathodic electrode the plastic clips are removed, a new activated cathode is installed and the package is again attached with plastic clips. This approach allows for easy replacement of the deactivated electrodes at the plant site without the need for sophisticated equipment or specially trained personnel. From the partial cutaway of two bipolar cell elements in Fig. 16 we get a better view of how the zero-gap package occupies the gap between the two cell elements and how the membrane is contained between the anodic and cathodic electrodes.

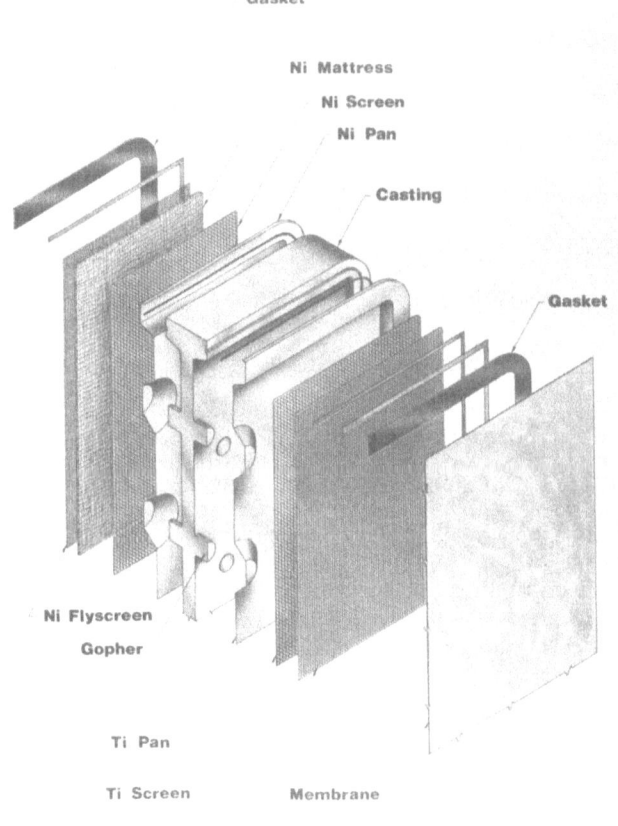

Fig. 15.   Exploded view of a DD-type bipolar cell showing the various components that make up the zero-gap package.

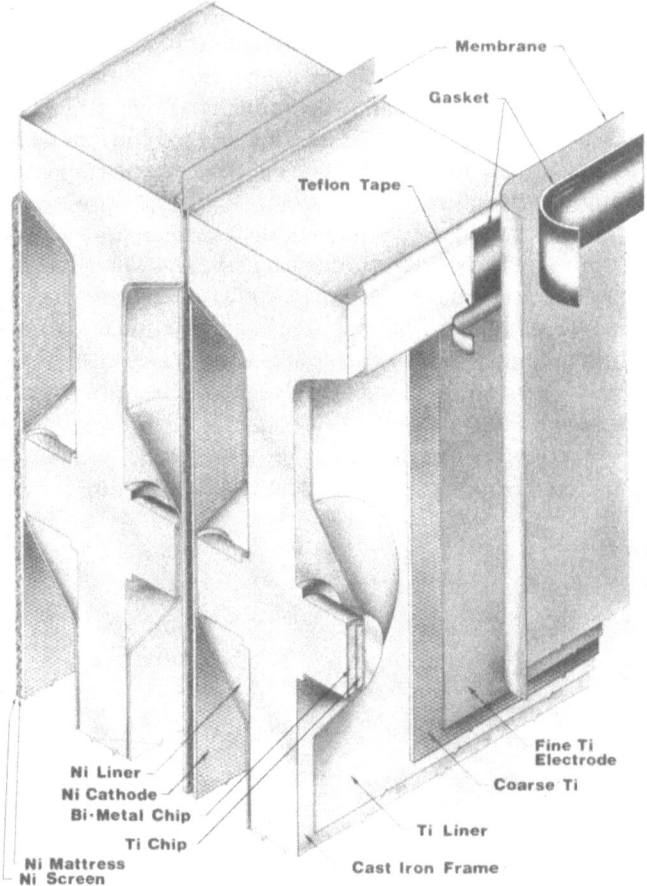

Fig. 16.   A partial bipolar cell cutaway with zero-gap electrode arrangement.

Membranes that are allowed to move or flutter during operation due to pressure fluctuations in the electrode chambers, or because of changes in electrolyte flow, can quickly become abraded by the electrodes, resulting in holes. Figure 17 shows a pictorial view of how pressure changes affect the movement of the membrane in the gap-type electrode configuration as compared to the De Nora Technologies' zero-gap electrode configuration. It can easily be seen that the zero-gap electrode holds the membrane in place during pressure fluctuations while the gap electrode does not.

To understand how the De Nora Technologies' zero-gap electrode configuration minimizes IR loss as compared with gap electrode configuration, an examination of the voltage losses of each component is needed. In the gap electrode configuration the anodic electrode and the cathodic electrode are spaced apart, allowing for a gap to exist between the membrane and the electrodes (a typical gap may measure up to 3 mm). Typically, the electrode voltage loss for a gap electrode configuration is in the range of 230 mV. It was this loss that inspired work by the Oronzio De Nora Group to develop a zero-gap electrode. In the zero-gap electrode configuration the gap that

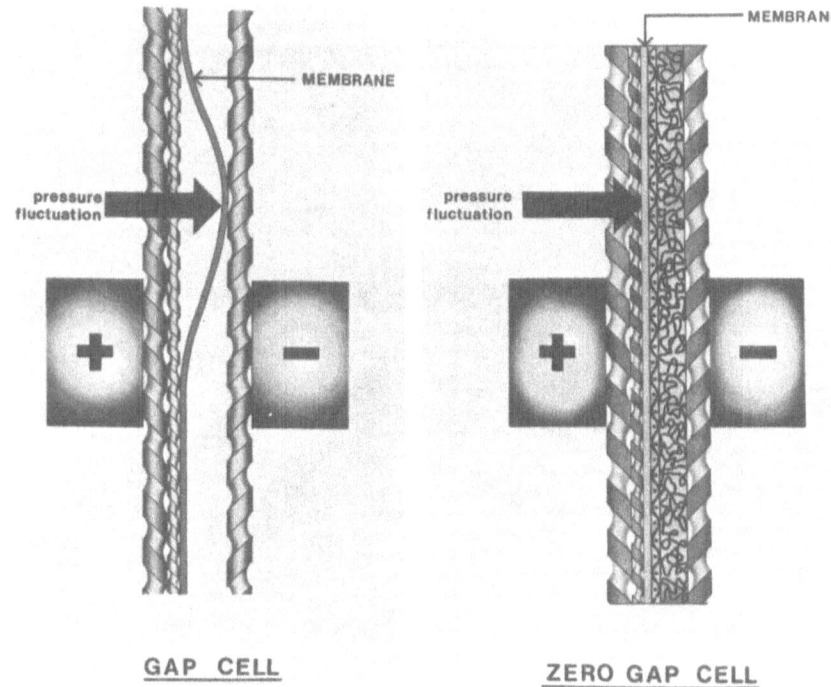

Fig. 17.   Effects of pressure fluctuations on membrane movement in a gap cell and a zero-gap cell.

existed in the gap electrode cell design is filled with a nickel mattress and an activated fine screen mesh (see Fig. 15). Because this low resistant material fills the gap and the activated fine screen is pressed firmly against the membrane, the IR loss has been substantially lowered. The net savings can exceed 150 mV.

Another key feature of the DD-type concept of construction and zero-gap electrode configuration is that it allows free circulation of electrolytes between the lined central cell element and the membrane. This mostly uninhibited circulation of the electrolytes combined with the fine screen electrodes (see Figs 18 and 19) assures the effective use of substantially all the membrane surface. This maximizes the renewal of the electrolyte on the membrane surface and therefore reduces the risk in the anodic side of the cell of depleting the chloride ions which would allow the evolution of oxygen. Such a side reaction, besides entailing a loss of current efficiency, has a detrimental effect on the active life of the anodes, which rapidly lose their activation.

On the other hand, membranes are also particularly sensitive to the uneven caustic concentration on the cathode side. This 'open area' electrical approach to the construction of the DD-type membrane cell also allows uniform circulation and therefore uniform caustic concentration in the cathodic side of the cell element.

Still another key consideration is the prevention of stagnant chlorine gas forming on the anodic electrode. This trapped chlorine gas can penetrate into the membrane and precipitate sodium chloride crystals within the structure of the membrane, causing small separations (blisters) which can eventually lead to pinholes, rendering

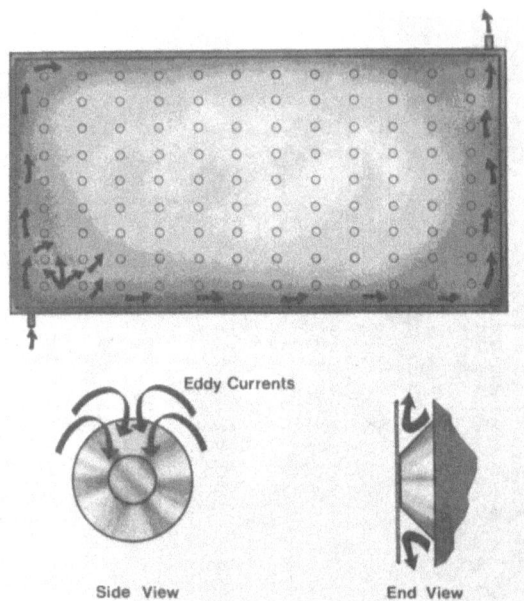

Fig. 18.   Pictorial view of electrolyte flow pattern in a DD-type cell.

the membrane less efficient or even inoperable. Since the electrolyte is allowed to circulate freely in the DD-type membrane cell, stagnation of chlorine is eliminated. Figure 19 shows the membrane area occupied by the anodic electrodes of the industry standard and by De Nora Technologies' 'fine electrode' design.

Lastly, for future higher current densities and high caustic strength membranes, the DD-type cell, with its versatility and dimensional flexibility of the unitized

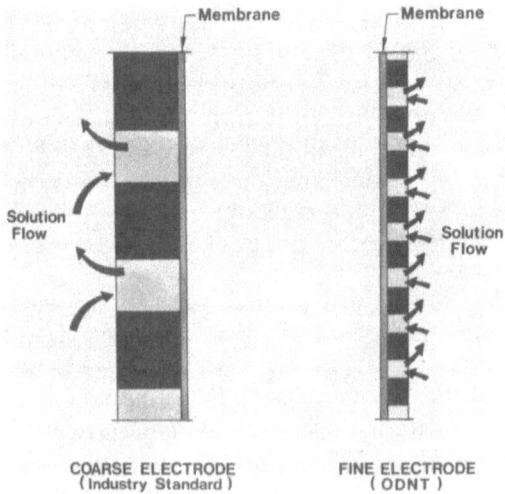

Fig. 19.   Solution flow between electrodes.

casting, can provide the necessary thickness and number of bossed electrical transmission and collection means required. Simply by increasing the thickness of the center board and adding bosses the existing $5000\,A/m^2$ rated call design can easily be adjusted to accommodate higher current densities.

CONCLUSION

The DD-type De Nora Technologies' membrane cell utilizes construction and assembly techniques that were established through many years of development of fabrication methods for mass-produced diaphragm and mercury cells, in both Dow and De Nora workshops.

In those early years sophisticated fabrication procedures were developed for the assembly of titanium, steel and nickel components. Incorporating this experience into the DD-type design has resulted in a membrane cell that operates with the same longevity and quality achieved in earlier diaphragm and mercury cells.

The DD-type zero-gap membrane cell exhibits the ruggedness and longevity required for both the membranes and cells of today, and those of the future.

Dow and De Nora researchers will continue to work in the areas of membranes, cells, electrode coatings and associated processes. Both companies have established a history of pioneering many new electrochemical technologies and remain dedicated to this effort.

ORONZIO DE NORA TECHNOLOGIES BV

The establishment of Oronzio De Nora Technologies BV, a joint venture of the Dow Chemical Company and the Oronzio De Nora Group for the development and commercialization of chlor-alkali membrane cell technology, brought together two of the world's leading chlor-alkali companies with more than a century of caustic and chlorine experience.

In the early days of chlor-alkali production, both companies established themselves as firm believers in the future industrial importance of chlorine and caustic soda. Dow's earliest chlorine plant was built in the early 1890s in Midland, Michigan. This was the beginning of what was to become the world's largest chlor-alkali producer. In 1923 Oronzio De Nora undertook the development and manufacture of innovative diaphragm and mercury cathode cells that found ever-increasing acceptance among chlor-alkali producers.

In pursuit of technical excellence and commercial viability, Oronzio De Nora joined their research efforts with Dow in the late 1970s for the development of an improved membrane electrolyzer for use in commercial-scale chlor-alkali plants. The result of this joint research effort is the development of the DD-type monopolar and bipolar membrane cells described in this paper. This innovative technology has been made exclusively available to Oronzio De Nora Technologies BV for design and construction of chlor-alkali plants all over the world.

19

'WHAT IFs...' IN MEMBRANE CELLROOM OPERATION

Donald J. Groszek and James A. Moomaw

OxyTech Systems Inc., USA

1 INTRODUCTION

Chlor-alkali plants based on membrane cells have been in operation for more than a decade. With rare exceptions new plants will be based on this relatively recent technology. Over the years you have heard, read and seen why membrane cells are the technology of choice. Now that the introduction phase is over, most of the questions from potential licensees or other interested parties are related to specific design or operational 'what ifs'. These are items that you may have heard about during plant visits or are considering in the design of your new plant. Included are holes in membranes, loss of flows, turndown rate considerations plus several others.

Our objective in this chapter is to address some of these 'what ifs' by sharing OxyTech's experience in developing and designing membrane cell electrolyzers for over 20 years, and engineering and operating chlor-alkali plants for over 50 years. The conclusions presented will be supported by actual plant data and observations.

Most of the conclusions are applicable for all types of cellrooms and a majority apply to cellrooms which use monopolar membrane electrolyzers and, in a few cases, specifically to the OxyTech MGC (membrane gap cell) electrolyzer. We believe that this does not limit the usefulness of the conclusions because at the very least you will be aware of some of the factors that differentiate the various types and brands of electrolyzers. The cases have been simplified to allow discussion of a number of them. Each case could be the subject of a paper itself.

2 HOLES IN MEMBRANES

One of the greatest concerns of membrane cell operators is holes in membranes. Holes lead to decreased current efficiency, impurities in the product caustic, and hydrogen in the chlorine.

To quantify the effects, two tests were run with holes intentionally placed in a membrane. In the first test, two 3 mm holes were punched, being careful to remove any 'flap' of membrane that could close the hole. One hole was positioned 7·5 cm

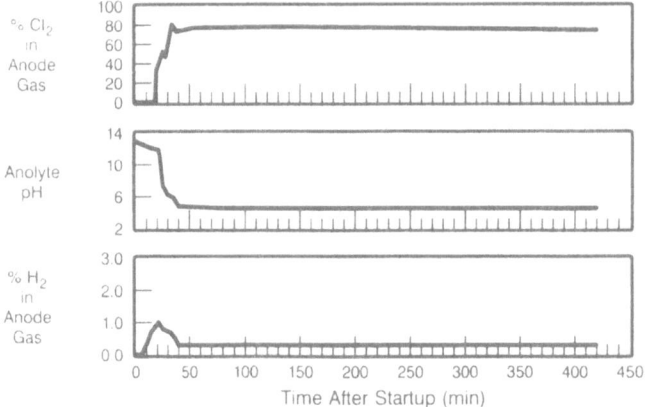

Fig. 1.   Hole in membrane test No. 1 performance results.

below the top flange while the other was positioned over a flange near the bottom of the cell. In the second test, only one 1·6 mm hole was punched. It was positioned 10 mm below the top flange. In terms of safety, holes in the top of the cell are of greater concern because if there was an accumulation of H_2 or Cl_2 gas it would be in this location.

The electrolyzers were started using a standard procedure—raising the load in 1 kA/m² increments until reaching 3·5 kA/m². These steps were done in 10–20 min intervals. During start-up the chlorine gas was analyzed for hydrogen, chlorine, and oxygen using an ORSAT. Samples were collected as fast as possible or approximately every 4–5 min. Several samples were taken for gas chromatograph analysis as a comparison to the values obtained by ORSAT. Depleted brine and caustic samples were also collected.

Following the tests, the electrolyzers were disassembled, being careful to observe the location of the holes to assure that the membrane had not moved from the intended location. The membrane was examined after each run to assure no additional holes were found.

Figures 1 and 2 show that no chlorine was detected during the first few minutes of either run. In fact, in test 1, it was almost 0·5 h after start-up before Cl_2 was detected. In test 2, chlorine was detected within a few minutes after start-up.

The reason for the difference was because of the caustic concentration in the anolyte prior to start-up. In the first test there was 44 g/liter NaOH while for the second test it was only 2·0 g/liter. During start-up most of the chlorine was absorbed until the caustic was converted to hypo. Indeed, no chlorine *gas* would be liberated until most of the caustic was reacted.

Test 1 showed that the H_2 level in Cl_2 stayed at or below 1% during the entire 7-h run. Oxygen levels, though, remained very high throughout the test—20–30%—undoubtedly caused by a significant amount of caustic entering the anolyte. Upon disassembly this was shown by severe anode coating loss directly under the holes.

Test 2 operated for approximately 2 h and reached full load before being shut down. Initial results during the first 45 min showed very low levels—less than 0·5% H_2. However, during the next hour the level increased to just below 2% H_2.

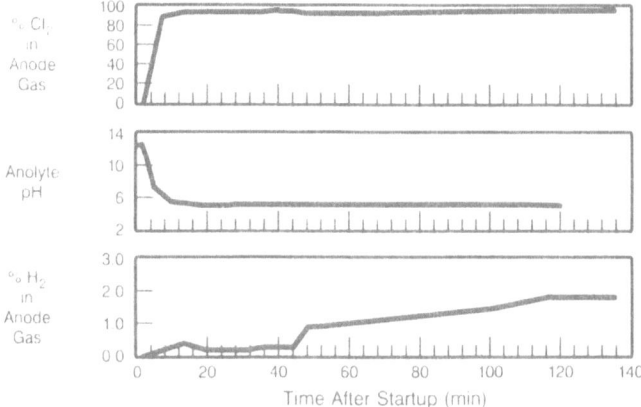

Fig. 2. Hole in membrane test No. 2 performance results.

Although in tests 1 and 2 the hydrogen levels were higher than would be desired in an operating membrane cell, the fact that they were less than 2% hydrogen indicated that they were still below the explosive limit. Typically, concentrations in a membrane cell are less than 0·05% H_2 on a dry basis. For comparison, some mercury plants will operate with hydrogen levels of 2% in the header as long as liquefaction can maintain less than 4% H_2 in the gas to sniff recovery. Certainly, however, no one would argue that something should be done to correct the problem since the potential is there for the problem to get worse. In addition, current efficiency and product quality would be affected.

The data demonstrate two points. The first is that holes in membranes are not a catastrophic problem. This is particularly true for multi-cell electrolyzers where the inefficiencies and impurities are diluted over a large number of cells. Even with the fairly large holes used in the test above, hydrogen concentrations are quite low as compared to other chlor-alkali technologies (i.e. mercury and diaphragm cells).

Secondly, it shows that the caustic concentration in anolyte prior to start-up is important. In the first test, the 44 g/liter NaOH in the anolyte is very high compared to the 2 g/liter NaOH in the second test. In the start-up of a new cellroom the anolyte of each electrolyzer can be analyzed for NaOH content. Electrolyzers with anolyte NaOH content (or pH) above certain values should be monitored more closely than others so that potential membrane holes can be detected before start-up. This technique is our standard procedure to confirm that the electrolyzer is ready to be placed in operation.

Two commercial examples illustrate this point. In the first case the pH of each electrolyzer was measured. A summary is shown in Table 1.

The lowest pH was only 0·4 units below the average whereas the highest cell was 1·0 units above. Two additional electrolyzers had high values and were scheduled for renewal. However, the renewals did not occur until after start-up. This represented only 2% of the electrolyzers.

In a second case the anolyte NaOH concentration was measured in g/liter. All but one of the 30 electrolyzers had concentrations of approximately 1·0 g/liter NaOH while one electrolyzer had 3·0 g/liter. Upon renewal one membrane had a small hole.

TABLE 1
Results of anolyte pH analysis before plant start-up

Average pH	10·7
Median pH	10·6
Highest pH	11·7
Lowest pH	10·3
Standard deviation	0·2

Had any electrolyzers had concentrations as high as in test 1 (i.e. 44 g/liter) the cell would have been renewed before start-up.

OxyTech recommends that this technique be used to assure the best possible performance.

3 LOSS OF FLOWS TO AN ELECTROYZER

Loss of flows to an electrolyzer can result in damage to the membranes. Because of this several tests were carried out on a one-cell MGC electrolyzer to determine the effect of loss of flow while the current remains on. These tests simulated loss of pumps or broken pipes and answer the question 'how much time do we have to restore flows before a shutdown is necessary?' The tests also help determine the best method to detect the problem.

3.1 Loss of Feed Brine Flow

The first test involved shutting off the feed brine flow for 5 min 40 s with caustic flow and electrical power maintained. Prior to shutting off the flow, the cell had been operating at 3·5 kA/m² with a feed and depleted brine concentration of approximately 300 and 200 g/liter NaCl, respectively. Test results are shown in Fig. 3.

Not unexpectedly, no sample could be collected from the top sample port after the loss of flow. However, the bottom concentration decreased from 301 to 239 g/liter

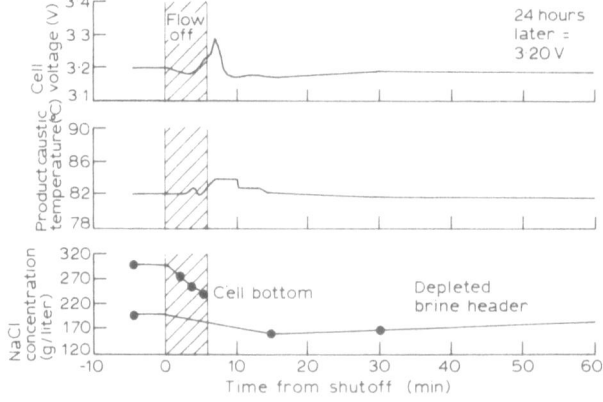

Fig. 3.   Brine flow shutoff test results: 1-cell MGC electrolyzer.

NaCl. If we assume the same decrease would occur at the top, the strength should have fallen from 191 (the concentration before flow shutoff) to approximately 130 g/liter NaCl. The first sample from the top was 6 min after the brine flow was restarted and had 146 g/liter NaCl—this was close to the estimate of 130 g/liter.

One interesting result was that, even after 55 min from restarting the brine flow, the depleted brine strength was only 178 g/liter NaCl. This indicates that after a cell loses brine flow the feed flow rate should be increased to raise the level back to normal as fast as possible.

Upon disassembly, which was several days after this test, no significant blisters were found on the membrane. Only at the very top could a row of small blisters be found. Significant blisters caused by low brine concentrations are known to cause performance decline. Of course, if blisters were to form it would be likely at the top since the strength probably decreased to 130 g/liter NaCl.

Based on these data there is a period of 3–5 min after feed brine loss before the rectifier should be tripped to prevent membrane damage. This amount of time allows the operator to restart the spare brine feed pump without need for a complete shutdown.

Obviously a brine flow alarm would detect the problem. However, the data also show that, besides anolyte NaCl concentration (which is not normally checked continuously), the next best value to monitor would be individual electrolyzer voltage. A rise of 100 mV occurred very shortly after loss of flow (Fig. 3). Thus a monitoring system should scan each electrolyzer in a cellroom at least every 5 min and preferably every 2 min.

It should be noted that the voltage response shown may not be as extreme in a multi-cell MGC electrolyzer. For example, if the feed brine inlet of one cell in a multi-cell electrolyzer is plugged, the voltage would rise in this cell causing the current to be redistributed through the other cells. This will increase the average voltage of the electrolyzer slightly. In addition, even if the inlet is totally plugged, brine feed will not completely stop because depleted brine will flow into the cell through the discharge manifold. Because the cell continues to be fed NaCl the voltage will not rise as rapidly or as much. On the other hand, if flow is cut off to an entire electrolyzer it would be expected to respond in the same manner as the results shown in the above test.

3.2 Loss of Feed Caustic Flow

Unlike a diaphragm cell, a membrane electrolyzer has two solutions fed to it. We previously discussed the loss of feed brine to the anolyte; now we consider the loss of feed to the catholyte. In a cellroom using MGC electrolyzers, a portion of the product caustic is recycled back to each electrolyzer after dilution with deionized water. Some electrolyzer licensors feed water directly to the catholyte compartments. We have evaluated only the case where diluted product caustic is recycled. Recycle systems recover faster after the flow is resumed.

The data collected is shown in Fig. 4. The caustic flow was turned off for 8·5 min while operating at 3·5 kA/m². Feed and depleted brine concentrations were 300 and 200 g/liter NaCl, respectively.

The first overflow occurred 30 s after restarting the flow and was 34·1 wt % NaOH.

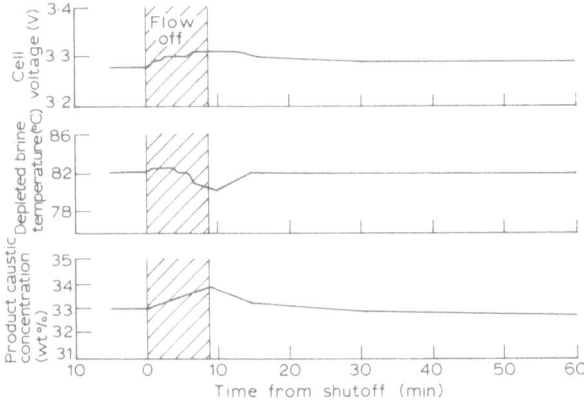

Fig. 4.   Caustic flow shutoff test results: 1-cell MGC electrolyzer.

This was an increase of 0·9 wt %, which averages to a rise of about 0·1 wt %/min after loss of caustic flow. If this average rate remained constant, the flow could be off for 30–40 min before the concentration would reach the specified maximum for irreversible damage (i.e. 35–37 wt % NaOH depending on the type of membrane). This shows that if caustic flow is lost a greater amount of time is available for correcting the problem (e.g. restarting a spare pump) as compared to loss of brine flow.

It should be noted that the amount of time it takes for the catholyte concentration to reach a danger point will be dependent on the anolyte NaCl concentration. This is because the amount of water going through the membrane can be affected by changes in concentration. Thus, at higher NaCl concentrations, water transport will be less and caustic concentration rise will be faster.

Locating the problem is again best done by monitoring individual electrolyzer voltages. Typically, cell voltage rises 15–25 mV for every 1·0 wt % increase in NaOH concentration, which would occur in about 10 min. As in the case of brine feed loss, a scan at least every 5 min would be suitable.

4 CURRENT, CAUSTIC, AND SALT DISTRIBUTION IN ELECTROLYZERS ADJACENT TO JUMPER SWITCH

A major concern during jumpering of an electrolyzer is the effect on current distribution in the electrolyzers adjacent to the jumpered-out electrolyzer.

With reference to Fig. 5, the question asked is: 'what is the effect on electrolyzers 1 and 3 when current is redirected through the jumper switch?' If the jumper switch arms and electrolyzer distributor/connecting bus are not properly designed and installed, the cells closest to the jumper switch arms may carry more current. This in turn could cause a corresponding maldistribution of anolyte NaCl and catholyte NaOH. Cells with higher current densities would have low NaCl and high NaOH concentrations—both of which could cause permanent membrane damage.

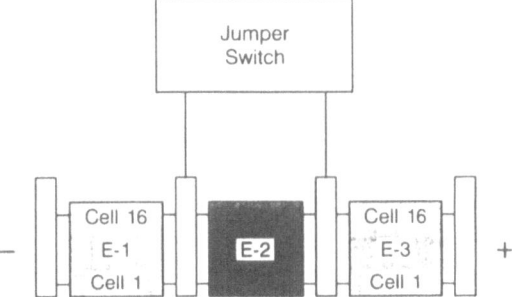

Fig. 5. What is the effect of current flowing through the jumper switch on E-1 and E-3?

The current distribution of electrolyzer 3 during jumpering out of electrolyzer 2 is presented in Fig. 6. A linear regression shows that the cells closest to the switch (i.e. cell 16) carry slightly more current than those further away. The distribution after electrolyzer 2 is placed back on line is shown in Fig. 7.

As can be seen the pattern before and after jumpering was not significantly changed—only the magnitude varied. The deviation is slightly greater during

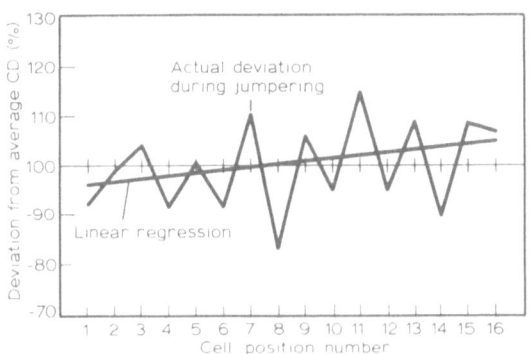

Fig. 6. Deviation from average current density versus cell position.

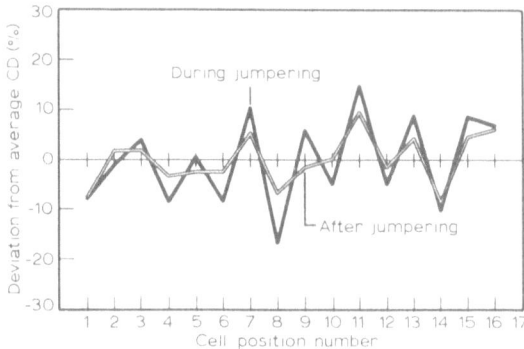

Fig. 7. Deviation from average current density versus cell position.

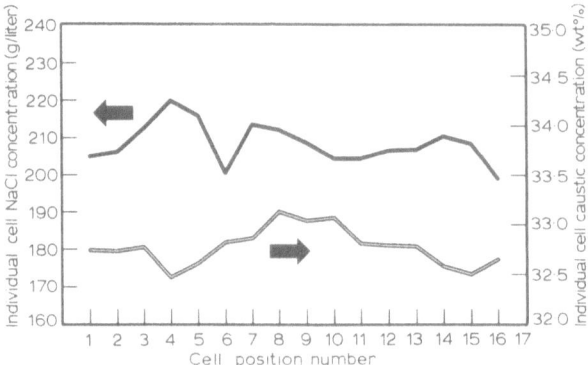

Fig. 8. NaOH and NaCl distribution in electrolyzer adjacent to jumpered electrolyzer versus cell position.

jumpering, probably because the current flow equilibrium was disturbed as current was shifted toward the jumper switch.

The resulting anolyte and catholyte concentrations during jumpering are shown in Fig. 8. The NaCl concentration range was 200–220 g/liter, which is at the design criteria of ±10 g/liter. A slight downward trend can be noted. This trend would be expected because of the slight positive slope of the current distribution; that is, the regression line shown in Fig. 6.

In conclusion, it is apparent that some distortion of the current does occur in electrolyzers adjacent to the one being jumpered, but the extent is minimal and presents no operational problems. Indeed the current distribution falls within the ±10% design criteria during steady-state operation.

5 CALCIUM IN THE ANOLYTE BEFORE START-UP

Users of membrane cells are aware of the need to keep calcium levels in the *feed brine* as low as possible to maintain high current efficiencies. It follows, then, that it is equally important to maintain low calcium concentrations in the anode compartment. This is particularly true in the start-up of an electrolyzer for the first time or restart after renewal. Calcium as well as other impurities are present in dust and dirt from the air which can contaminate the electrolyzer components during renewal.

An example of this is shown in Fig. 9. The calcium concentration in the first overflow from the cells following the initial filling was nearly 100 ppb—well over the limit specified by membrane suppliers. Depending on the membrane it may take as much as 2–4 h to reduce the concentration to an acceptable level with feed brine calcium of approximately 5 ppb.

Although the example cited above was ready for start-up in a relatively short period of time, other cases have been observed in which as much as 10–15 h of circulation were required before a safe concentration could be obtained. Reasons for an extended time include the pretreatment of membranes in a solution with a

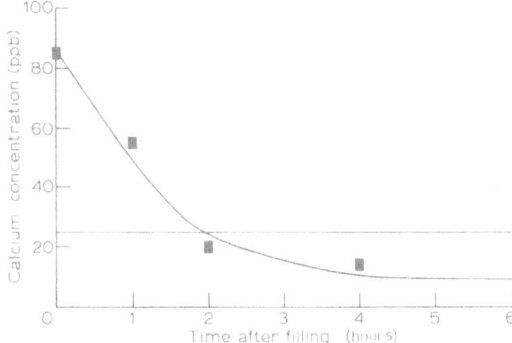

Fig. 9.   Calcium in feed brine versus time after filling.

high level of contaminants, washing of the electrodes with 'hard' water, or an extremely dirty cell renewal atmosphere. It is necessary for the cell renewal staff to recognize that these items can cause problems and take necessary precautions to prevent them. The precautions are not always 100% successful, therefore it is important to analyze the initial overflow after filling the electrolyzer.

If the calcium value is very high, increased brine flow rates may be used to help reduce the level to acceptable limits. Only when the brine purity from each electrolyzer is within specification should the operator begin heating the electrolyzer in preparation for start-up.

6 TURNDOWN RATE CONSIDERATIONS

The limitation on plant turndown is usually given as the lowest allowable electrolyzer current density. This limit, typically $1.5 \, kA/m^2$, is set by the membrane supplier to avoid damage to the membranes and minimize the effects of operation at lower current densities. As you will see there are other factors to consider in answering the question.

Some of the effects of operating at low current densities are: an increase in sodium chloride and chlorate content in the caustic soda product, lower current efficiency, and higher heat addition to the feed stream. Any of these changes could limit a specific plant's turndown rate depending on the plant's product caustic soda quality requirements, etc. Some of these effects can be minimized by adjusting cellroom operating conditions. For example, lowering the cell temperature while at reduced load will offset the sodium chloride in caustic soda to some extent.

Multiple circuits allow for stepdown production cuts by shutting down one circuit without lowering the current density of the remaining operating cell line(s). However, even with multiple circuits the minimum acceptable current density eventually has to be considered.

Thus the lowest acceptable current density along with the cellroom transformer/ rectifier and other plant process and utility equipment and controls, and product quality requirements, determine the lowest operating rate.

One item frequently not given enough consideration in plant economic studies and risk analysis evaluations is the amount of heat required relative to the operating rate of the cellroom. Newer membrane cells use electricity more efficiently, therefore less heat is produced within the cell at any given current density. In order to operate the cells at the optimum cell temperature (typically 90°C), the feed streams usually have to be heated instead of cooled.

Figure 10 shows the operating region on the heat required versus current density curve between 80°C, the lowest allowable operating temperature, and 90°C, the typical optimum operating temperature. In this operating region, either (1) steam must be added to the feed brine and/or caustic soda recycle stream (right of the vertical axis) or (2) heat must be removed from these streams (left of the vertical axis) to maintain the desired cell temperature. Note that steam has to be added to maintain the 90°C temperature at all current densities up to nearly 5·0 kA/m² and at all current densities below 2·8 kA/m² to maintain 80°C.

Most cellrooms are designed to operate in this region (i.e. heat added) during standard operation based on economical reasons which take into account the cost of power, steam, cooling water and capital. Cells can be operated with cell temperatures as low as 80°C but the resulting voltage rise increases power consumption.

Line 'A' on Fig. 10 illustrates the heating/cooling required at temperatures between 80 and 90°C at a constant operating rate of 4·0 kA/m². The cell temperature can be varied along this line based on the specific plant economics and design as discussed above. The ability to operate along this line permits more flexibility in operating the plant.

Line 'B' illustrates the case where the operating rate is cut from 4·0 kA/m² at a cell temperature of 90°C and the heat input remains constant. As a result the cell temperature will drop as the current density is lowered. This could illustrate a case where more steam may not be available to keep the temperature on the 90°C temperature curve. When designing a plant a major point to consider is the ability to add adequate steam to allow the operating rate to be reduced to the lowest required rate without the cell temperature dropping below 80°C. Example—if the cellroom is designed to operate at a temperature of approximately 85°C at 4·0 kA/m² but without the ability to add more steam, the minimum operating rate will be

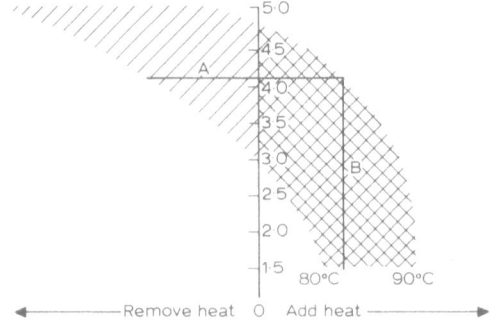

Fig. 10.   Heat input/removal versus current density (current density in kA/m²).

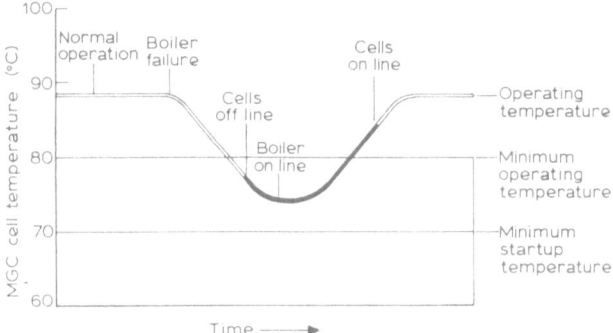

Fig. 11.   Effect of boiler shutdown on cell temperature.

approximately $2.8 \, kA/m^2$. This is the point where the cell temperature reaches $80°C$ (where the $80°C$ curve crosses the current density axis).

Insulating the cells and cross-heat exchanging the feed brine stream with the depleted brine and hydrogen streams can reduce the net heat input somewhat. The net heat input will decrease during the lives of the membranes and cathode coatings as the current resistance of the membranes and the hydrogen overvoltage of the cathodes increase. These produce more heat in the cells, thus reducing the external heat requirements.

Assuming that the plant has been designed to be operated at the minimum acceptable current density, the question of onstream factors for boilers, heat exchangers, etc., comes into question. Figure 11 illustrates what happened at one licensee when a boiler failed.

The cellroom stayed on-line for several hours after the boiler failure, during which time the cell temperature dropped from the standard $88°C$ to $77.5°C$. This temperature was slightly below the minimum recommended operating temperature but still above the minimum start-up temperature. At this point the cells were taken off-line. During the shutdown the cell temperature continued to drop until it bottomed out at $74°C$ at the end of the boiler shutdown. After the boiler was restarted the brine and caustic soda feed streams were heated up until the cell temperature reached $85°C$ approximately $3.5 \, h$ later. The cells were then restarted and the cell temperature was back to normal $1 \, h$ later.

A number of factors that determine the lowest acceptable turndown rate have been reviewed. These include physical, product quality and economic factors which must be considered during the plant design period. Equipment failures during operation, such as the boiler failure case that was discussed, can also limit the turndown rate.

7 CONSIDERATIONS DURING SHUTDOWNS

Procedures during shutdowns are as important to follow as those during operation. Failure to do so can create permanent damage to the anode and cathode coatings as well as the membrane. Additionally, good shutdown practices minimize the off-spec caustic that typically occurs during a shutdown. Our discussion will focus on the anodes, cathodes and product purity.

7.1 Anolyte and Catholyte Available Cl_2 during Shutdowns

An inherent problem in the shutdown of membrane cells is the transport of available Cl_2 (e.g. OCl^- ion) into the catholyte. The transport increases after the current is off because the voltage gradient is reduced, thus the available Cl_2 ion is no longer 'attracted' to the anode. Furthermore, during operation available Cl_2 entering the catholyte is reduced at the cathode. But during a shutdown the available Cl_2 in the catholyte can corrode the cathode and/or the active coating, reducing its life. In addition, this reduction of available Cl_2 during shutdowns creates a reverse current which can damage both membrane and cathodes or cathode coatings. For the above reasons the elimination of available Cl_2 from the cell is important for optimum long-term operation.

Figure 12 shows the concentrations of available Cl_2 in the anolyte and catholyte as a function of time after shutdown. In the first case (labeled 'no circulation'), the brine and caustic flows were stopped after loss of the electrolyzing current. This would simulate a complete power failure. Note that the anolyte concentration decreased during the first 15 min and then rose. The initial decrease occurred as the acidic brine dechlorinated. The increase after that period was caused by the reabsorption of the Cl_2 from the header by the OH^- diffusing into the anolyte. As can be seen from the catholyte analysis, the hypochlorite in the caustic increased and remained constant.

In the second case shown in Fig. 12 (labeled 'circulation'), the flow of brine and caustic remained on after the d.c. power failure. Note that the anolyte concentration began to fall, as in the first case, but did not rise again. Correspondingly, the available Cl_2 in the catholyte also rose but decreased to nondetectable limits within a short period.

Two points are important:

(1)   In lieu of other methods to reduce available Cl_2 in the catholyte, the brine and caustic flows should remain on during a d.c. power failure. To accomplish this emergency power should be provided for the feed brine and caustic pumps. An alternative would be the use of head tanks to provide fresh solutions.

(2)   The Cl_2 header should be purged with air or N_2 to remove the Cl_2 during a shutdown. This prevents the reabsorption of Cl_2 into the anolyte and consequent transport into the catholyte. In the case of an individual electrolyzer being jumpered from a circuit, a method to isolate the electrolyzer from the header must be made before the flows are stopped and the electrolyzer is drained.

7.2 Caustic Concentration Decrease during Shutdowns

A potential problem during shutdowns in membrane cells is the dilution of the caustic. It occurs because of the water concentration gradient between the anolyte and catholyte (i.e. osmosis). This is a problem for three reasons. First, during long

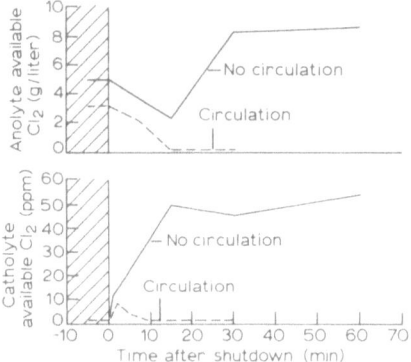

Fig. 12.   Anolyte and catholyte available Cl₂ concentrations during shutdowns versus time after shutdown.

shutdowns it is possible for the strength to decrease low enough to cause damage to the membranes by affecting the water transport. Second, it represents a loss of product through dilution. Finally, when restarting a cellroom with low caustic concentrations the efficiency remains low because current efficiency is related to caustic strength. Optimum operation will not be achieved until the concentration rises as a result of continued operation.

Caustic analyses were performed during a 1-day shutdown and are shown in Fig. 13. As can be seen the concentration declined from the operating level of 33 wt% NaOH to 28·5% during this period. The rate of decline was about 4·5 wt% per day. However, the rate was not constant over the entire period. Analyzing the data closely shows the rates as presented in Table 2.

The reason for the change in slope is cell temperature. A plot of cell temperature during the shutdown is also shown in Fig. 13. The temperature declined rapidly because the feed caustic was cooled. A 50°C temperature was maintained since operation much below this can result in reaching the brine NaCl saturation level.

It is important to note that the rates of decline listed above will be different for each cellroom and possibly for each shutdown. There are several reasons for this: (1)

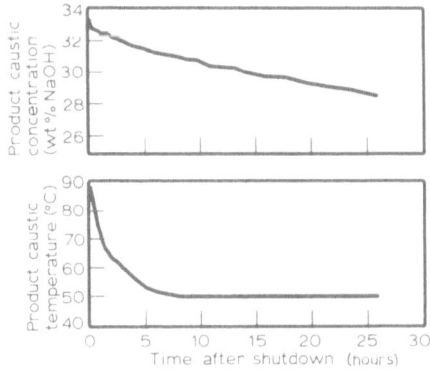

Fig. 13.   Caustic concentration and temperature versus time after shutdown (continuous circulation).

TABLE 2
Rate of decrease in caustic concentration

Time after shutdown (h)	Rate of concentration decrease (wt %/h)
0–1	0·72
1–3	0·28
4–6	0·23
8–11	0·19

the rate of cooling may change because of seasonal variations in temperature; (2) the ratio of caustic volume to membrane area will vary for each cellroom and, possibly, for each shutdown depending on the level in the caustic circulation tank; (3) the time it takes to turn cooling water off the caustic may not be the same unless it is automated; (4) the type and condition of the membranes can cause the diffusion rate to vary.

The main conclusion is that operators should be aware of this occurrence and be prepared to add concentrated caustic during extended shutdowns. During a plant's initial start-up phase this means procuring at least 2–3 times the minimum amount of caustic or have the caustic evaporation system operational before filling the electrolyzers.

7.3 Caustic Impurities during Shutdowns

Membrane cells typically produce very high quality caustic with regard to impurities; however, what happens during shutdowns can affect this purity. Two of the major contaminants in caustic are iron and salt.

Figure 14 shows actual plant results of the concentration of iron after a shutdown. The iron level prior to shutdown was 0·5 ppm. Within 5 min after loss of power the concentration had already risen to 1 ppm. Within 1 h it leveled to slightly under 2 ppm.

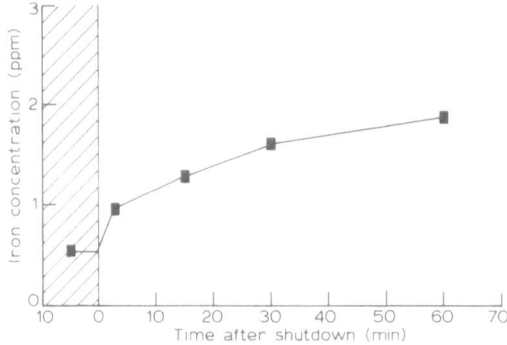

Fig. 14. Iron concentration in caustic soda versus time after shutdown.

The reason for the increase is because of the cathodic plating of iron onto the cathodes from the caustic solution during operation. Once the current is off, however, the iron corrodes rapidly. Tests have shown that iron is not transported through the membrane from the brine unless there is a hole. Therefore it must come from the caustic used to initially fill the cells, corrosion of materials used in the caustic circulation system, and/or the water used as the reactant/diluent to the catholyte.

Because the iron deposits (i.e. plates) onto the cathode the process is cumulative. Even very low levels of iron in the caustic (e.g. 0·02 ppm) during normal operation will eventually build up on the cathode as long as there is a continuous source. Thus, during each successive shutdown the iron level will increase unless the system is flushed with new caustic to remove it.

One advantage of the fact that the iron cathodically plates is that even without flushing the contaminated caustic can be recovered by allowing it to be recirculated to the cells after operation is restarted. It must be remembered, however, that it will dissolve again during the next shutdown.

Figure 15 shows how the NaCl concentration in the caustic soda increased after the shutdown. Salt contamination is slow by comparison to iron. The increase to 200 ppm took 6 h as compared to the iron contamination which only took minutes. Many shutdowns, therefore, will be short enough so that NaCl contamination will not be a problem.

Contamination of the caustic soda by NaCl is caused by transportation through the membrane from the anolyte to the catholyte. In a membrane cell it is controlled primarily by diffusion, electropotential gradient, membrane rejection ability and water transport flow through the membrane. During operation diffusion is insignificant and overcome by the electropotential gradient created by the current flow. The higher the voltage the less transport will occur. However, during a shutdown, when the current is zero, diffusion takes on greater importance because only membrane rejection ability resists the flow. Cell temperature then becomes very important during shutdowns since diffusion is strongly a function of temperature. This means that cells that are left hot during shutdowns will have more NaCl contamination than those that are cooled.

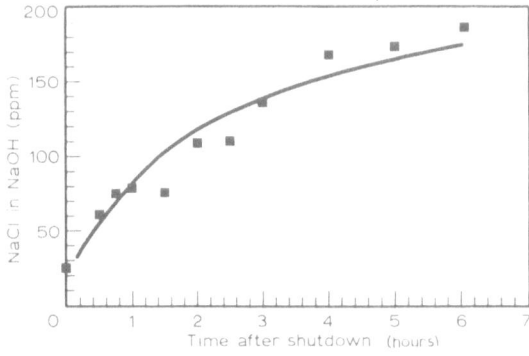

Fig. 15. Sodium chloride in caustic soda versus time after shutdown.

In conclusion, to reduce and minimize impurities in the caustic during shutdowns the following steps should be taken.

(1) Monitor the Fe and NaCl concentrations on a regular basis during shutdowns. This will keep you aware of a potential problem and give you time to make plans for 'off-spec' caustic soda.

(2) Design caustic systems with nonferrous piping. This will eliminate one major source of iron.

(3) Use deionized water for caustic makeup and employ nonferrous materials of construction for this system also. Deionized water is low in both NaCl and Fe as well as other undesirable impurities.

(4) Make provisions for the inevitable time when 'off-spec' caustic is made during shutdowns. This could include separate storage until it can be blended with good material or sold to customers that do not require high purity.

(5) Reduce the temperature of the electrolyzers as soon as possible after loss of current to reduce the transport of NaCl from the brine. (This will also help reduce the rate of caustic dilution.)

8 MEMBRANE INSTALLED BACKWARDS

One of the most important aspects of cell assembly is installing a membrane in the correct direction. Failure to install them properly leads to permanent damage of the membranes. Further details and reasons for this can be discussed with membrane manufacturers. However, as hardware suppliers, we are interested in the effect on operation if and when a membrane is installed backwards.

One effect that can be studied is current distribution. When membranes are installed backwards they will have greater resistance which results in higher voltage. In a monopolar electrolyzer the voltage of all cells must be equal, therefore current will be redistributed throughout the *electrolyzer* inversely related to individual *cell* resistance (i.e. as the resistance of one cell goes up the current to that cell will go down while the current to the other cells will go up). In addition, the impact of a current maldistribution can be checked by analyzing the depleted brine

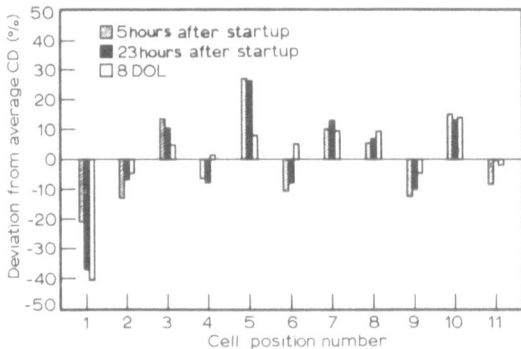

Fig. 16.   Current distribution versus cell position (cell 1 membrane installed backwards).

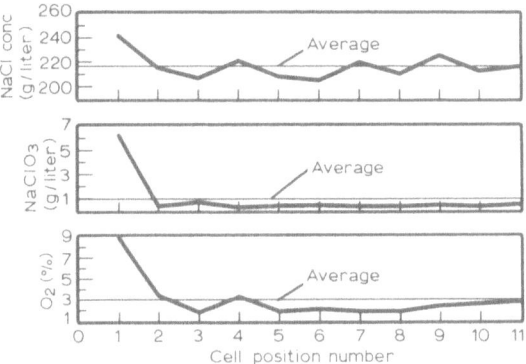

Fig. 17.   Anolyte distribution in individual cells (cell 1 membrane installed backwards).

concentration from each cell. Those cells with higher current should have lower NaCl concentrations and vice versa.

For this test the current distribution was measured in an 11-cell monopolar electrolyzer at three different times. The first measurement, taken 5 h after start-up (Fig. 16), shows a wide current distribution. It was not, at this time, wide enough to indicate a major problem.

Also shown is the current distribution after only 23 h of operation. Note the difference particularly in cell 1. Eight days later the results were even more dramatic. By this time it was obvious that the cell 1 membrane was installed backwards because it was carrying only 60% of rated current. Note the remaining 10 cells; all except three were carrying current above the ideal value showing the redistribution of current from cell 1 to the rest of the cells in the electrolyzer.

Analyses of depleted brine from each cell clearly shows the problem (Fig. 17). NaCl, $NaClO_3$ and O_2 were all much higher for cell 1 than for the rest of the cells. The average values (also shown) that would be reported for the electrolyzer might not be poor enough to cause removal. However, if cell 1 data were removed, the average performance would improve to 214 g/liter NaCl, 0·4 g/liter $NaClO_3$ and 2·4% O_2.

The point of this test is that installing a membrane backwards will not cause a major problem which could curtail production significantly. This electrolyzer operated for 9 days before it could be scheduled for renewal. Some damage may have been done to the anode coating, because of the poor efficiency, but the damage would be limited to only one cell.

9 INCREASED BRINE FLOW AT START-UP

It should be noted in Fig. 16 at 5 and 23 h after start-up that the current distribution was much wider than desired. Typically we expect a deviation of ±10% from the average value at the steady state. The data at 23 h after start-up still showed a variation as high as 26% above the average, but after 8 days on-line the variation was within the desired limit. The reason for the variation is not understood completely but is probably a result of slight variations in such things as electrode potentials, membrane lot variations, etc. As this 'break-in' period passes all components become 'equal' and the load distributes itself evenly—unless a perturbation occurs, such as the case with a membrane installed backwards.

This 'break-in' phenomenon points out the need to increase brine flow rates during the early stages of start-up. As we also saw from the previous data, current distribution affects NaCl concentration in the anolyte. If there was a temporary shift such that 1 or 2 of the cells were carrying a much higher level of current, the NaCl concentration could fall below the minimum required to prevent membrane damage. Typically OxyTech recommends a 10% increase over the standard operating value.

10 CONCLUSIONS

We have provided you with data and our conclusions on a few of the 'what ifs' that are frequently asked about membrane cell operation. You may not have been aware of some of these and the fact that these potential problems exist may raise some new concerns. The fact is that all of these can be planned for in the design and the operation of a plant. The real proof can be found in the over one hundred cellrooms which use membrane technology. This represents over 10% of the world's capacity. These plants are:

—safe,
—environmentally sound,
—efficient,
—reliable,
—economical, and
—proven.

For these reasons and more, membrane cells continue to be the technology of choice.

20

HOECHST–UHDE SINGLE ELEMENT MEMBRANE ELECTROLYZER: CONCEPT—EXPERIENCES—APPLICATIONS

W. Kramer and B. Lüke

Uhde GmbH, FRG

1 INTRODUCTION

Figure 1 shows a cell room which is equipped with Hoechst–Uhde single element electrolyzers. This picture introduces the membrane cell technology of Hoechst and Uhde, which is the subject of this chapter.

Today's market for chlor-alkali electrolysis plants consists of new plants as well as of conversion of mercury or diaphragm cell rooms to membrane cell technology. Membrane cells consume up to 30% less electrical energy than mercury cells. Therefore a conversion is sensible in areas with high power prices.

Political reasons for conversion can be saving of electrical power as a target of national policy or environmental considerations, which cause legislative actions. This was the case in Japan and later in the Republic of Taiwan.

Fig. 1

New grass roots membrane cell plants are often on-site units to overcome transportation risks and cost.

The Hoechst–Uhde single element membrane electrolyzer concept is preferably designed to meet these different requirements of new installations. A flexible system made of standard basic components fits smoothly into existing space and capacities.

Our two recent contracts for conversions from mercury to membrane technology with India (200 t/day NaOH) and Taiwan (525 t/day NaOH) shows the confidence in the Hoechst–Uhde cell technology.

2 SINGLE ELEMENT CONCEPT

The single element mainly consists of three components (Fig. 2):

—anode pan with anode,
—membrane, and
—cathode pan with cathode.

These three components form an electrochemical cell with the anode chamber, where chlorine is produced from NaCl–brine, and the cathode chamber, where caustic and hydrogen are generated. The two electrode pans are clamped together by circumferential flanges (Fig. 3) to seal the electrolyte chambers against each other and prevent leakage to the outside. The electrolytic current is conducted from the

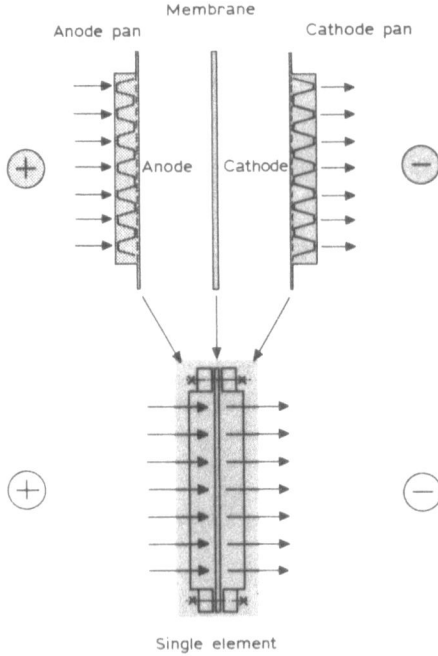

Fig. 2.   Single element concept.

Fig. 3

anodic backwall made from titanium into the cell element and leaves it through the stainless steel or titanium backwall.

Inside the element the current passes along a plurality of corrugated strips and is evenly distributed (Fig. 4). The figure shows the electrode pan before assembly of the electrode.

The single element, which is completed now by an electrode (Fig. 5), has individual inlet and outlet connections for electrolytes and products. Therefore it is an independent electrolysis unit. Insert pipes ensure the filling of the anolyte and catholyte chambers with electrolyte. The figure shows such a pipe being inserted for initial assembly.

Plastic distance strips, which are buttoned onto the electrodes, fix the membrane inside the element and define the electrode gap.

The basic components (Fig. 2), which form a single element, have according to their design several degrees of freedom:

Fig. 4

Fig. 5

—The electrode area can be varied in size, step by step, from 0·9 to 2·7 m²
  according to demand.
—Each membrane type, finite and zero gap, is applicable. Adaption is done only
  by different gaskets and distance strips.
—For cathode and cathode pan manufacture stainless steel can be used if power
  cost is low. Lowest power consumption is achieved by using activated nickel
  cathodes within a nickel cathode pan.

These single elements and electrolytical units, which are described roughly, can now
be produced to specific requirements. The configuration is carried out by pressing
the elements together. The current is then conducted from element to element via
contact strips, which are welded onto the anode backwalls (Fig. 6). Long series of
single elements can be realized in this way, which cause low electrolytic current and
in turn low investment and operating cost for rectifiers and bus bars.

If a high current has to be utilized from existing cell installations, single elements
can be arranged in parallel (Fig. 7). This can be done with each element connected in
parallel as shown on the figure as well as with groups of elements forming a package
in series, where several of these packages are connected in parallel.

Figure 8 shows a series of elements being suspended into a rack, connected to the

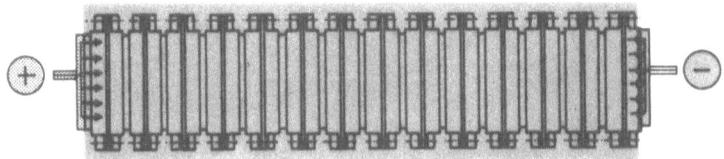

Arrangement in series
Low current / High voltage
(5 - 12 kA)   (100 - 400V)

Fig. 6.   Single element concept.

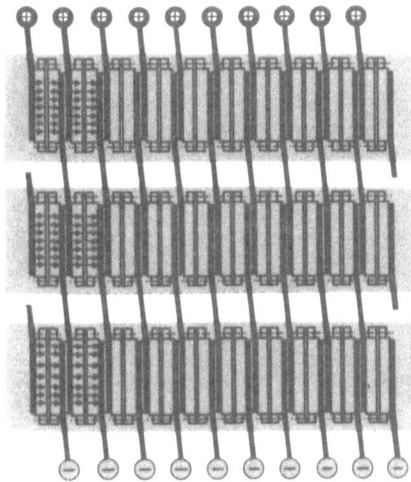

Arrangement in parallel
High current / Low voltage
(100-300kA) (3-15V)

Fig. 7.   Single element concept.

feed and collector pipes, and pressed together by a plurality of screws for electrical contact. One compression arrangement is called an electrolyzer. Two electrolyzers are combined in one steel rack. Several of these so-called double electrolyzers can be connected in series to form a stack which has a common feed and collecting header system. Figure 9 shows a single stack which is constructed by combining three double electrolyzers, which can contain up to 40 elements each. This stack was started recently at Hoechst.

Now will be shown the reliability of the single element concept to clarify how optimal adaption can be achieved using the same basic components.

The figures that follow show various cell room arrangements for the same capacity of 265 t/day NaOH.

Fig. 8

Fig. 9

2.1 Grass Roots Plants (Fig. 10)

A very important criterion for optimal cell room design is keeping the electrolysis current as low as possible. This leads to a low investment cost for rectifiers, switches and bus bars. Furthermore, low energy losses within the structures result in low operating costs.

If we have a free choice for a grass roots plant design, we will arrange stacks consisting of three double electrolyzers with 40 elements, each resulting in 120 elements per stack connected in series. Two groups of rectifiers supply the current to three stacks each, thus forming two independent circuits. For our pattern of a 265 t/day NaOH cell room only two 32-kA rectifiers should be used with a total voltage of 400 V.

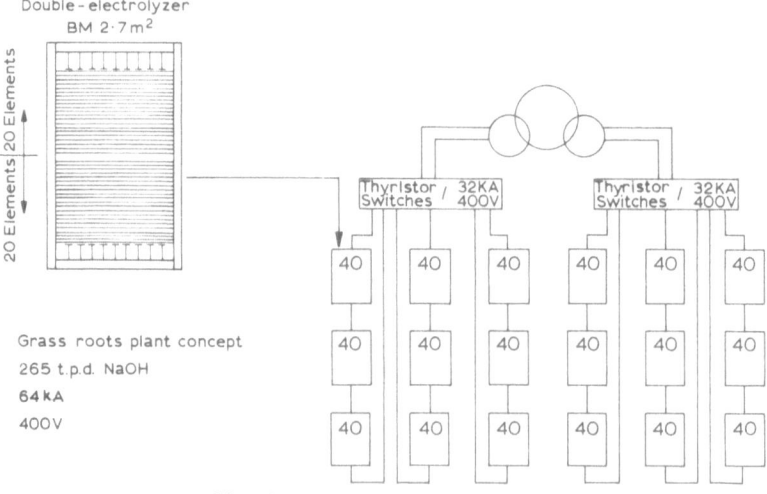

Fig. 10.   Cell room arrangement.

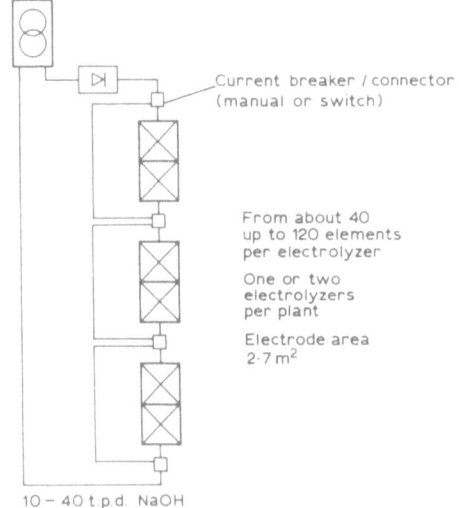

Fig. 11. Small capacity concept H–U cells.

The electrode area should be designed as large as possible to decrease cost for gaskets, membranes, electrode coating and instrumentation.

It is also possible to use the same type of circuit for small plants without loss of flexibility (Fig. 11). Within this stack each double electrolyzer can be switched out for maintenance while the others keep on operating. With one single stack series connection of 40–120 elements can be realized, giving capacities between 10 and 40 t/day NaOH.

2.2 Cell Room Conversions

The optimization criteria are different in this case. The common objective is to utilize existing power supply equipment and cell buildings. Production loss should

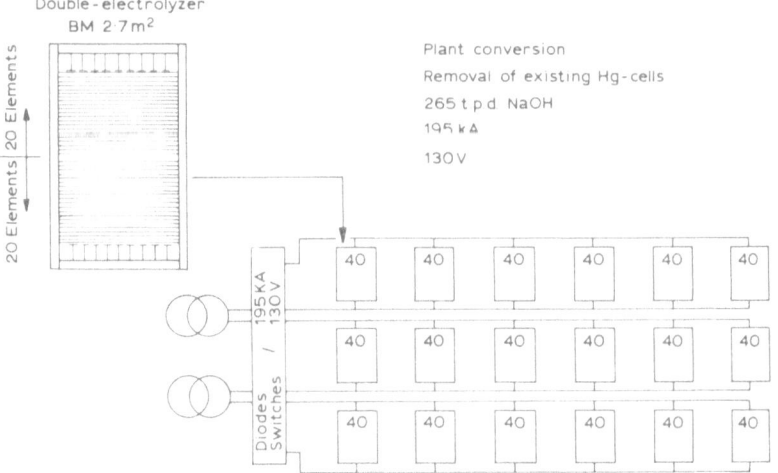

Fig. 12. Cell room arrangement.

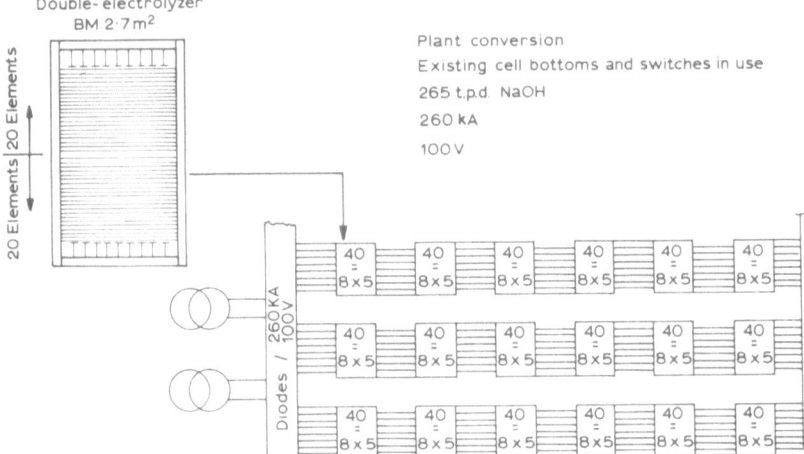

Fig. 13.   Cell room arrangement.

be minimized during the period of conversion. Nevertheless, the cell room should remain flexible for operation and easy to maintain.

The first example for plant conversion shows (Fig. 12) that the identical double electrolyzers as used for grass roots plants can be connected to an existing mercury cell rectifier by arranging all double electrolyzers in parallel. Each double electrolyzer can be switched in the rectifier building. Three double electrolyzers have common electrolyte feed and product discharge. The existing mercury cells and bus bars have to be removed completely. The production loss during the conversion

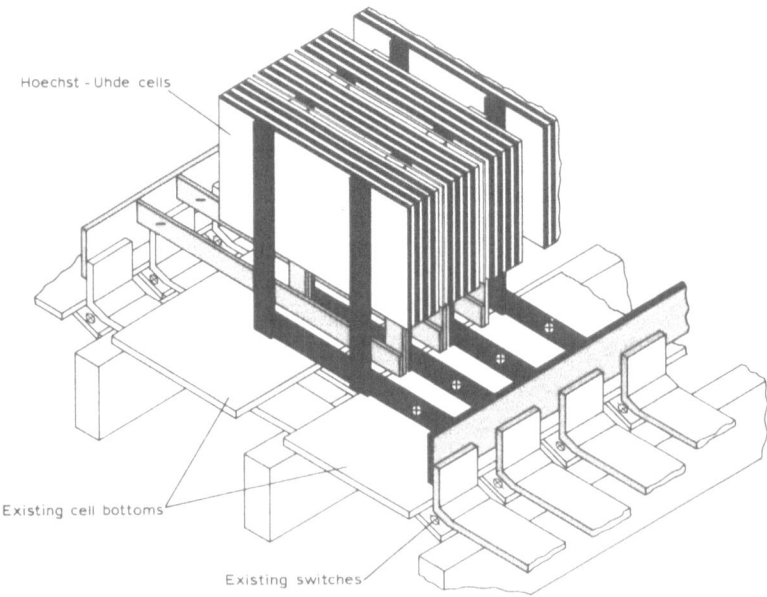

Fig. 14.   H–U conversion of Hg-cell room.

period is small, because the membrane cells need only 30% of cell room area for the full replacement capacity.

If the production loss has to be minimized, the conversion has to be carried out Hg-cell by Hg-cell on top of the existing cell bottoms, re-using also the bus bars and switches. Such a conversion procedure is also possible with the single element concept (Fig. 13): 120 elements are arranged in 24 groups connected in parallel consisting of five elements in series on top of two 30-m^2 cell bottoms. Here (Fig. 14) the described arrangement is shown in a perspective figure. From the existing Hg-cell switches with a distribution bar the current enters the element packages ($\oplus$ bars), passes through the five elements in series and is conducted via the $\ominus$ bus bars to the next distribution plate. Using the existing cell bottoms the stack can be short-circuited for maintenance.

Experiences and optimization studies have found that the high copper requirement makes this alternative of secondary importance.

3 EXPERIENCES AND FEATURES

The single element concept was initially invented by Hoechst AG. Since 1977 commercial-scale pilot membrane cells have been in operation.

First there were four single elements with 1 m^2 electrode area; later this number was increased to 10 elements. In a next step the electrode area was enlarged from 1 to 2·7 m^2. Here five elements were tested, later 15 in series.

With this background the cooperation between Hoechst and Uhde was started in 1983. Both companies used their combined knowledge for further development of membrane cells.

The milestones of this joint development will now be presented in more detail (Table 1).

The first two milestones show directly a very important property of the single element system. After testing of three elements (1 m^2) of the new Hoechst–Uhde design a big step was taken by the installation of 280 elements (1·7 m^2) in the electrolysis plant at Tofte, Norway, where Hooker-plastic, filterpress technology

TABLE 1
Milestones of Hoechst–Uhde single element development

Start-up	*Electrode area (m^2)*	*Number of elements*	*Location*	*Membrane gap*	*Cathode*	*DOL*
May 1983	1	3	Hoechst	Finite	SS	273
Dec. 1983	1·7	280	Tofte, Norway	Finite	SS	>1 540
Apr. 1984	1	10	Hoechst	Zero	Ni + act.	1 250
June 1985	2·7	18	Hoechst	Finite	Ni + act.	>910
Dec. 1987	1·8	544	GEP, The Netherlands	Finite	Ni + act.	>140
May 1988	2·7	96	Hoechst	Finite	Ni + act.	>20
June 1989	2·7	2 384	FPC/GACL	Finite	Ni + act.	

was replaced. The success of this operation showed that, using the single element concept, escalation from small quantities to big ones is possible with very limited risk. In other words, the property of one unit cell is almost the same as that of a large complex.

The third milestone was the testing of the zero gap technology in combination with activated cathodes using a 1 m² pilot cell with ten elements. In the fourth step the electrode area was increased from 1 to 2·7 m². Relying on these preparations, the important fifth milestone was carried out: it was the erection and start-up of the electrolysis plant of General Electric Plastics, The Netherlands (Fig. 1). In this case the application of activated cathodes was carried out on a large scale.

The sixth milestone is the latest one. It is, as already shown, the large stack containing three double electrolyzers and 96 elements in series at Hoechst AG, Frankfurt, which produces 10 000 t/year Cl_2 and realizes for the first time a high voltage circuit with 2·7 m² elements. This electrolyzer will be the basis for 2384 elements which have to be installed in India and Taiwan within the first half of 1989.

The milestones listed here are only those relevant for development of technology. This is used in the 17 Hoechst–Uhde membrane cell plants which are now being planned or erected, or are in current operation. This number of plants represents a total capacity of approximately 700 000 t/year NaOH.

As a result of our development the operating data of the different single element combinations are given in Table 2. The most realistic combinations listed are activated nickel cathodes combined with zero and finite gap membranes, and stainless steel cathodes with finite gap membranes. The data are independent of electrode area and single element arrangement.

Naturally, the lowest power consumption is achieved with zero gap technology and activated cathodes (< 2100 kWh/t NaOH). However, the economical lifetime of these special membranes, which are much more sensitive to brine impurities and shutdown conditions, seems to be not clearly determinable.

TABLE 2
Performance data of Hoechst–Uhde membrane cells

Cathode	Membrane gap		First month		After 2 years		After 3 years	
		i	3	4	3	4	3	4
Ni + act.	Zero	U	2·95	3·13	3·05	3·23		
		CE	95	95	93	93		
		E	2080	2210	2200	2325		
Ni + act.	Finite	U	3·05	3·26			3·20	3·41
		CE	96	96	⟶		95	95
		E	2130	2275			2255	2405
Stainless steel or nickel	Finite	U	3·40	3·65			3·40	3·65
		CE	96	96	⟶		95	95
		E	2375	2550			2400	2575

U (V); E (kWh/t NaOH); CE (%); i (kA/m²).

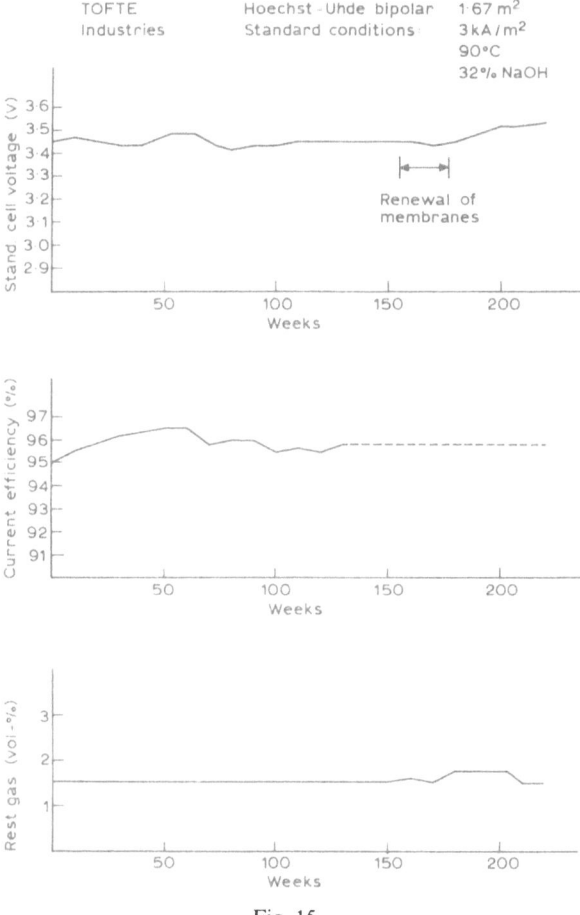

Fig. 15

Finite gap membranes have in our electrolyzers a 100 mV higher cell voltage than the zero gap ones. However, the current efficiency is much more stable.

Compared to the stainless steel cathodes the cell voltage with activated cathodes is approximately 350 mV lower; this works out to 250 kWh/t NaOH.

The long-term stability of finite gap membranes was proven in the Tofte plant in Norway (Fig. 15), where a current efficiency of greater than 95% was obtained for more than 3 years. The middle graph shows the current efficiency versus time.

For activated cathodes an increase of voltage of 50 mV/year is to be expected. However, the electrolyzers of GEP have now been in operation for almost half a year and do not yet show an increase of voltage (Fig. 16). The standardized cell voltage ranges around 3·0 V (upper graph).

Before applications are discussed, the most important properties of the Hoechst–Uhde single element concept and the experiences gained are summarized (Table 3):

1. The single element design applies individual sealing of each element using only metal and corrosion-resistant plastics.

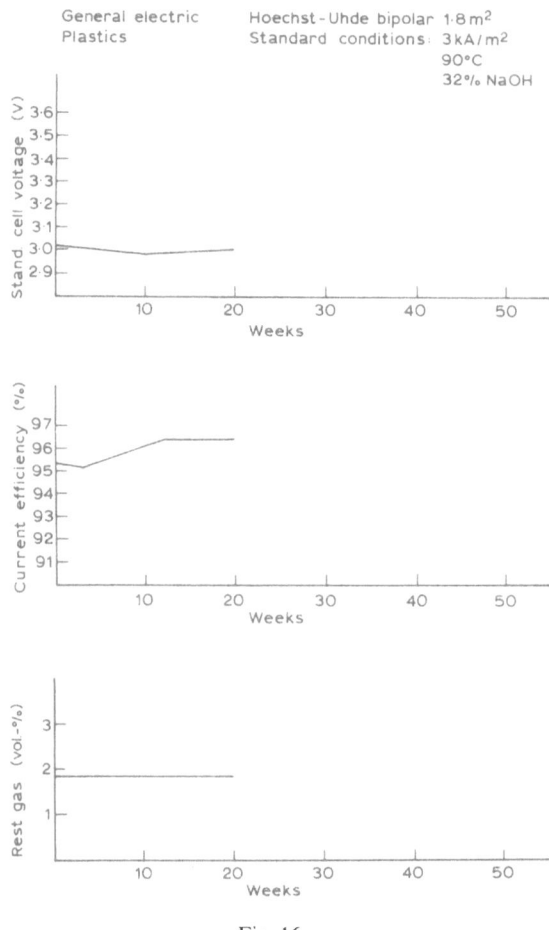

Fig. 16

TABLE 3
Characteristics of the Hoechst–Uhde membrane cell

---

1. Single element design with individual sealing of each element; only metal and corrosion-resistant plastics used
2. Flexible arrangement of basic components suitable for all types of membranes and electrode activations
3. Flexible cell room layout: connection in series for low structural voltage losses and investment cost; connection in series and parallel for optimal adaption to existing cell rooms
4. Large electrode area for reduction of investment and maintenance costs
5. High safety by large electrolyte volume, voltage and flow control of each element
6. Easy assembly/disassembly of single elements; quick exchange of preassembled elements
7. Ten years' operating experience in various combinations

---

2. The flexible arrangement of basic components is suitable for all types of membranes and electrode activations.
3. The cell room layout is flexible using element connection in series for low structural voltage losses and investment cost. Optimal adaption to existing cell rooms is obtained through the combination of series and parallel connection of single elements.
4. Large electrode area reduces investment and maintenance costs.
5. High safety is achieved by large electrolyte volume and by voltage and flow control of each element.
6. The assembly and disassembly of single elements is easy. Preassembled elements can be changed quickly.
7. Hoechst–Uhde relies on 10 years' operating experience in various combinations.

4 APPLICATIONS

Because of the fact that the design of grass roots plants does not have more characteristic issues than those mentioned already, two examples of mercury cell room conversions to membrane cell technology will be presented here.

The two electrolysis plants of Gujarat Alkalies (GACL), India, and Formosa Plastics Corp. (FPC), Taiwan, will be dealt with.

In India the high power prices made a plant conversion economical while in Taiwan statutory requirements demanded the conversion. Both mercury electrolysis plants were planned, erected and started up by Uhde. The technical data are summarized in Table 4.

FPC operates a total of 48 amalgam cells with an electrode area of $30.2\,m^2$ each, supplied by three rectifiers with a total current of 322 kA. The task is to equip the entire cell room with membrane electrolyzers while keeping the rectifier unit in use. The plant capacity of maximum 525 t/day NaOH is to be maintained.

GACL is operating two separate electrolyzer plants (Fig. 17) in one complex. The older of the two cell rooms, with a capacity of 200 t/day NaOH, is to be completely

TABLE 4

Technical data	GACL—India	FPC—Taiwan
NaOH capacity	200 t/day	525 t/day
Number of Hg-cells	38	48
Electrode area per cell	$18\,m^2$	$30.24\,m^2$
Load	$8.6\,kA/m^2$	$10.6\,kA/m^2$
Start-up	1976	1975
Total current	155 kA	322 kA
Rectifier current	$2 \times$ max. 90 kA	$3 \times$ max. 116 kA
Rectifier voltage	60–180 V	max. 225 V

Target: conversion with membrane technology for constant capacity

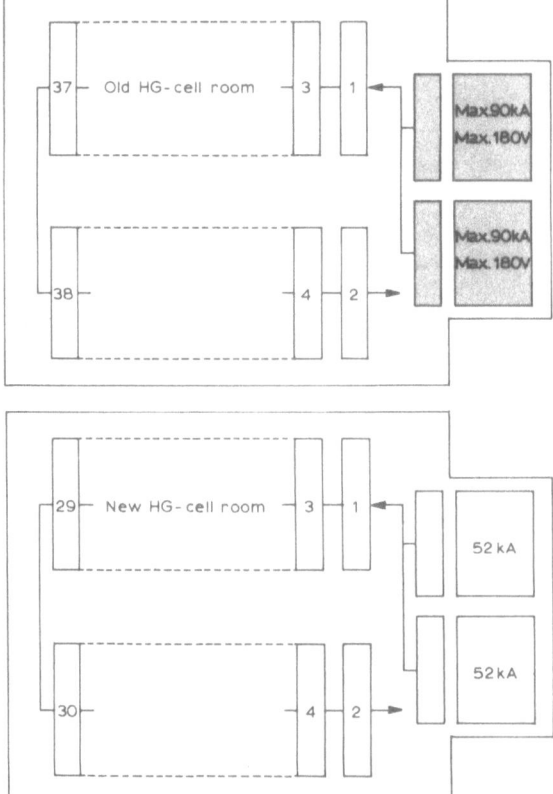

Fig. 17.   Conversion GACL (200 t/day NaOH).

converted to membrane technology while, as is the case with FPC, maintaining the
capacity and employing the two existing rectifiers.

In GACL's old Hg-cell room the 38 amalgam cells are arranged in two rows. Both
rectifiers feed one power circuit. The new Hg-cell room, which will be operated with
a separate brine system, is also shown here.

For planning the conversion, it is important to mention that GACL is prepared to
accept a capacity reduction of the old cell room of not more than 50% while the
plant is being converted to membrane technology. This prerequisite makes it
possible to convert the cell room to membrane electrolyzers in one go.

The diagram in Fig. 18 shows a plan view of the cell room in the GACL plant after
the conversion. In order to provide room for the membrane cells part of the existing
Hg-cells will be dismantled. The space occupied by 16 old Hg-cells will be taken by
16 large bipolar membrane electrolyzers which will be mounted on the existing cell
foundations. Groups of eight electrolyzers are each connected to a rectifier in
parallel. The installation of load switches and breakers allows each electrolyzer to be
switched on and off individually. The technical data of the new membrane
electrolyzers are summarized in Table 5 in the form of a comparison with the data of
the old amalgam cells. As all units are connected in parallel, each electrolyzer can

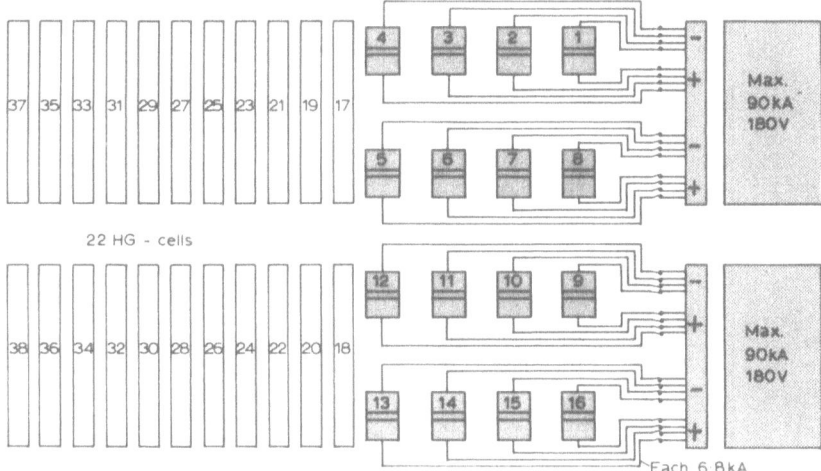

Fig. 18. GACL: cell room with bipolar membrane electrolyzers.

employ the voltage range of the rectifier to the full extent. For this purpose a total of 53 Hoechst–Uhde elements are connected in series in one electrolyzer. For the production of 200 t/day NaOH the required current flow is only 109 kA.

In order to achieve favourable power consumption figures, the cells are to be operated with 2·5 kA/m² only. As we are using elements with a large surface area, namely 2·7 m², in order to reduce the required investment and operating costs, the resulting number of electrolyzers is eight per rectifier. This means that a total of 16 × 53 = 848 Hoechst–Uhde elements will be required.

Due to the low cell load and thanks to the use of nickel cathodes with activation, the energy consumption can be reduced from 2800 (in the case of Hg) to less than 2100 kWh/t NaOH. A considerable advantage of the selected arrangement is the low space requirement of the new electrolyzers of only approximately 42% of the cell room. Even if 16 Hg-cells are dismantled in one go and the membrane electrolyzers

TABLE 5
Cell room conversion GACL—India (chosen arrangement: bipolar)

	Amalgam	*Membrane*
NaOH capacity	200 t/day	200 t/day
Number of cells	38	16 electrolyzers
		53 elements each
		= 848 H–U elements
Electrode area/cell	18 m²	2·7 m²
Specific load	8·6 kA/m²	2·5 kA/m²
Total current	155 kA	109 kA
Rectifier	2 × max. 90 kA	16 × 6·8 kA
	max. 180 V	< 160 V
Energy consumption	> 2 800 kWh/t NaOH	2 090 kWh/t NaOH
Space requirement	100%	42%

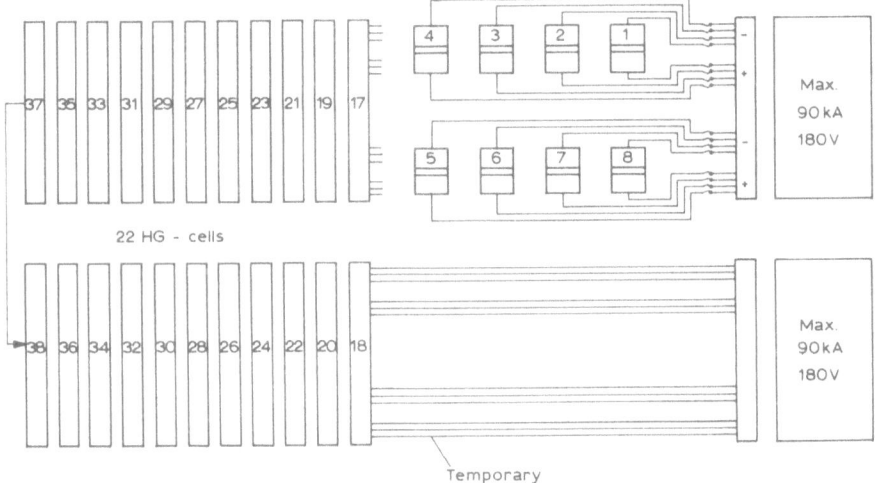

Fig. 19.   GACL: cell room with bipolar membrane electrolyzers.

mounted in their place, the temporary bus bar as shown in Fig. 19 permits the capacity of the Hg-cell room operating with the remaining 22 cells to be maintained at 58% at least. In other words, GACL's requirement to keep at least 50% of the capacity during the conversion period can be fulfilled in one step even if complete conversion is to take place.

The conversion will therefore proceed as follows. First, temporary bus bars will be laid underneath eight Hg-cells on both sides of the cell room. The reconnection to these bus bars will take place during a shutdown. In the course of the following months production will continue with the aid of the remaining 22 Hg-cells. The 16 closed-down Hg-cells will be dismantled, including bottoms and switches, in order to render the membrane cells accessible to the best possible extent, and the new

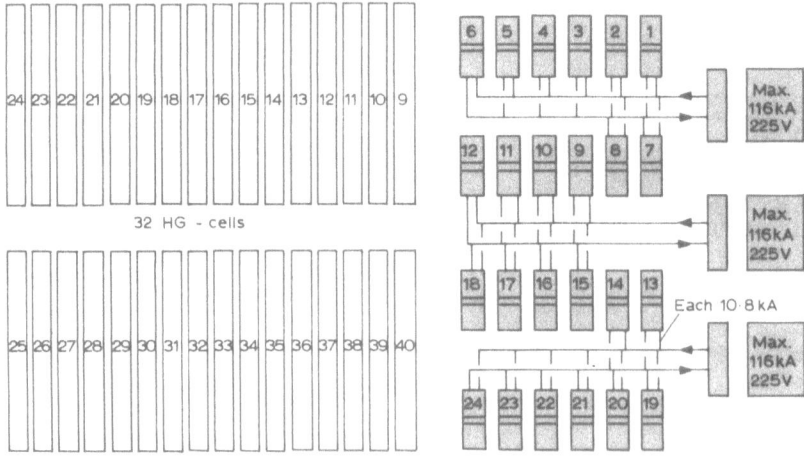

Fig. 20.   FPC: cell room with bipolar membrane electrolyzers.

TABLE 6
Cell room conversion FPC—Taiwan (chosen arrangement: bipolar)

	Amalgam	Membrane
NaOH capacity	525 t/day	525 t/day
Number of cells	48	24 electrolyzers
		64 elements each
		= 1 536 H–U elements
Electrode area/cell	30·24 m²	2·7 m²
Specific load	10·6 kA/m²	3·7 kA/m²
Total current	322 kA	242 kA
Rectifier	3 × max. 116 kA	24 × 10·1 kA
	max. 225 V	< 210 V
Energy consumption	> 3 000 kWh/t NaOH	2 260 kWh/t NaOH
Space requirement	100%	33%

membrane electrolyzers will be mounted together with all the required piping and bus bars. In addition, one set of switches will be installed for each electrolyzer. After completion of erection, the reconnection from the amalgam to the membrane cells can be performed. Branches will be laid from the temporary bus bars to the individual electrolyzers at the same time. The final state can be seen in the upper part of Fig. 19. A total of eight electrolyzers is connected to each rectifier, each electrolyzer having its own bus bars with load switches and breakers.

The 525 t/day plant owned by FPC will be converted according to the same principle as just explained.

Figure 20 shows here, too, the final state of the cell room equipped with Hoechst–Uhde membrane cells: 24 electrolyzers are connected to three rectifiers in parallel to end up with one single current circuit. As the 30 m² Hg-cells installed here are a little larger, the space requirement for the membrane electrolyzers in this conversion project is only 33%.

As can be seen in the comparison (Table 6), each of the 24 electrolyzers is equipped

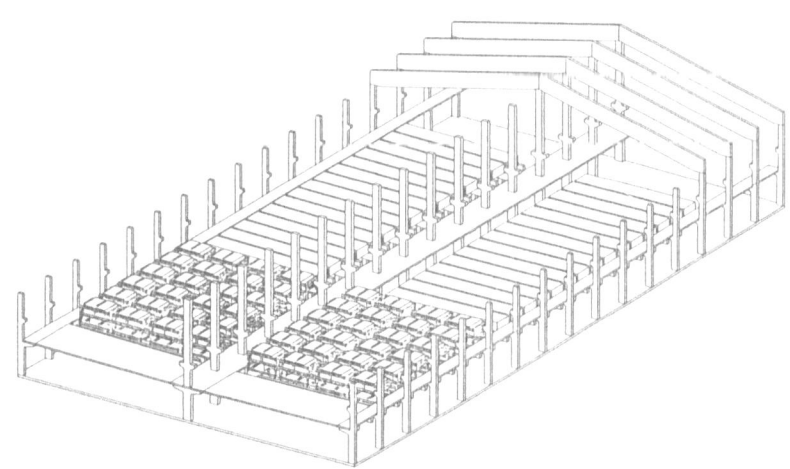

Fig. 21. FPC: cell room with bipolar membrane electrolyzers.

with 64 individual series-connected elements of $2.7\,m^2$ electrode area each and operated with maximum $10.1\,kA$ corresponding to $3.7\,kA/m^2$. The energy consumption can thus be reduced from more than 3000 (Hg) to approximately $2260\,kWh/t$ NaOH.

A total of 1536 Hoechst–Uhde elements are required in the 24 electrolyzers for the production of $525\,t/day$ NaOH.

Figure 21 shows a perspective graphic of the FPC cell room after conversion and the small space requirement of the membrane electrolyzers.

5 SUMMARY

The most important facts of this chapter are summarized as follows:

—Flexible membrane cell technology applying the single element concept is available for small- and large-scale grass roots plants and cell room conversions of any kind.
—Ten years' operational experience with Hoechst–Uhde single elements forms a reliable basis for executing further projects in the future.
—The two contracts for mercury plant conversion with India and Taiwan are evidence that a new era has begun for the Hoechst–Uhde membrane cell technology.

Very recently three new customers have decided to use our technology:

—200 000 t/year NaOH for the VC/PVC producer Salgema CQR in Brazil.
—36 000 t/year NaOH for the pulp mill Aracruz, also in Brazil.
—A 9000 t/year NaOH mercury cell conversion for Cellulose Attisholz, Switzerland, which will be the first plant conversion in Western Europe.

21

DECHLORINATION OF BRINES FOR MEMBRANE CELL OPERATION

T. F. O'Brien

United Engineers and Constructors Inc., USA

1 INTRODUCTION

One feature shared by mercury- and membrane-type chlor-alkali cells is the production of a depleted brine which is saturated with dissolved chlorine. The increased use and developing predominance of membrane cells have revived interest in processes for its removal.

With membrane cells dechlorination must be more thorough than has been the case with mercury-cell technology. The ion-exchange resins which produce brine with μg/litre levels of hardness are attacked by dissolved chlorine. A typical specification for the circulating brine therefore is 0.1 ppm Cl_2. Even in conversion of an existing mercury-cell plant additional treatment for scrupulous removal of chlorine usually will have to be added [1].

Accordingly, processes for dechlorination of brine can be divided into those aimed at bulk recovery of chlorine and those intended for thorough removal or destruction of free chlorine. (This chapter is concerned with the practical aspects of dechlorination of concentrated brines. The terms 'free chlorine' and 'dissolved chlorine' are used somewhat loosely to refer to chlorine and its active hydrolyzed forms. The latter may be taken to be HOCl and OCl^-. Other forms which may be important in water treatment are less so here. Reactions will usually be presented as reactions of Cl_2, unless there is a specific point to be made about the actual form. This is done in part for consistency. It also serves to distinguish among reducing agents which differ in their net consumption of other reactants.) In a mercury-cell brine process, only the first type may be practised. In a membrane-cell plant, the second type becomes necessary, and so these processes are of particular interest to membrane-cell technologists.

In very small membrane-cell plants, this latter type may be the only dechlorination process used. In larger plants, economy will require that both types be used in sequence. We can then refer to them as primary and secondary methods of dechlorination. The discussion which follows will treat these methods in that order.

The data presented refer only to NaCl brines. While the principles are similar with

KCl brines, chlorine is more soluble, and primary dechlorination, at least, is more difficult.

2 PRIMARY DECHLORINATION

Primary dechlorination has two objectives, reduction of the amount of free chlorine in solution and recovery of a major fraction of the dissolved chlorine as the gas. Since removal will not be complete, economics and good practice will determine how much residual chlorine is to be accepted.

The normal technique is to create or provide a gas phase in contact with the brine. Dissolved chlorine will tend to transfer into the gas phase until it exerts an equilibrium partial pressure. In chemical engineering terms, the process can be studied as a normal vapor–liquid transfer operation.

2.1 Thermodynamics

When gaseous chlorine dissolves in water or an aqueous solution, it hydrolyzes to HCl and hypochlorous acid:

$$Cl_2 + H_2O \rightleftharpoons H^+ + Cl^- + HOCl \tag{1}$$

The apparent solubility is therefore higher than the true solubility of molecular chlorine, and under some conditions it is many times higher. When we speak of dechlorination of brine, we mean the reduction of the total amount of dissolved chlorine, whether molecular or combined as hypochlorous acid and its derivatives. Physical chemical considerations must include the chemical equilibrium of eqn (1) as well as the usual vapor/liquid equilibrium. The way in which these equilibria interact is treated below.

Two phases in contact are in physical equilibrium when each chemically independent component has equal fugacities in both phases. In the case of brine dechlorination, the fugacity of interest is that of chlorine, and at equilibrium we have

$$f_g = f_l \tag{2}$$

where the meanings of the symbols are obvious. Gas-phase fugacity is usually given by

$$f_g = \pi \gamma_g y \tag{3}$$

where π = total pressure
γ_g = gas-phase activity coefficient
y = mole fraction in gas phase

while liquid-phase fugacity is given by

$$f_l = P^o \gamma_l x \alpha \tag{4}$$

where P^o = vapor pressure

γ_1 = liquid-phase activity coefficient of chlorine

x = mole fraction of available chlorine in liquid phase

α = fraction of total available chlorine in liquid phase which is present as Cl_2.

At equilibrium

$$x = \frac{\pi \gamma_g y}{P^o \gamma_1 \alpha} \tag{5}$$

The objective of dechlorination is the reduction of x. Consideration of the terms in eqn (5) shows how this can be done.

The gas-phase activity coefficient and the vapor pressure can be disposed of rather quickly. The activity coefficient is usually taken to be close to unity and is in any case beyond our practical control. The brine temperature should be kept high, but heating the brine to increase P^o even further is of doubtful value. Leaving those two quantities aside, we have

$$x \propto \frac{\pi y}{\gamma_1 \alpha} \tag{6}$$

2.2 Acidification

The denominator of eqn (6) shows that we can improve dechlorination by increasing the activity coefficient of dissolved chlorine or the fraction present as Cl_2. Equation (1) indicates a simple and highly effective approach. HCl is almost universally added to the brine to reverse the reaction and Cl_2 is sprung from solution.

The solubility behavior of chlorine in acidic brine is quite complex. To illustrate the importance of acidification, the chlorine–water–HCl system can be considered first. Whitney & Vivian [2] have shown from eqn (1) that the total solubility of chlorine in water can be expressed by

$$S = Hp + (KHp)^{1/3} \tag{7}$$

where S = total solubility

H = proportionality constant for molecular solubility

p = partial pressure of chlorine in gas phase

K = equilibrium constant for eqn (1).

The term Hp represents physical solubility of molecular chlorine. The second term shows how much dissolved chlorine is present as the products of the hydrolysis of eqn (1). At 80°C, with solubility expressed in mg/litre and pressure in mm Hg, tabulated data can be fitted by

$$S = 1 \cdot 51p + 201p^{1/3} \tag{8}$$

At typical vacuum dechlorination pressures, this indicates that at least 75% of the dissolved chlorine has been hydrolyzed. At low partial pressures, this fraction increases. At 10 mm Hg, for example, the amount hydrolyzed will be close to 97% of

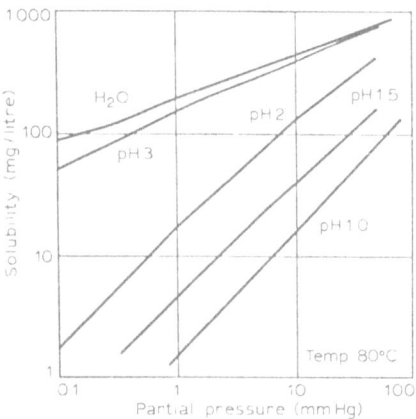

Fig. 1.   Solubility of chlorine in water and dilute HCl.

the total. Reduction of chlorine pressure from 100 to 10 mm Hg, a factor of 10, would reduce the total solubility of chlorine only by a factor of 2·4.

Addition of acid changes the situation dramatically. When chlorine is dissolved into pure water, the concentrations of H^+, Cl^- and HOCl are equal. This fact, along with the definition of the equilibrium constant of eqn (1), leads to the 1/3 power in the second term of eqns (7) or (8). When HCl is added to the dissolving water, the three concentrations are no longer equal. The contribution of hydrolysis to the final concentrations of H^+ and Cl^- now is less important, and the solubility more closely approaches Henry's law. Figure 1 illustrates this fact as well as the major reduction in total solubility. The following tabulation also shows the extent of reversal of hydrolysis at 80°C and a chlorine partial pressure of 10 mm Hg:

	Molar ratio,
Starting pH	*hydrolyzed/molecular*
7	33
3	30
2	10·7
1·5	1·6
1	0·16

Henry's law is a limiting case at low pH of the dissolving water because the contribution of H^+ and Cl^- ions by hydrolysis of the Cl_2 is small compared to the levels already present. In other words, $[H^+]$ and $[Cl^-]$ are nearly constant, and only $[Cl_2]$ and $[HOCl]$ are affected by pressure. Similarly, if data are expressed at a given final pH (below that at which ionization of HOCl must be considered), $[H^+] = [Cl^-] = constant$. The amount of chlorine in solution then becomes directly proportional to its partial pressure in the gas and Henry's law holds.

We should note in passing that the extension of physical data given above relies on calculations which assume a simple common-ion effect. As HCl concentration increases, other factors come into play and the solubility of chlorine soon begins actually to rise.

In brine, the effects of acidification are similar but less pronounced. Addition of HCl in the usual amounts has little effect on the total concentration of chloride ion. We must now be more precise in representing the equilibrium and use activity rather than concentration of Cl^-:

$$[Cl_2] = \frac{1}{K} [H^+][HOCl] \frac{a_{Cl^-}}{a_{H_2O}} \tag{9}$$

For brine of a given salt concentration, we incorporate the activities into the constant. If the brine before exposure to the chlorine gas is neutral, $[H^+] = [HOCl]$, and so

$$[Cl_2] = \frac{1}{K'} [HOCl]^2 \tag{10}$$

This leads to

$$S = Hp + (K'Hp)^{1/2} \tag{11}$$

This differs from the form of eqn (7) only by the higher exponent in the second term, which indicates that the deviations from Henry's law are less extreme in brine than in water.

Again, if we look at data which apply at constant acidic final pH, we shall find that $S = H'p$, where H' is the sum of a pseudo-Henry's law constant for physical solubility and a coefficient accounting for the presence of hydrolyzed chlorine. This assumes only that the activity coefficients of eqn (9) are constant and predicts that the total solubility will be proportional to the partial pressure of chlorine. The data of Bott & Schulz [3] at pH 2 confirm this.

When pressure is held constant, along with temperature and brine strength, acidity is the only variable to consider. Equation (9) predicts that $[HOCl]$ will be proportional to $[Cl_2]$ and inversely proportional to $[H^+]$. Data obtained as a function of pH then should follow the equation

$$S = a + \frac{b}{[H^+]} \tag{12}$$

Typical data in nearly saturated brine are shown in Fig. 2a. The solubility of chlorine increases linearly with the reciprocal of the hydrogen-ion concentration, as predicted by eqn (12). Figure 2b compares the responses of chlorine solubility in water and in brine to changes in pH. First, this shows that the solubility of molecular chlorine is reduced by the presence of NaCl. Less acid is required to reverse the hydrolysis, because of the presence of a large excess of chloride ion. The total solubility of chlorine in brine is substantially depressed at pH 3 and is close to its minimum value at pH 2. It is much easier to release chlorine from brine than it is from water.

As a final note on this subject, it can be seen that the hydrolysis eqn (1) and the derived solubility eqns (7) and (11) ignored the possibility of such reactions as $Cl_2 + Cl^- \rightleftharpoons Cl_3^-$. This does not affect any of our conclusions. In very dilute solution, as when chlorine is dissolved in water, Sherrill & Izard [4] indicate that less than 1% of the unhydrolyzed chlorine will form Cl_3^-. Its concentration is therefore low enough

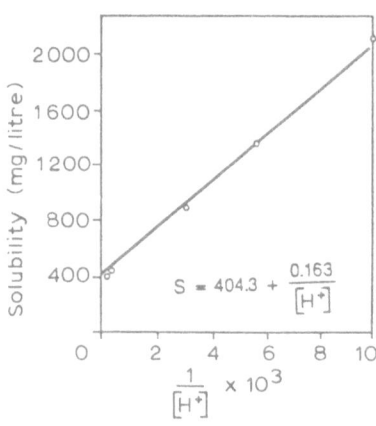

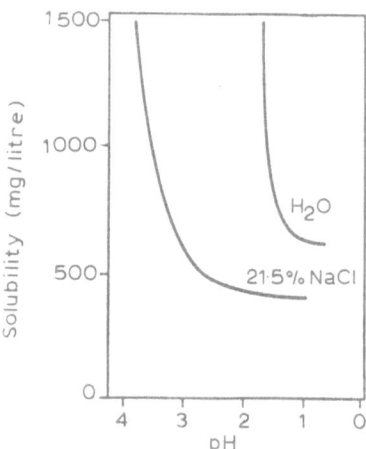

Fig. 2a. Solubility of chlorine in brine versus   Fig. 2b. Solubility of chlorine in brine and
reciprocal of hydrogen ion concentration.   water versus pH. Temperature = 80°C.

to be ignored. In concentrated brine, the fraction which forms Cl_3^- can be much higher, but at a given brine strength it is independent of pressure. We can write $[Cl_3^-] = K_c[Cl_2]a_{Cl^-}$. In strong brine, the activity of chloride ion is substantially independent of the amount of chlorine which dissolves and can be regarded as constant. Thus we have

$$[Cl_3^-] = K_c'[Cl_2] = K_c'Hp$$
$$[Cl_2] + [Cl_3^-] = (1 + K_c')Hp = H'p$$

For our purposes this is mathematically indistinguishable from the case where $[Cl_3^-] = 0$. The 'Henry's law' part of the solubility expression is simply taken to apply to the sum of the concentrations of Cl_2 and Cl_3^- (plus higher complexes if necessary).

2.3 Reduction of Partial Pressure

The use of acidification was suggested by eqn (6). This increased the magnitude of the denominator and so reduced x, the concentration of dissolved chlorine. Equally well and simultaneously, eqn (6) suggests that the partial pressure of chlorine in the gas phase should be reduced (i.e. the value of the numerator should be decreased). This can be done by lowering either the mole fraction of chlorine or the total pressure at the point of separation.

Lowering the mole fraction implies addition of a diluent gas. In practice, this means stripping with air or steam. Section 2.3.1 deals with this process.

Total pressure usually is reduced by application of a vacuum. This is probably the most widely used primary method of dechlorination, at least in large mercury-cell plants. It is treated in Section 2.3.2.

2.3.1 Diluent Stripping

Deliberate addition of another gas will reduce the vapor-phase mole fraction of chlorine. In practice, this means addition of air or steam, and air stripping is in fact a

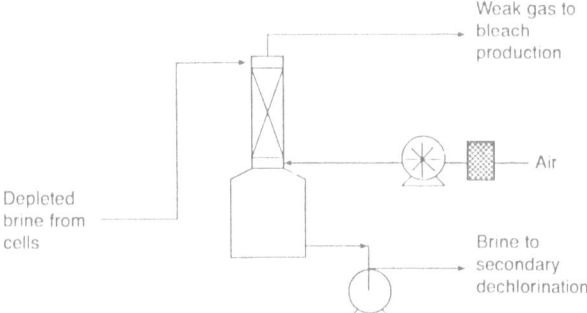

Fig. 3. Dechlorination by air stripping.

common approach to dechlorination. It is usually carried out in packed columns (Fig. 3).

In this process, there is no equilibrium limitation on the efficiency of chlorine removal. The slow rate of desorption and the practical aspects of column design, however, usually dictate a residual concentration of 5–50 mg/liter chlorine in the stripped brine.

In most cases, air stripping is the simplest and cheapest primary method of dechlorination. Its major disadvantage is the dilution of the recovered chlorine with large quantities of air. Sending such a gas to a liquefaction process may actually have a negative effect on the amount of chlorine recovered. It is most suitable for scrubbing with an alkaline liquor and recovery as a hypochlorite, usually NaOCl or $Ca(OCl)_2$.

The use of steam can help separation by increasing the temperature of the brine, and the water vapor which accompanies the stripped chlorine will be removed in the normal cooling and drying process and so not interfere in chlorine liquefaction. However, the brine must be raised almost to its boiling point to provide a useful amount of vapor, excess sensible heat must be removed during recycle of the brine, and the condensate from the stripped vapor is itself saturated with dissolved chlorine. Steam stripping has not found widespread use on a large scale.

2.3.2 Vacuum Stripping

In this case, chlorinated brine is injected into an evacuated vessel. Common designs are simple horizontal tanks and vertical packed columns. The former are empty vessels, perhaps baffled to spread the brine into a thin film, in which the chlorine escapes from the surface of the brine and is evacuated from the vapor space. Packed columns are conventional in design. The packing spreads the brine and increases the interfacial area; there is little countercurrent flow unless some air or steam is added to improve stripping. Because the vapor traffic in a column is so low in simple vacuum dechlorination, even a small amount of vapor addition can make a substantial difference in the mole fraction of chlorine, and the process will reduce the partial pressure of chlorine in two different ways.

This is a highly corrosive service, and the vessels must be designed for some pressure and full vacuum. Rubber-lined steel is frequently used but will have a limited life. The designer will do well to consider the problems of replacement or in-place repair. Titanium is an alternative material of construction, particularly on a small scale.

The stripper is usually mounted in a structure at barometric height in order to allow the brine to flow to a conventional horizontal pump for transfer. The chlorinated feed brine may be pumped to the stripper. If the plant arrangement permits, it can also be moved by the vacuum. The transfer line then must be designed with flashing in mind.

Vacuum is produced by a pump or a steam jet. There is a familiar tradeoff between the two sources. While the jet has a greater energy consumption, it provides a cheaper installation. While it is inherently noisier, it is perhaps more easily muffled. While it dilutes the recovered gas with steam, it is less likely by itself to introduce air.

Two-stage systems sometimes are used for superior dechlorination, but a single-stage jet usually will suffice. The second stage may also be useful in developing the higher discharge pressures which may be called for by newer membrane-cell operations. A precondenser can be used to reduce the operating load on the jet (and therefore its steam consumption). The discharge may be sent to an aftercondenser or directly to the chlorine header, before the cooler. At least that amount of water removed in the precondenser must then be re-used or treated for disposal.

In a membrane-cell process with solid salt feed, the anolyte water balance usually will allow this condensate, along with that from the chlorine coolers, to be returned to the brine loop. The simple flowsheet which results is shown in Fig. 4. The chlorinated condensate is shown returning directly to the depleted brine collecting tank. The layout of the plant may make the addition of a pump necessary.

If a vacuum pump is selected for brine dechlorination, it most probably will be of the liquid-ring type and sealed with water. The choice facing the designer is one of materials. Both titanium and ceramic-lined pumps have been used. The latter may have titanium or ceramic-coated impellers. These pumps may be less familiar to some chemical plant operators, but they have given excellent service with wet chlorine. The available vacuum is limited by the vapor pressure of the water used to seal the pump, but this is not a practical limitation.

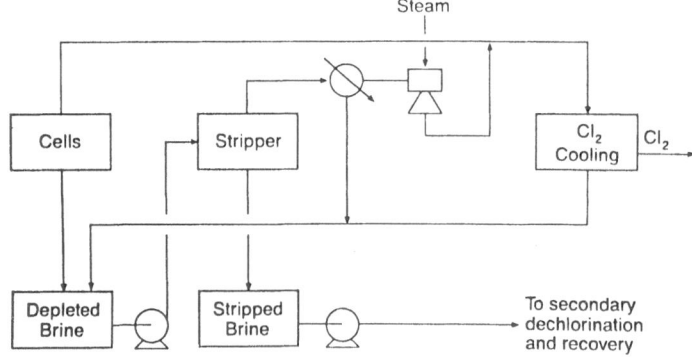

Fig. 4.   Vacuum dechlorination—steam jet.

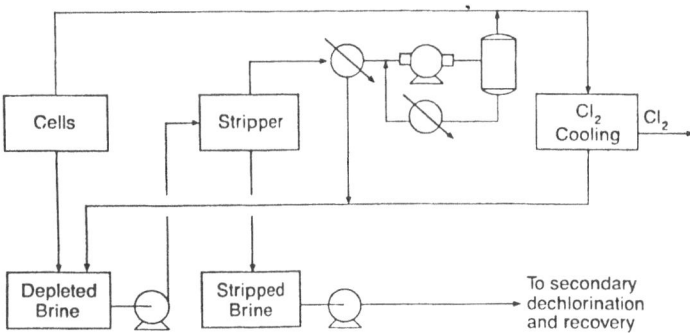

Fig. 5. Vacuum dechlorination—pump.

The dechlorination process using a vacuum pump is shown in Fig. 5. Again, a precondenser relieves the duty of the pump, which becomes equivalent to that of the jet in Fig. 4. The water content of the chlorine going to the first cooler will be less when the vacuum pump is used, and so the load on the cooler will be lower. Its size can therefore be reduced slightly, although not in proportion to the reduction in duty. This reflects the fact that the additional cooling load forced by the steam jet is all at the hot, wet end of the condensing curve, where both the temperature driving force and the heat transfer coefficient are at their highest.

3 SECONDARY DECHLORINATION

Secondary methods of dechlorination are fundamentally different from those of Section 2, where the emphasis was on the reversal of the hydrolysis reaction. Here, the free chlorine is 'destroyed' by conversion to the chloride ion. The methods which we shall consider are chemical reduction, chemisorption, and destruction by metallic catalysts.

Chemical reduction, treated in Section 3.1, is carried out by adding an easily oxidized reagent to the brine. This is an offshoot of wastewater treatment technology and often involves sulfur compounds. These invariably lead to the formation of sulfates, which must not be allowed to accumulate beyond a specified limit. If they are, membrane damage results. If the salt supply contains enough sulfate, its separate removal may be justified. Many techniques are available to do this [5], and the additional load due to a sulfur-based reducing agent may be of little concern. Alternatively, the addition of sulfur can be avoided by the use of other agents (e.g. H_2O_2).

Removal of free chlorine by chemisorption usually involves coal- or coconut shell-based activated carbon [6]. Again, this follows water treatment technology, where activated carbon is used to purify drinking water or water used in the manufacture of beverages. This technique is described in Section 3.2.

The hypochlorites formed by hydrolysis of dissolved chlorine are decomposed catalytically by certain metals. The discussion of Section 3.3 centers on cobalt-based catalysts manufactured for this specific application.

3.1 Chemical Reduction

Dissolved chlorine is an oxidant because of its tendency to form hypochlorous acid and its derivatives (which contain Cl^+). It will therefore react with many reducing agents, and this fact is used in wastewater treatment and in corrosion inhibition. Since reducing agents are almost by definition subject to oxidation during storage or handling, some may be supplied with stabilizers, antioxidants or promoters. The user must ensure that none of these is present in concentrations which can affect performance of the cells.

3.1.1 Sulfur-based

The reducing agents most widely used for removal of chlorine probably are the sulfur-containing compounds, and the most familiar of these are based on S^{4+}. The specific form applied might be SO_2, SO_3^{2-} or HSO_3^-. Within normal hydrolytic equilibrium at a given final pH, these are equivalent. SO_2 is used when large volumes are required or when it is readily available on site. It is widely used in water treatment, one of its advantages being the similarity of some of its properties to those of chlorine. It can therefore be applied in the same equipment used for chlorination.

A choice between the two reacted forms, sulfite and bisulfite, will usually be determined by local availability and price. If a commercial product is to be used and there is a source nearby, aqueous HSO_3^- will be favored. This is the most easily prepared form, and the sodium salt is produced as a solution containing about 23–27% SO_2 equivalent. If a solid reagent is favored, sulfite is the more easily prepared form. Solid bisulfite is actually supplied as its anhydrite, metabisulfite ($S_2O_5^{2-}$). Without prejudging local variables, it would seem that the relatively modest amounts of reagent used in brine dechlorination would not often justify use of a dry form simply to save transportation cost.

In solution, the following equilibria apply:

$$SO_2 + H_2O \rightleftharpoons H_2SO_3 \rightleftharpoons H^+ + HSO_3^- \rightleftharpoons 2H^+ + SO_3^{2-} \tag{13}$$

The two steps of ionization have pK_a values, respectively, of 1·81 and 6·91. Solutions of $NaHSO_3$ and Na_2SO_3, as supplied, would have pH values of about 4·5 and 10.

Figure 6 shows calculated ionic distributions as a function of solution pH. The first ionization is quite similar to the second step with S^{6+}:

$$H_2SO_4 \rightarrow H^+ + HSO_4^- \rightleftharpoons 2H^+ + SO_4^{2-} \tag{14}$$

Here, the first ionization is essentially complete, and the second has a pK_a of 1·92. There is a difference of about 0·1 pH unit between the ionization curves of H_2SO_3 and HSO_4^-. At a given pH, H_2SO_3 is slightly more highly ionized, such that

$$\frac{[HSO_3^-]}{[H_2SO_3]} = 1 \cdot 29 \frac{[SO_4^{2-}]}{[HSO_4^-]} \tag{15}$$

In any event, both acids are completely ionized at high pH, where the dechlorination reaction is simply

$$SO_3^{2-} + OCl^- \rightleftharpoons SO_4^{2-} + Cl^- \tag{16}$$

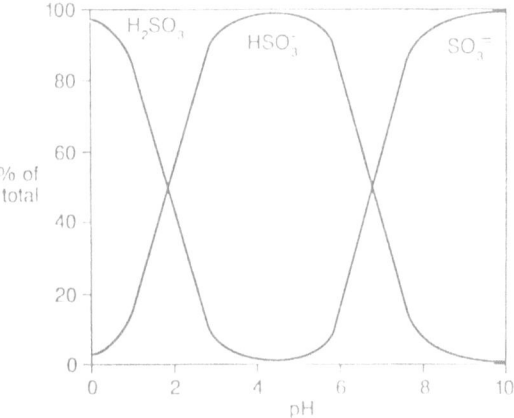

Fig. 6. Distribution of S^{4+} species versus pH.

Theoretical requirements of various forms of S^{4+} for reaction with a unit weight of free chlorine are tabulated below. The numbers in parentheses represent moles of reagent per mole of chlorine.

Reagent	Theoretical usage	NaOH required[a]
SO_2	0·90 (1)	1·13 (2)
$Na_2S_2O_5$	1·34 ($\frac{1}{2}$)	0·56 (1)
$NaHSO_3$	1·46 (1)	0·56 (1)
Na_2SO_3	1·77 (1)	—

[a] For conversion to SO_3^{2-}

When a brine has a high chlorine content, there is an advantage to be gained by matching the form of the reducing agent to the pH of the treatment process. For example, the pH of the brine is usually adjusted before recycle to the saturators. This may be done along with secondary dechlorination or in close conjunction with it. Common practice is to control the addition of a chemical reducing agent to maintain a given oxidation–reduction potential in the treated brine. Since this control signal is itself a function of pH, the control loops will interact.

Suppose that $NaHSO_3$ is to be used for dechlorination and that the brine is to be made alkaline. The caustic soda required to convert HSO_3^- to SO_3^{2-} would be a substantial fraction of that required to decompose HOCl and adjust the pH. Fluctuations in reducing agent supply would be reflected directly in OH^- demand, and the stability of the control system would suffer.

Another form of sulfur compound which may be used is the thiosulfate, $S_2O_3^{2-}$. The apparent oxidation state of $+2$ is the average of $+6$ and -2, with sulfur replacing one of the oxygen atoms in the sulfate radical. This would appear to be an advantage for thiosulfates. If all sulfur is oxidized to $+6$, each mole will remove twice as much chlorine as will S^{4+}. For example:

$$S_2O_3^{2-} + 4Cl_2 + 10OH^- \rightleftharpoons 2SO_4^{2-} + 8Cl^- + 5H_2O \tag{17}$$

In this case, only 0·56 weights of $Na_2S_2O_3$ would be required to deactivate a unit weight of chlorine. In practice, behavior of thiosulfates is not so simple. In neutral or slightly acidic solution, for example, they decompose to sulfur and sulfite [7]:

$$S_2O_3^{2-} \rightleftharpoons S + SO_3^{2-} \tag{18}$$

The sulfite (or bisulfite, depending on pH) then reacts as in eqn (16). The overall chemistry becomes

$$S_2O_3^{2-} + Cl_2 + 2OH^- \rightleftharpoons SO_4^{2-} + S + 2Cl^- + H_2O \tag{19}$$

This is only 25% as efficient as reaction (17) and so requires 2·23 parts of $Na_2S_2O_3$ for each weight of chlorine. Another 1·13 parts of NaOH are required to replace the hydroxyl consumed in the reaction. Thiosulfate must therefore be used with discretion and at high pH to avoid the formation of colloidal sulfur and to approach the efficiency promised by eqn (17).

On an equal-sulfur basis, thiosulfate is in fact less effective than the S^{4+} reagents below a pH of about 10·5. Supplier data [8] show the following requirements for removing one part of free chlorine from water:

pH	$Na_2S_2O_3$ required
11	1·00
9	1·60
6·5	2·23
4	2·67

Throughout this range of pH, requirements for the S^{4+} compounds are listed as

$NaHSO_3$	1·61
$Na_2S_2O_3$	1·96

Each of these is 10% greater than the stoichiometric quantity and has the same sulfur content as 1·22 parts of $Na_2S_2O_3$.

Sulfides, too, can theoretically be very efficient scavengers of chlorine. If oxidized to sulfate, 1 mole of S^{2-} would account for 4 moles of chlorine. The reaction would be similar to eqn (17). Sulfur formation, once again, presents the problem encountered with the use of $S_2O_3^{2-}$. Data given by White [9] on the oxidation of dissolved H_2S show that complete reaction to sulfate, even with an excess of chlorine, was achieved only at low pH.

3.1.2 Non-Sulfur-based

The reducing agents covered in the preceding section form an important class because of their widespread use in water treatment. This experience and their ease of application have led to their use in brine treatment as well. They have, in fact, been the standard for chemical dechlorination of brine.

Reduced oxides of other nonmetals could also be used, e.g.

$$NO_2^- + Cl_2 + 2OH^- \rightleftharpoons NO_3^- + 2Cl^- + H_2O \tag{20}$$

This, however, offers no economic incentive, and there is a general reluctance to add nitrogen compounds to a brine circuit.

More generally, any oxidation with the proper free energy might be coupled with the reduction of OCl^-. One reducing agent suggested by this criterion and producing no troublesome byproduct in solution is hydrogen peroxide. This is now used in several plants and will be discussed below. First, we shall consider other potential reducing agents.

(a) Miscellaneous. Alternatives to sulfate-forming compounds include:

(1) nitrogen compounds,
(2) ferrous compounds, and
(3) simple organic acids.

Certain of these leave undesirable byproducts. With nitrogen compounds (e.g. ammonia and urea), there will always be concern over the possible formation and accumulation of nitrogen trichloride. Since the dechlorination process would be controlled to avoid a large excess of reducing agent, this possibility as well as the possibility of formation of chloramines should not be discounted. The rate of the reaction must also be considered. When ammonia is used, White [10] recommends that at least 20 min be allowed in neutral water and more than 30 min when water is slightly alkaline. Control of the system then becomes more complicated, and holding volume may have to be added at additional cost.

Ferrous compounds in reaction with dissolved chlorine will be oxidized to the insoluble Fe^{3+}. While similar reactions have been used in conjunction with mercury removal to promote coagulation and sedimentation, the usual processing sequence considered for dechlorination does not favor this approach.

The acids which might be considered are formic and oxalic, which require only an atom of oxygen to produce carbonic acid or CO_2 and water. Formic acid would be the more practical choice. If carbonate is formed, as at a pH somewhere above 10, the reaction becomes

$$HCOOH + Cl_2 + 4OH^- \rightleftharpoons 2Cl^- + CO_3^{2-} + 3H_2O \tag{21}$$

While only 0·65 parts of HCOOH are needed to remove one part of chlorine, the hydroxyl requirement seems quite high. Four moles (or 2·26 parts NaOH) are needed, against 2 in most previous examples. In a brine recycle process, however, the net alkali consumption is normal. The 'extra' 2 moles form a mole of carbonate. This may be useful in brine treatment when a crude salt is being used, so long as rejection of calcium does not cause fouling in the dissolver.

(b) Hydrogen peroxide. Hydrogen peroxide, with its excess of oxygen in the molecule, is usually thought of as an oxidizing agent and in fact competes with chlorine in some bleaching applications. It can also serve as a reducing agent when in the presence of a stronger oxidizer, such as free chlorine. Oxygen then leaves as the gas rather than being reduced to O^{2-}:

$$Cl_2 + H_2O_2 + 2OH^- \rightleftharpoons 2Cl^- + 2H_2O + O_2 \tag{22}$$

The reducing agents considered previously share a problem in their formation of undesirable byproducts in solution. H_2O_2 has the advantage that it forms only

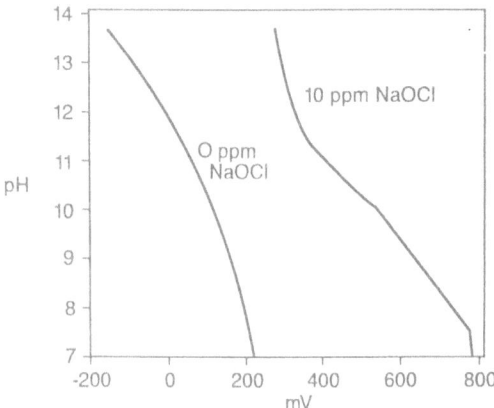

Fig. 7. ORP response of hypochlorite solution (data of Mannig & Scherer [11]).

water and oxygen. It is therefore being used to dechlorinate brine in a number of membrane-cell plants.

Mannig & Scherer [11] recently discussed this application. They recommend that the peroxide be added to brine in response to the oxidation–reduction potential (ORP) of the treated solution. Figure 7 repeats their data on the effect of pH on ORP. The shape of the curve itself depends on residual chlorine content. Good pH control therefore is essential to proper control of any process based on chemical reduction of free chlorine. If the pH is not at its target value, the oxidation–reduction potential will shift and its response to addition of peroxide will vary. The comments made in Section 3.1.1 on the use of $NaHSO_3$ at high pH reflect this situation.

Hydrogen peroxide, by its dual nature in oxidation–reduction reactions, presents a different sort of problem. Except in the impracticable case where H_2O_2 is always fed in precisely the right quantity, treated brine will contain some concentration of an oxidant. While H_2O_2 is weaker than chlorine, it still can damage brine softening resins. The designer must ensure that the resin is protected. Oxidizable material in the brine circuit (for example in the salt) will consume a certain amount of peroxide. A method described in the next section uses activated carbon, in what might be called tertiary reduction, to remove either excess peroxide or unreduced free chlorine.

Dechlorination by H_2O_2 is used in several of the plants designed by Stearns Catalytic. Figure 8 shows the approach taken in one of these. First, 35% H_2O_2 is added by metering pump to brine made alkaline after vacuum dechlorination. The control point is slightly on the free chlorine side of equivalence. The primary control measurement, as shown, is close to the peroxide addition point. This minimizes dead time in the control loop. Because the reaction takes some time to reach completion, a second probe was installed and a cascade control system established. Retention time between the two probes is more than 10 min. A more stable reading, nearer the true equilibrium ORP, was expected from the second probe. Its output could then be used to reset the main controller. In practice, with the chemical dechlorination at high pH, the first probe gave quite a stable reading, and it is not necessary to use the cascaded system to achieve good control and successful dechlorination.

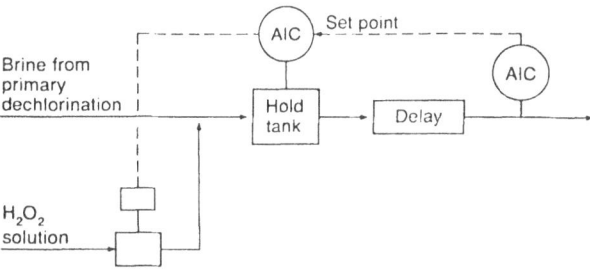

Fig. 8. Reduction by hydrogen peroxide.

3.2 Carbon Chemisorption

Free chlorine can be destroyed by contact with activated carbon. This process has been used widely in treatment of potable water and is now being adapted to chlorinated brines [6]. Coal- or coconut shell-based carbons with surface areas of $700–1200 \, m^2/g$ are normally used.

The process involves chemisorption of chlorine, in any of its active forms, followed by decomposition of the hydrolyzed species. For HOCl, the reactions can be summarized as below, with C^* representing an active site:

$$C^* + HOCl \rightleftharpoons C^*HOCl \tag{23}$$
$$C^*HOCl \rightarrow C^*O + HCl \tag{24}$$
$$2C^*O \rightleftharpoons C^* + CO_2 \tag{25}$$

The first step is a reversible adsorption, shown by eqn (23). This is followed by the irreversible decomposition of eqn (24), which is the rate-limiting step in the destruction of free chlorine.

Release of oxygen is not complete or immediate. Reaction (24) leaves a once-active site oxygenated. This can remain so, in any of various forms (e.g. —COOH). Alternatively, CO_2 will be evolved as in eqn (25). This restores one active site, with another being consumed.

If only the ideal reactions listed above took place, the overall chemistry would be

$$2Cl_2 + C + 6OH^- \rightleftharpoons 3H_2O + 4Cl^- + CO_3^{2-} \tag{26}$$

The theoretical consumption of carbon would be about 0·085 parts per part of chlorine, and hydroxyl consumption would be 3 moles per mole of chlorine, one more than the 2 required by stoichiometry. As was the case with reduction by formic acid, the excess OH^- goes into the formation of carbonate. This may be of value in brine treatment or may contribute to operating problems in a dissolver.

In practice, the life of a carbon bed is much less than projected. There are several reasons:

(1) Reaction (24) is slow, so that there is something resembling a steady-state fraction of the active sites which is not available.

(2) Reaction (25) is not complete; the off-gas from a column has a high concentration of CO [12], from a reaction which consumes twice as much carbon.

(3) To prevent breakthrough in an ordinary fixed bed, one must at the end of the bed's life still retain enough available carbon for the full process duty.

(4) Gradual reaction of a given particle will cause fines to break off and so reduce the particle to one which is small enough to affect the permeability of the bed.

(5) Because of this accumulation of fines, occasional backwashing of the bed is necessary, removing carbon before it can be completely reacted.

(6) As the extent of reaction of the carbon increases, a point is reached at which some of the products of reaction are released, contaminating the effluent from the bed.

The first five of these effects are at least qualitatively predictable. The last becomes the real limitation on the useful life of the carbon. After reaction of about $2\,kg\,Cl_2/kg\,C$, a brown or black discharge appears in the effluent [13]. A substantial safety margin must be allowed to avoid this. Activated carbon thereby becomes more expensive as a means of chlorine removal.

An advantage of a fixed bed of carbon as a chlorine trap is that, until the capacity of the bed is being approached, carbon is 'present' in substantial excess. Its supply is not dependent on another piece of equipment, and there is no control system involved to regulate its rate of addition to the system.

Carbon adsorption is most attractive when used as an adjunct to other methods. When combined with chemical dechlorination, for example, it provides insurance in case of upset or fluctuation in flow of the reducing agent. At the same time, it does its more familiar job of adsorption of organic contaminants, which with some brines may be important. Carbon also has some capacity for the removal of mercury compounds.

In the particular case in which hydrogen peroxide is used as the reducing agent, activated carbon serves as a backup regardless of the direction of the error in the H_2O_2 addition rate. While the peroxide is adsorbed less strongly than chlorine, the added degree of protection afforded by carbon's ability to remove either of the reactants is a very useful feature. The lower capacity of the carbon for peroxide removal explains the use of a control point on the chlorine side of equivalence in Section 3.1.2.

If reaction (24) is the rate-limiting step, we can expect dechlorination to follow first-order kinetics. Magee [14] has shown this to be the case. Rates of reaction follow the sequence $Cl_2 > HOCl > OCl^-$. The process thus becomes less efficient as pH is raised. Magee's work demonstrated this and showed that the effect of pH on reaction rate was due to the changing distribution of free chlorine species. Suidan et al. [15] confirmed this and found the activation energy to be about $10.5\,kcal/mol$.

Generally, dechlorination rates are lower in brine than in water. The decrease of rate with increasing pH is less pronounced, however, so that performance in brine and in water begins to converge at high pH. Figure 9 shows typical data. The rate constant for dechlorination of water shows two areas of rapid change, corresponding broadly to changes from Cl_2 to HOCl to OCl^-. The rate constant in brine drops more slowly, and the two become equal at about pH 9.

Within the normal brine process, carbon treatment might be applied at low or at high pH. While low pH would give a faster reaction, it also increases the rate of

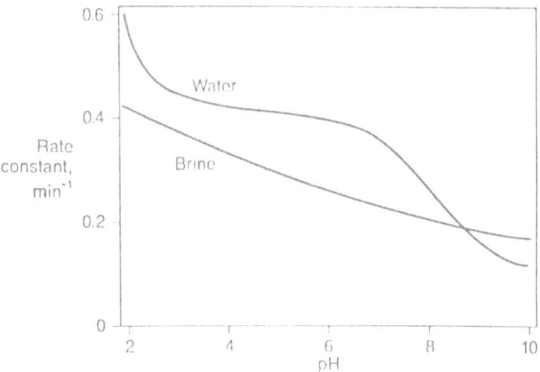

Fig. 9.   Effect of pH on rate of dechlorination by carbon (data of Forster [6]).

decomposition of the chlorates which inevitably are present in brine after electrolysis. This can cause severe gassing which limits the flow capacity of the bed, leads to instability, and quickly abrades the carbon. The safer approach is to use carbon on alkaline brine.

Mechanical design of an activated carbon dechlorinator is not a highly demanding task. The same considerations used in the design of packed-bed filters, and indeed of brine ion-exchange systems, apply here. Dechlorination is often a simpler application because of the relatively infrequent need to take a column off-line. The automatic sequencing common to the other processes may not be appropriate.

In a downflow system, the superficial velocity will be 5–25 m/h. The pressure drop will depend on the flow rate and the particle size of the carbon, which usually will average between 1 and 2 mm. In the mid-range, it should be about 20 mbar/m when the column is clean. On a regular basis (approximately weekly), a column should be backwashed. For good results, the bed should be expanded by 10–50%. A superficial upward velocity of 40–50 m/h should put one safely in the upper part of this range. A vessel, unless very small, should be designed with flow distributors and collectors. When designed for downflow, it will also have a false bottom and a bed support device. The latter must be designed to permit maximum flow in either direction without excessive pressure drop or sufficient stress to deform any component at the temperatures met during operation.

Backwashing should require less than 30 min. The on-line attainment of a bed is therefore very high, and it is not necessary to design a system which can continue to operate at full design rate even when one bed is off-line. In a small plant, a single column will serve if there is enough storage. In larger facilities, multiple beds with little reserve capacity might be used. While a bed was being backwashed, there would be little or no forward flow of brine through this part of the process. Feed brine could be used as the backwash fluid and returned to a point in the treatment process where any entrained solids would be removed without causing new problems.

With a system so designed, it is important to consider the problem of removal of carbon from an exhausted bed. While scheduling can alleviate the problem, the time will come when one must replace a charge of carbon while the plant is operating. Design should make this reasonably convenient, to avoid or minimize production losses. The simplest procedure, a common one with activated carbon systems, is hydraulic. Most of the carbon can be removed simply and easily by flushing it out of the vessel with water. A small connection near the bottom support plate should be provided. There should be space enough near the bed(s) for a portable vessel or bin, which would be used to catch the carbon. Not all the carbon would be expected to be released. The space near the support plate and opposite the point of removal is the most difficult to clean. Supplementary backwash or a small flow through nozzles added for this purpose will help, but scrupulous removal probably will require entry into the vessel. It is not necessary to be scrupulous each time a bed is changed, and so one might expect to be able to unload and recharge a vessel within a few hours. The more elaborate procedures can be left to the time of a scheduled shutdown.

3.3 Heterogeneous Catalysis

Free chlorine can also be destroyed by certain metals or their salts. This is a catalytic process in which oxygen is evolved. Starting with the dissolved chlorine itself, this can be written as

$$Cl_2 + 2NaOH \overset{cat}{\rightarrow} 2NaCl + H_2O + \tfrac{1}{2}O_2 \tag{27}$$

Representative catalytic metals are nickel, copper and cobalt.

Waste streams are sometimes treated in a homogeneous process by addition of dissolved salts of these metals (e.g. $NiSO_4$). This would not be appropriate in a recycle brine process. Heterogeneous catalysis is more attractive, and catalysts based on cobalt spinels have been applied commercially. The major practical problem in their development has been to find a solid composition robust enough to resist the stresses in an operating bed.

Dow Chemical [16] has developed a simple composition in which cobalt is deposited on fairly large (6×10 mm) cylinders of dense ceramic after etching the surface with acid. A catalyst developed by Pennwalt Corporation [17] has cobalt deposited on silica with a poly(vinylidene fluoride) resin binder. This is a porous material with good activity. The binder provides a degree of strength and durability which other products have lacked. Both these catalysts have been used by their developers to dechlorinate wastewaters as well as more concentrated solutions of hypochlorites. The latter have included scrubber liquors as well as depleted brines.

Hodges [18] has discussed the Pennwalt catalyst and described the construction and operation of a test unit as well as the full-scale installation now being used to dechlorinate brine in Pennwalt's plant in Tacoma, Washington. He gives an equation which shows that the temperature effect is normal and the pH effect is similar to that seen with activated carbon. This was obtained by linear regression of the measured first-order rate constants and is modified here for ease of comparison with the equation to follow:

$$k = 0.103\,33 + 0.001\,36T - 0.012\,13H \tag{28}$$

where k = rate constant (s^{-1})
T = temperature $(°C)$
H = pH.

The recommended pH range is 9–11. While a lower pH would increase the rate of reaction, a minimum of 9 is recommended with automatic diversion of flow below that point.

In tests made during plant operation, brine containing more than 1000 ppm Cl_2 was handled successfully. With pH near 10, an average of more than 98·5% of the free chlorine was destroyed.

Caldwell [19] has described development and commercial use of the Dow catalyst. Temperature and pH effects are qualitatively similar to those above. The form of the regression expressed by the Dow design equation is different. The half-life of chlorine, an inverse of the reaction rate constant, is given by

$$t_{1/2} = 7·011 \times 10^{(0·232H - 0·0312T)} \tag{29}$$

where $t_{1/2}$ = half-life of free chlorine (min) (contact time defined as volumetric rate of feed divided by total volume of catalyst bed)
T, H = same as in eqn (28).

In a later version of this catalyst [20], some of the cobalt is replaced by other metals. In one example using zinc and zirconium in the ratio $10Co:5Zn:1Zr$, activity increased by 64%.

Commercial experience in the treatment of tail gas scrubber effluent with the unmodified catalyst is said to be consistent with eqn (29). The preferred pH is close to neutral. This reduces the half-life while avoiding the release of chlorine from the solution. Overacidification (to a pH of about 2) again can lead to deterioration of the catalyst.

Table 1 compares the results of eqns (28) and (29). The calculated half-life is used

TABLE 1

pH	Temperature (°C)	Half-life (s)	
		Pennwalt	*Dow*
7	50	—	488
	70	—	116
	90	—	28
9	50	11·2	1419
	70	7·8	337
	90	6·0	80
10	50	13·9	2420
	70	9·0	575
	90	6·6	137
11	50	18·3	4130
	70	10·6	982
	90	7·5	233

to show the activity of each catalyst. The responses to temperature and pH are similar. The relative importance of these two variables is much the same for the two catalysts. A change of one unit in pH has as much effect as a change of about 7·5°C in one case and 9°C in the other.

Temperature effects are more pronounced with the Dow catalyst. This would be expected from its relatively simple structure. In the temperature range of interest here, the calculated activation energy is about 18 kcal/mol. For comparison, the activation energy with the Pennwalt catalyst is only about 4 kcal/mol. It should be noted that the activation energy for carbon, referred to in the previous section, is about 10·5 kcal/mol.

Catalytic destruction of free chlorine should be regarded as a real option which has been demonstrated technically. Its commercial value will depend on the price of a catalyst, its activity, and its expected service life. Developments are continuing, and improved versions will continue to be made available [21].

REFERENCES

1. O'Brien, T. F., *Modern Chlor-Alkali Technology*, Vol. 2, ed. C. Jackson. Ellis Horwood Ltd, Chichester, 1983, pp. 190–212.
2. Whitney, R. P. & Vivian, J. E., *Ind. Engng Chem.*, **33** (1941) 741.
3. Bott, T. R. & Schulz, S., *J. Appl. Chem.*, **17** (1967) 356.
4. Sherrill, M. S. & Izard, E. F., , *J. Am. Chem. Soc.*, **53** (1931) 1667.
5. O'Brien, T. F., *Modern Chlor-Alkali Technology*, Vol. 3, ed. K. Wall. Ellis Horwood Ltd, Chichester, 1986, pp. 326–49.
6. Forster, G., Brine dechlorination by activated carbon. *Proceedings 29th Chlorine Plant Operations Seminar*, Tampa, 1986.
7. *Kirk–Othmer Encyclopedia of Chemical Technology*, 3rd edn, Vol. 22. John Wiley, New York, 1983, p. 979.
8. Allied Corporation product bulletin, *Water Treatment Uses and Applications—Dechlorination*. Syracuse, New York, 1985.
9. White, G. C., *Handbook of Chlorination*. Van Nostrand Reinhold, New York, 1972, pp. 208–9.
10. Ibid., p. 348.
11. Mannig, D. & Scherer, G., Hydrogen peroxide in the chlor-alkali industry. *Proceedings 30th Chlorine Plant Operations Seminar*, Washington, 1987.
12. Puri, B. R. *et al.*, *J. Indian Chem. Soc.*, **37** (1960) 171.
13. Snoeyink, V. L. *et al.*, *Environ. Sci. Technol.*, **15** (1981) 188.
14. Magee, V., *Proc. Soc. Water Treatment*, **5** (1952) 17.
15. Suidan, M. T. *et al.*, *Environ. Sci. Technol.*, **11** (1977) 785.
16. Caldwell, D. L. & Fuchs, R. J. Jr (to Dow Chemical Co.), US Patent 4 073 873, Feb. 14, 1978.
17. Clark, R. T. & Gardner, D. M. (to Pennwalt Corp.), US Patent 4 400 304, Aug. 23, 1983.
18. Hodges, J. R., K-Cat process for brine/water dechlorination. *Proceedings 30th Chlorine Plant Operations Seminar*, Washington, 1987.
19. Caldwell, D. L., Catalytic destruction of hypochlorite in chlorine plant effluents. Electrochemical Society Meeting, Seattle, 1978.
20. Caldwell, D. L. (to Dow Chemical Co.), US Patent 4 442 227, Apr. 10, 1984.
21. Wagner, L. A., Pennwalt Corporation, personal communication, June 1988.

22

SECONDARY BRINE TREATMENT: ION-EXCHANGE PURIFICATION OF BRINE

Ian F. White

Badger Catalytic Ltd,* UK

and

T. F. O'Brien

United Engineers and Constructors Inc., USA

1 HISTORY

Electrolysis of brine for production of caustic soda (or potash) and chlorine is a process with a history of over 100 years. For most of that period brine quality has been an important factor in plant operability and economics.

Until recent years the industry has been dominated by mercury cells and diaphragm cells. In terms of hardness (principally calcium and magnesium) these technologies have required impurity levels which are measured in ppm (mg/litre). As a rule, brine of this purity has been produced by use of a modified lime/soda technique in which the alkaline earth cations are reacted with appropriate anionic species to form precipitates which are then removed by a combination of settlers and filters. Using this process it is quite possible to achieve a hardness level in low single figures of ppm, this being the level generally regarded as necessary for good operation of diaphragm cells. Mercury cells are often run with somewhat higher impurity levels, although this can require careful control of pH around the brine circuit.

Other means are employed to control the hardness of brine feed to the traditional electrolysis cells. These can include the use of unconventional dissolving techniques allowing rejection of a portion of the impurities as undissolved solids. Even so, some form of chemical purification will still be required. A further alternative has been to use a salt with very low impurities, such as vacuum salt or salt recovered from diaphragm cell liquor evaporation.

* Formerly Stearns Catalytic International Ltd.

In the 1970s a new technology began to create interest in the chlor-alkali industry. Early membrane cells were fitted with low efficiency (by today's standards) sulphonic membrane exemplified by 300 series NAFION®. At the outset diaphragm grade brine was considered adequate for these electrolysers [1], although the deterioration of efficiency with time could be correlated with brine quality. Some mitigation of this effect was realised by the addition of phosphates to sequester the hardness ions.

Higher efficiency membranes were developed by the manufacturers. It was found that these rapidly deteriorated with even the purest of diaphragm cell brines and phosphate dosing. It became necessary to respecify hardness contamination in the ppb (μg/litre) range [2, 3]. Today's membrane electrolysers typically require a brine with 50 ppb total hardness or better. Such purity is not achievable with conventional purification techniques.

2 ION EXCHANGE

In order to obtain brine of the purity necessary for modern high efficiency membrane cells a well-established technique has been adapted from the water treatment industry, that of ion exchange. This technique has long been used for the removal of unwanted ionic species from water streams, for example to produce boiler feed water or where high purity water is required for process use. The technique is even familiar in the domestic environment, finding an application in the water softener in the humble dishwasher, for instance.

For brine purification it is not necessary to achieve the extremely low level of hardness that is required in some forms of water treatment. The pharmaceutical industry, for instance, or the electronic components manufacturing industry, need exceptionally pure deionized water. The requirements of secondary brine treatment are much more onerous in another respect, however, in that by far the greatest ionic concentration present is the desired sodium (or potassium) chloride. The problem is one of removing unwanted cations and yet retaining the far larger concentration of wanted species.

2.1 Resins

The chemistry involved in ion exchange softening of brines is similar to the softening of water:

$$2\,R^-Na^+ + Ca^{2+} \rightleftharpoons (R^-)_2Ca^{2+} + 2\,Na^+$$

This is not to say that water softening resins can be used to prepare membrane cell brine. The final hardness level must not only be very low but must also be achieved against a very high concentration of alkali metal ions, which will tend to drive the reaction to the left. The selectivity of the resin for multivalent cations must be very much higher than in more conventional operations.

Two types of resin meeting this requirement are in active use. The first is patterned after the familiar use of EDTA to chelate certain metal impurities; the active group is an iminodiacetate (Fig. 1(a)). The second type has aminophosphonic acid end

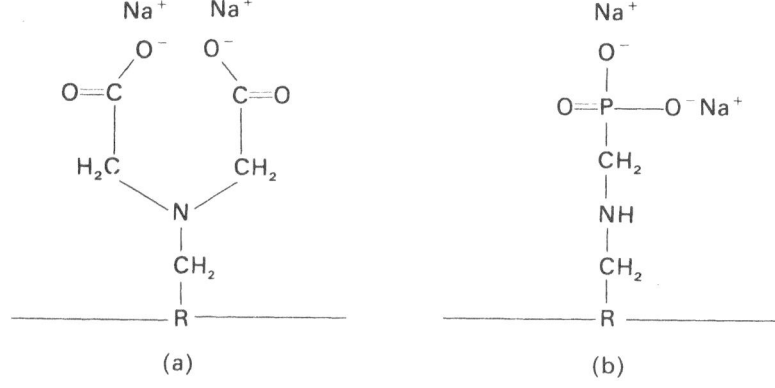

Fig. 1

groups (Fig. 1(b)), which also can chelate hardness cations, and in fact has a greater capacity.

Several manufacturers now supply resins for this application, and both types are in use and available. The designer of an ion exchange unit can find himself confronted with a bewildering stack of product literature. It is essential, when trying to determine the breakthrough capacity of a resin, not only to understand the conditions under which test data were obtained but also to define exactly what is meant by 'breakthrough'.

As membranes have become more selective and at the same time more sensitive, and as analytical techniques have become more sophisticated, the target for the ion exchange process has become more daunting. Over the past few years the commonly referred to level of hardness has gone from about 100 to 50 ppb and now is often 20 ppb or less. The designer who wishes to compare resin performance must be careful not to mix data obtained against different bases.

When data reliable below 50 ppb are analysed, it appears that, at normal rates of brine flow, the breakthrough capacity for 50 ppb is about 10% greater than that obtained at 20 ppb. A similar increase is found if breakthrough is taken to be at 100 ppb.

Important design variables are pH, temperature and brine flow rate or space velocity. The resins are best applied to alkaline brine, and in this respect are a good match with conventional primary or chemical brine treatment. The capacity of the typical resin rises continuously with pH. A typical minimum pH for design is 8 or 9. Below this range capacity begins to drop rapidly as pH is lowered. At high pH there is a danger of precipitation of hydroxides, and so a good range for operation is from 9 to about 11.

The capacity of a resin also increases continuously with temperature. The only constraints are the materials of construction of the system (including piping, valves, gaskets, etc.) and the stability of the resin. The maximum temperature for the resin should be that defined by the manufacturer. There would seem to be no overall advantage in operating the ion exchange system at a temperature above that desired for feed to the cells. This is probably no more than 80°C at full load. Resins are

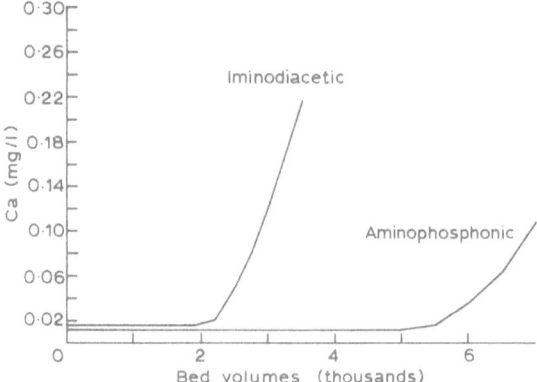

Fig. 2.   Brine purification breakthrough profile: 20 BV/H, 70°C, 1–2 mg/litre Ca^{2+} plus Mg^{2+}.

usually stable up to a temperature of 80 or 85°C—so thermal stability does not constrain design.

The optimum temperature for operation can be chosen only after considering the rest of the process and the ways in which the brine will finally be adjusted to the cell feed temperature. If there is a need to heat the brine from primary treatment before it reaches the cells, there is a case for doing much of this prior to ion exchange. The capacity at 60°C, for example, is about one-third higher than the capacity at 40°C. We will usually strive for a temperature no lower than 60°C.

The design flow rate, or the size of a resin bed, will depend on temperature, pH and type of resin used. We shall consider the effects of the type of resin below and so only note here that hourly space velocities will usually be in the range 10–25. This statement will be qualified later when comparing two- and three-bed systems.

Figure 2 compares operating results with iminodiacetic and aminophosphonic resins. Breakthrough is defined at 20 ppb. In both cases the space velocity was 20/h, the temperature 70°C and the pH approximately 10. The feed brine contained 1–2 mg/litre Ca^{2+} plus Mg^{2+}. The superiority of the aminophosphonate in this

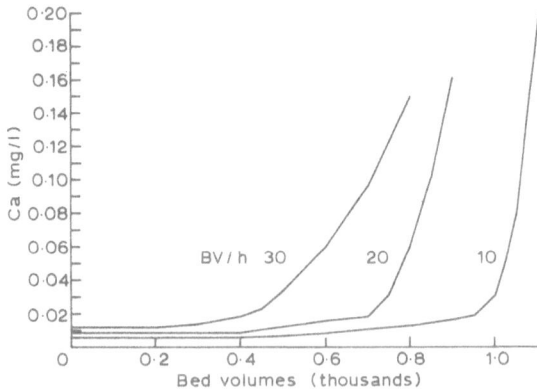

Fig. 3.   Aminophosphonic breakthrough profiles: 15 mg/litre Ca, 60°C, pH 10.

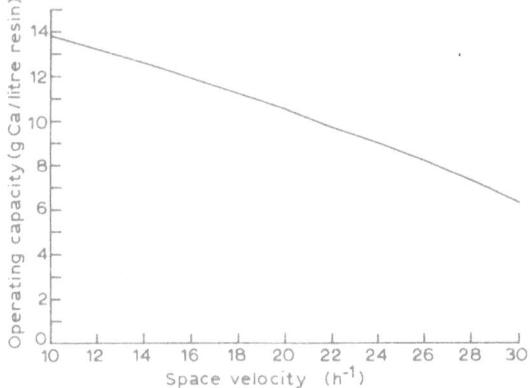

Fig. 4. Aminophosphonic operating capacity: 0·02 mg/litre end point, 60°C, pH > 9.

particular case is obvious and is due both to a higher equilibrium capacity and to superior dynamic properties.

The effects of space velocity are illustrated by Fig. 3. These were accelerated tests, using brine with 15 mg/litre Ca^{2+} and a temperature of 60°C. At a space velocity of 30/h the breakthrough capacity of the resin is only half the value at 10/h. The slope of the breakthrough curve is reduced, as is normal with a less favourable equilibrium or with a breakthrough zone which occupies a larger fraction of the bed. Figure 4 shows the operating capacity, expressed as ionic loading, as a continuous function of space velocity. Such data can be used by the designer to optimise the system. Higher space velocities allow the use of smaller beds; they also reduce the operating capacity of the resin and require more frequent regeneration. In this application security is also a very important consideration.

2.2 Resin Consumption

Predictions of resin consumption were initially based on water treatment experience with a conservative downrating for the more arduous service. This resulted in an expectation of a 3-year life with the requirement to top up the bed by 10% after each of the first 2 years' service to compensate for loss of resin by attrition and the deactivation of active sites.

Whilst these figures remain the basis of Stearns Catalytic's quotations for expected consumption we have found, in practice, that better performance has been achieved. The addition of 5% has proved sufficient to compensate for annual losses and overall lifetimes appreciably longer than 3 years have been realised.

3 EQUIPMENT AND OPERATION

3.1 System Configuration

Many configurations have been used or proposed. Most of these are based on two- or three-bed systems, although we have used a single bed unit on one small facility. This is operated on a semi-batch basis with intermediate storage of purified brine sufficient to cover the regeneration period.

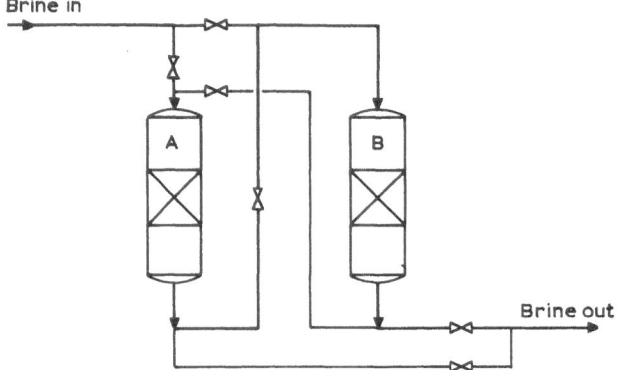

Fig. 5.   Two-bed system.

Two-bed systems are generally designed on a lead-lag basis (Fig. 5). This means that either bed can be first in the process flow. Our practice has always been to specify the unit such that each bed is sized for the full process duty. Operation will be such that the bed in the second position is the most recently regenerated so that it can act as a guard in the event of hardness breaking through the first bed.

When a three-bed system is desired there is a very much wider selection of process schemes available. The third bed can be added as a guard bed (Fig. 6), always operating in the final position. In this case the choice must be made as to whether the guard bed should be the same size as the two main beds, or whether a smaller bed will suffice. It must also be decided whether to incorporate the guard bed in the automatic regeneration sequence or not. Our preference has been to provide a bed rather smaller than the main beds and to have the regeneration automatically controlled but manually initiated.

Alternatively, a third bed can be incorporated in a 'merry-go-round' system where all three beds are identical from a processing standpoint (Fig. 7). If this route is selected there are still further decisions to be made. First the designer should decide

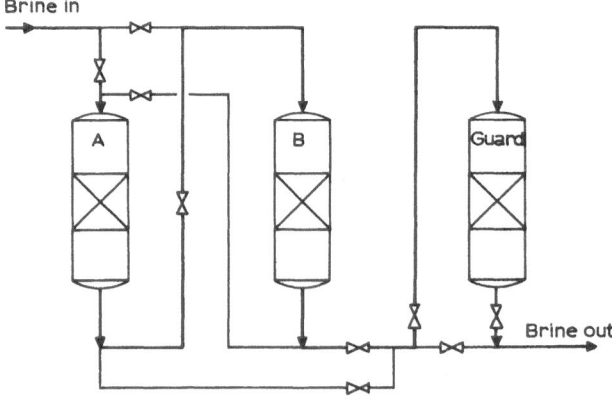

Fig. 6.   Two beds + guard bed.

Brine in

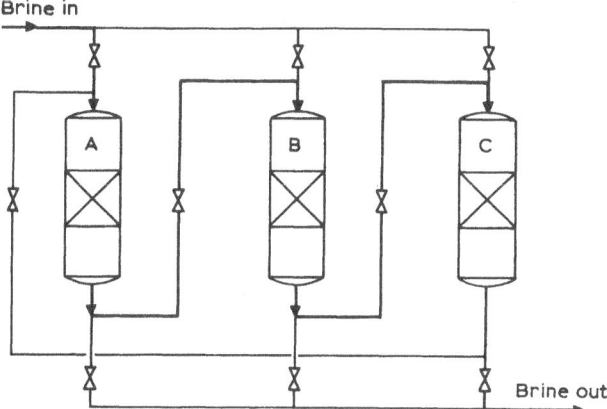

Fig. 7.   Three-bed system.

whether he requires the brine to pass through the beds in any order or whether one way round is sufficient. In other words, if the beds are designated A, B and C, whether he is satisfied with the orders ABC, BCA and CAB as in Fig. 7, or whether ACB, BAC and CBA are also necessary. The additional complexity arising from the latter requirement is, in our view, not justified by the marginal benefit of extra flexibility.

Secondly, the size of the individual beds must be chosen. Whereas with two beds there is really no case for anything less than 100% capacity for each bed, when three are selected an argument can be made for each bed having a lower capacity. We have recently commissioned a plant with three 50% beds in a 'merry-go-round' arrangement.

The feasibility of this approach depends on two beds always being in active service, even during regeneration of the third. The amount of resin which would otherwise be provided in a single bed can now be contained in two. Such a three-bed system can thus be built with a smaller resin inventory than the standard two-bed system. Since the resin is rather expensive, the cost is quite a bit lower.

Disadvantages include the more frequent need for regeneration and the need to compromise the joint operating goals of security and efficient utilisation of the resin. If the space velocity desired in a single bed is 15–20/h, the space velocity in this case will be 30–40/h in each bed. Figure 4 suggests that the capacity of the resin in the smaller bed will be reduced by about 40%. If breakthrough beyond 20 ppb is to be avoided, the frequency of regeneration will be much higher and the amount of hardness removed by a unit of resin will be reduced. Since the second bed is still fresh at this point of incipient breakthrough, however, and since the third bed is also on line, the service cycle of the first bed could be extended to increase its resin utilisation. Doing this necessarily imposes a load on the second bed. The relatively flat breakthrough curve characteristic of high space velocity operation (see Fig. 3) means that an appreciable fraction of the second bed's capacity could be used in this

assume that two half-sized beds are in all ways equivalent to a full-sized bed. A specific analysis should be made of the state of the resin inventory throughout the proposed cycle. Only then can the security of the system be judged and the trade-offs evaluated opposite the lower cost. We find in general that a resin utility comparable to the two-bed system can theoretically be achieved with beds 55–60% of the 'standard' size. This allows a 10–15% saving in resin inventory.

3.2 Regeneration

Once a resin bed has been loaded with hardness it must be regenerated before re-use. Basically this is achieved by eluting the impurities with acid as the resin's affinity for divalent ions is very low in low pH conditions. This use of acid converts the active sites to the hydrogen form. After regeneration the resin must be reconverted to the active sodium form before it can be returned to service. A typical regeneration sequence is described in Table 1.

The bed to be regenerated is taken off line and drained of brine. This brine can normally be recovered by recycling to an appropriate point in the brine circuit. The bed is then rinsed with water. Resin fines and any material collected by the resin's tendency to act as a filter can then be removed by backwashing, again with water. The flow rate here must be sufficient to expand the bed to allow release of the fine material but not so great that whole beads are carried out of the bed.

Next comes the regeneration stage itself when dilute hydrochloric acid is injected into the bed. A further water rinse to remove residual acid is followed by injection of dilute caustic soda to reactivate the resin. Caustic soda is rinsed from the bed which is then drained prior to refilling with brine. After a final rinse with brine the bed is ready to be returned to service.

If regenerant solutions are at ambient temperature, the resin will undergo both thermal and osmotic shocks. Nevertheless this is the routine approach, the resins being quite sturdy. However, the designer should be aware of and follow the resin supplier's advice.

Clearly, the control of this regeneration sequence is complex. Equally the possibility of operator error is high and the consequences can be very expensive. It is

TABLE 1
Regeneration sequence

Step	Space velocity (h^{-1})	Duration (h)	Quantity, bed volume	Direction of flow	Composition
1. Drain	—	—	1	Down	25% NaCl
2. Displacement	4	1–1·5	4–6	Down	Water
3. Backwash	10	0·5	5	Up	Water
4. Regeneration	2–6	0·5	[a]	Down	4–15% HCl
5. Rinse	2–4	0·5–1	2	Down	Water
6. Conversion	2–4	0·5	[b]	Up	4–15% NaOH
7. Rinse	2–4	0·5–1	2	Down	Water
8. Rinse/fill	—	—	—	Down	Brine

[a] Total acid dosage 100–200 g/litre of resin.
[b] Total alkali dosage equivalent to 80–150 g NaOH/litre of resin.

quite conceivable that the membranes from an entire cellroom could be written off as a result of high hardness being allowed to bypass the ion exchange unit. For this reason we specify that the regeneration should be automatically controlled, preferably by a dedicated programmable logic controller but alternatively by a process control computer if such is available on the plant.

A feature of the weak base resins used for brine purification is the substantial volume changes between their different states. The sodium form can be 40% less dense than the hydrogen form. Therefore substantial pressures can be built up within beads during the resin reactivation, leading to bead fracture. It is important that the system and the regeneration sequence are designed to alleviate this by ensuring that the beads are given room and the opportunity to expand. The directions of flow suggested in Table 1 are chosen with this in mind.

In order to achieve satisfactory regeneration it is plainly necessary that the regenerants used should be of adequate quality. The hydrochloric acid must not contain hardness ions or free chlorine above very low levels. Other contamination is also to be avoided. Similarly a pure grade of caustic soda is required. Fortunately a typical membrane cell product is suitable. The regenerants are used at concentrations as low as about 4% whereas they will normally be received at substantially higher strengths. The water used for dilution as well as that used in the various rinse stages should also be very pure, at least softened and preferably demineralised.

Pumps may be used to inject the acid and alkali for the regeneration sequence. An economic alternative can be to use eductors. These can make use of the pressure of the demineralised water and at the same time produce the necessary dilution of the regenerant liquors. We have found that consistent results can be obtained provided the eductor systems are properly set up at start-up. If pumps are favoured, again the acid and alkali can be used as supplied and injected into diluent streams of water. Alternatively, regenerant solutions can be prepared and stored at the concentration of use. They would then simply be pumped to the beds at the desired rates. This approach adds to the labour and space requirements but provides a margin of comfort by allowing routine checks of reagent concentrations.

The number of steps involved in the regeneration process and the use of intermediate rinses mean that large volumes of waste solution will result. The operator may be faced with the disposal of more than 20 bed volumes of fluid from each regeneration. Most of this can be usefully recovered for recycle. The brine drained at the beginning of the process, and even part of the first rinse, can be recycled for concentration and treatment. The acid and caustic can be recovered, at least as the equivalent salt value.

The simplest technique is to blend them as they issue from the system, and then to neutralise the resulting acidic solution and return it to the brine plant. This presupposes that there is primary chemical treatment of the brine to provide an outlet for the hardness elements. The acid effluent contains all the hardness but could still be used more effectively in dechlorination of depleted brine, for example. Here again a primary treatment system is necessary, unless a once-through brine system is employed. The caustic stream may also be recovered for high grade use. On the whole, this is only diluted by exchanging sodium for hydrogen ions. Thus it should be suitable for blending with concentrated solution in the next batch of regenerant, for example.

Most plants have some system for recapturing parts of the regeneration effluent, but we are not aware of any as elaborate as outlined above.

3.3 Prefabrication

The similarity in concept of a brine ion exchange unit to a deionised water unit immediately leads one to the idea of a prefabricated modular unit. This is a standard approach which is widely accepted in the water treatment industry and in the construction of deionisers for use in the chemical process industries. It allows a unit to be assembled in the fabricator's workshop in a relatively clean and controlled environment. It is usually much easier to maintain quality control and to detect and correct errors under such circumstances.

The unit will be divided into a logical number of subassemblies, based upon size and function. The maximum size of such an assembly will be determined by transport and site clearance considerations. With a two-bed system of fairly modest size, for example, two skids might be assembled, one holding the beds and the other the regeneration system and controls (see Fig. 8).

All the units that Stearns Catalytic has installed have adopted this approach, and we believe that it is in fact the industry standard. The advantages of prefabrication, even more widely applied, have been discussed by Dibble [4].

Prefabrication also presents opportunities for more effective precommissioning. This aspect is discussed in Section 4.

Fig. 8

3.4 Prefiltration

It is difficult to overemphasise the importance of adequately filtering the brine upstream of the ion exchange unit. Besides the obvious danger of plugging the resin bed with such materials as filter aid or suspended brine solids there is the potentially greater risk of hardness components passing through the bed in the solid phase, for which the resin has no selectivity. The solids would then redissolve in the purified brine, thus nullifying the effect of the ion exchange system.

3.5 Chlorine Removal

The resins used for secondary brine treatment are typically based on a styrene–divinylbenzene copolymer and as such are subject to oxidative attack. It is important, therefore, to ensure that free chlorine concentrations are reduced to minimal levels, bearing in mind that most membrane cell plants utilise a recirculating brine system. A maximum concentration of 0·1 ppm is generally specified to ensure reasonable resin life. It can be equally important to avoid the presence of other strong oxidising agents such as hydrogen peroxide which is sometimes used as a chlorine scavenger.

A further potential cause of bead oxidation occurs during the regeneration sequence. After being taken off line the bed is full of brine which, in most cases, will contain several grammes per litre of chlorate ion. If the sequence commences immediately with acid regeneration there is the potential for producing free chlorine by decomposition of that ion. Therefore it is important that the regeneration sequence be arranged to separate brine from acid.

TABLE 2
Characteristics of systems

Plant	A	B	C	D
Number of beds	2	2	3	1
Resin	Duolite C-467	Duolite C-467	Sunpearl SC-401	Duolite C-467
Operation				
Bed height (m) (Na$^+$ form)	1·2	1·4	1·2	0·61
Space velocity (h^{-1}) (per bed)	20	20	35	5
Design run time (h)	24	24	24	42
Design regeneration time (h)	5	6	12	4·5
Direction of flow	Down	Down	Down	Down
Regeneration				
Acid—direction of flow	Down	Down	Down	Down
—concentration (wt %)	4	4	8	5·5
—number of bed volumes	2·5	2·0	1·6	2·4
—how diluted	Eductor	Metering pump	Premixed	Eductor
Caustic—direction of flow	Up	Up	Down	Up
—concentration (wt %)	4	4	5	4·6
—number of bed volumes	2·0	2·0	3·0	1·9
—how diluted	Eductor	Metering pump	Premixed	Eductor
Disposal of waste	Collect in buffer tanks; neutralise and recycle	Mix with water treatment effluent; neutralise and recycle	Collect and mix; pump to waste stream treat tank	Route to drain and works waste treatment

3.6 Resin Trap

In the early days of secondary brine treatment units we experienced losses of resin after start-up that were considerably greater than expected. This was noticed at more than one plant. It was determined that most of this loss was occurring during the backwash phase of the regeneration sequence, despite the fact that backwash flow rates were within recommended limits. This loss was found to be a consequence of inconsistencies in bead size and density. Smaller and lighter beads were being backwashed from the bed at quite low flow rates. Quality control has since been improved so that this phenomenon is no longer a problem. Nevertheless we continue to specify a resin trap, a simple screen to catch any resin that is backwashed from the bed. This serves to guard against inadvertent maloperation during the regeneration sequence.

3.7 Characteristics

Table 2 gives a listing of the characteristics of some of the secondary brine treatment systems designed by Stearns Catalytic.

4 COMMISSIONING

Effective precommissioning is the best way to guarantee a successful start-up. This is a subject familiar to designers and operators, and so the generalities of commissioning programmes will not be discussed here.

The arguments for a highly automated, computerised regeneration sequence are given in Section 3.2. The use of such a system makes thorough precommissioning even more important. The constructor must be sure that the controller and the process hardware are in fact properly linked and that the correct sequence of operations is obtained in all reasonably likely circumstances. This should specifically include response of the system to variations which may be called for by the operator.

While the extent of manual back-up and the possibilities for operator intervention must in the end reflect the philosophy of the operating organisation, there is a minimum required for flexibility and response to problems or changing circumstances. The operator certainly must have the ability to initiate regeneration at his own discretion. Someone in the organisation should also have the means of changing some of the basic parameters (flow rates, durations, etc.). At the other extreme the ability to control the entire sequence from a panel or keyboard will also be required by some owners. Precommissioning checks should demonstrate that none of these operator actions can disturb the proper operation of the controller and that it can return to normal operation after such 'upsets'.

This checking of the controller and its relationship with the unit itself can be accomplished after final assembly and along with testing of the hardware.

In Section 3.3 we mentioned the use of prefabricated skid-mounted units. This approach allows a maximum of pretesting to be done in the fabricator's shop. This is especially advantageous so far as the automatic sequencing of valve operations is concerned. Once the sequencing has been demonstrated satisfactorily, the entire unit can be tested on water. Only after this is the unit shipped to site, where it should require only hooking up and final testing on process fluids.

Only in rare circumstances, however, will the fabricator of the columns and skid be the original supplier of the unit's programmable controller. This may impose a time constraint on project management which interferes with the wish to test the system complete with controller before it leaves the shop, or it may prove inefficient to make the first trials of the controller far from its own manufacturer's facility.

In such cases a certain amount of independent testing should be considered. There are many ways in which this might be done. Stearns Catalytic, for example, maintains a process simulator which can be used to test the logic of the programme. Circuit boards on which lights can represent valves or events are wired to and driven by the controller. Real-time tests can be made in which, for example, hold times can also be verified. This allows early discovery and correction of 'bugs' and is an excellent device for familiarising the commissioning team with the workings of the controller.

A system installed after such thorough checking would be expected to go into successful operation with the minimum of field adjustments.

In most cases the ion exchange system can be put on order fairly early in the design process, and its delivery and installation are not on the critical path. In many conversion projects it would be useful to put the system into operation even before the existing cells are replaced with membrane cells. Diaphragm cells, for example, could benefit for a while from the improved quality of the brine. At the same time the period available for shakedown of the system and training of the operators is extended. There is less incentive in the case of a mercury cell conversion and also the danger of contamination of the resin by mercury, unless combined operation is planned and mercury removal facilities are also available.

5 OTHER APPLICATIONS

5.1 Potassium Chloride Electrolysis

We are not aware of any published data concerning the performance of secondary brine treatment resins in potassium chloride brines. It would be expected that performance would be similar to, or even marginally better than, in sodium chloride. This is borne out in our experience.

5.2 Diaphragm Cells

As discussed previously, diaphragm cells are generally operated with substantially higher concentrations of hardness in the feed brine than is acceptable for modern membrane cells. However, as a major cause of diaphragm efficiency decay is known to be diaphragm plugging caused by deposition of alkaline earth compounds it would be expected that improved performance would be realised by the application of ion exchange technology to the feed brine in this case also. We are not aware of any diaphragm producer who is using the technique on a routine basis but experimental work applied to one or two cells has been reported by Langeland [5]. Although he concentrates on the ion exchange process rather than on the effects realised in the diaphragm cells, he does report a very positive effect on the current efficiency of the cell.

5.3 Chlorate Cells

Chlorate cells are usually operated on chemically softened brine. Treatment is similar to that given to diaphragm cell brine but is sometimes less thorough or less closely monitored. This is tolerable because chlorate cell operators can revel in the absence of an anolyte/catholyte separator. Problems still arise with hardness elements, however, one being the deposition of calcium sulphate on the anodes. This is most often a problem in plants which have no convenient outlet for solution but are dedicated to solid products.

Softer brine can relieve some of these operating problems. Ion-exchange brine, in particular, is said to increase current efficiency and to prevent the gradual increase in voltage which is caused by the sulphate deposit [6].

One should note that in the typical chlorate process brine is produced separately and then added to the cellroom circulation loop. Ion exchange is practised on a salt solution and not on a liquor rich in chlorate. The resin oxidation problem referred to in Section 3.5 therefore does not arise.

6 REMOVAL OF OTHER CONTAMINANTS

6.1 Strontium

Some brines contain strontium at levels greater than the limits specified by membrane suppliers. Owing to the similar chemical nature of strontium to the lighter alkaline earths it would be expected that the resins would be effective in removal of strontium also. This is, in fact, the case although the capacity for strontium is significantly lower than for calcium. Thus a large vessel volume would be required for a brine with a high strontium concentration. Fortunately membranes can tolerate a higher concentration of strontium than of calcium, so this rarely creates a problem.

In one case the brine, as analysed prior to plant specification, showed a very high strontium concentration. In order to cope with this we specified a three-bed ion-exchange unit, designed with the third bed in a guard role. This provides specification brine, though in the event the high strontium concentration in the feed brine is rarely experienced.

In other cases, systems designed for removal of calcium and magnesium were performing so well that the on-line part of the cycle was continually increased. This has actually resulted in strontium breakthrough and a retreat towards shorter cycles. It is important during design clearly to establish the quality of the feed brine and to determine which factors may limit operating results.

6.2 Barium

Proceeding further down the periodic table we come to barium. The trends noted with strontium continue; the membrane is less sensitive to the impurity, but the capacity of the resin for its removal is also lower. We are aware of one plant which has been operating its secondary brine treatment unit to control barium contamination of its membrane cell feed brine since the end of 1985. The plant uses a three-bed 'merry-go-round' arrangement with a feed rate of 18 bed volumes per hour. The resin used is Duolite® C467. The unit has been running with a barium end point of 400 ppb. Feed concentration of barium is in the range 800–900 ppb [7].

6.3 Aluminium

Increasing awareness of the effects of impurities on membranes has led to tightened specifications on aluminium concentrations, particularly where there is a high level of soluble silica present in the brine [8]. In the normal operating mode of secondary brine treatment systems, that is at a pH of 10 or higher, the capacity of typical resins for aluminium is not great enough to consider ion exchange as a practicable means of reducing aluminium concentrations. However, Rohm and Haas have shown that Duolite C467's capacity is increased by almost an order of magnitude if the pH is reduced to 4. Thus it becomes practicable to use ion exchange for aluminium control, albeit at the cost of an additional unit and possibly further consumption of acid and alkali for pH adjustments around the brine circuit.

6.4 Iron

There is clear evidence that the resins pick up iron from brine. This iron is not readily removed owing to the amphoteric nature of the resin. It probably reaches an equilibrium loading and is not seen as a problem, having no serious effect on the resin's capacity for calcium [7].

7 EFFECT OF MERCURY

A high proportion of completed and projected membrane cell plants are conversions from older technologies. In the case of conversions from mercury cells there is likely to be contamination of the circulating brine with mercury. In a total conversion this contamination will be temporary and can be removed by physical cleaning followed by short-term use of chemical precipitation methods. With partial conversions, however, where a common brine circuit is used for the remaining mercury cells and the new membrane electrolysers, the contamination is constantly replenished. The effect on the membrane cells themselves is well understood and in most cases it is necessary to provide some continuous mercury removal system. Candidates include chemical reaction to precipitate mercury in an insoluble form such as mercurous sulphide and use of an ion-exchange resin specific to mercury. In order to optimise this mercury removal it is desirable to know how the secondary brine treatment resin is affected by mercury.

Unfortunately there is little available information on the effects of mercury on the resins generally used for brine purification. Clearly any effects are likely to be complicated by the amphoteric nature of the resins and by the various ionic forms of mercury which exist in equilibrium in the circulating brine stream. We are aware of some work which has been carried out on Duolite C467 which clearly indicates that mercury is picked up by the resin, more readily than by the iminodiacetate resin, Duolite ES466. There is no significant removal from either resin during the normal regeneration procedures. It is possible, of course, that periodic additional treatment would remove sufficient mercury to enable an equilibrium set of conditions to be attained. However, it is worth noting that at a loading of 10 g/litre no effect on the C467's capacity for calcium is observed and also that the effective pick-up rate is very dependent on pH, being some five times greater at pH 11 than in the range 7–9 [7].

It is possible that the loss of activity associated with mercury contamination reaches a stable level and that provision of a larger bed volume will be sufficient to accommodate the mercury concentrations found in a typical mercury cell brine circuit. If this is so, then this course of action could prove the most economical solution to mercury problems, particularly where activated cathodes which can be contaminated by mercury are not employed. Further work is necessary in this area.

8 ANALYSIS

In order to monitor the unit's performance it is important that suitable analytical procedures are incorporated. Our practice and strong recommendation is to include an on-line analyser for the detection of unacceptably high hardness concentrations. Such an analyser must be robust and reliable. We have experience and equipment designed for this service by different suppliers. Whilst none has been totally free of problems, improvements have been made and we are now confident that a unit fulfilling the requirements is available.

The best location for an on-line analyser is a matter of some debate. We have installed them on the final outlet line and in the interbed position. One unit even had the analyser set up so that samples could be taken from either position, either on a prearranged sequence or at the decision of the operator. In practice it has proved better to leave it sampling from a single point. As a general rule, we recommend that, wherever the analyser is located, both interbed and product brine be regularly monitored. This allows the operator ample time to take appropriate action in the event of a breakthrough being detected.

On-line analysers for the detection of calcium and magnesium in brine work on a colorimetric principle. Coloured complexes of the alkaline earths are generated automatically during the operation of the analytical sequence. The density of the colour produced gives a good indication of the concentration of contaminant present.

It must be stressed that an on-line analyser cannot be relied upon in isolation to determine whether the ion-exchange equipment is performing correctly. It must always be backed up by sampling and off-line analysis. As always, when very low levels of contamination are to be looked for it is important that sampling techniques are properly developed. It is only too easy to take a sample of brine containing, say, 20 ppb of hardness in a sample bottle which has been washed with tap water, itself containing many times the acceptable degree of contamination. Analysis of a few such contaminated samples is enough to destroy confidence in the analytical technique or in the ion-exchange unit itself.

Sample lines must be of carefully chosen materials which not only are resistant to corrosion but also contain no compounding ingredients which can be leached into the sample. Reagents for the most part should be of the highest quality (usually 'Analar' grade).

It is at least as important not to contaminate the sample in its container or conversely to allow the impurity of interest to drop out of solution or to be adsorbed on the walls of the container. Plastic bottles should be used (e.g. polypropylene). When there is no danger of interfering with the analysis, the bottle should first be soaked in dilute HCl solution and the sample itself should be adjusted to low pH.

There are two schools of thought regarding off-line analysis for plant control. One holds that speed of response is paramount and that the analysis should be carried out on the plant, possibly by the process operators. The other contends that accuracy is the main criterion. This suggests the use of central laboratory facilities with sophisticated analytical instruments. Inevitably, however, this latter results in a slower response. Our view is that there is a place for both. Trained process operators are quite capable of carrying out simple analytical procedures which will give results of sufficient accuracy for immediate response to operating variables. A colorimetric technique, similar in principle to that employed by the on-line analyser, allows a rapid decision as to whether the unit is performing its functions adequately. More accurate assessment of the unit's performance, and also of that of the plant laboratory, can be obtained by periodic analysis off the plant in a fully equipped laboratory using atomic absorption spectrophotometry.

The total hardness should be checked by laboratory analysis on each shift. Barium and strontium should be checked approximately weekly, or more frequently if they are determinants of the length of the operating cycle. Those who supply membranes or electrolysers under warranty may also define minimum acceptable frequencies of analysis.

9 RECONTAMINATION

Having achieved a brine of sufficient purity for feeding to the membrane cells it is clearly important that no contamination be introduced subsequent to the ion-exchange unit. Because of the very low level of impurities required it only needs a small quantity of contaminant hardness to undo the good work of the secondary brine treatment unit.

Possible sources of recontamination include airborne material, any hardness present in such added streams as the acid used for brine pH adjustment and the equipment and piping between the ion-exchange unit and the electrolysers.

Particularly in the vicinity of power generation plants there can be appreciable quantities of airborne fly ash which can have a significant concentration of calcium. Naturally it is imperative that this ash be kept out of the purified brine. In many locations it may be sufficient merely to ensure that all vessels containing purified brine are covered. In others it may be necessary to consider blanketing the vessels with filtered air or nitrogen.

In many instances purified brine is acidified before it enters the membrane cells in order to improve the apparent anode current efficiency. Hydrochloric acid is used and must obviously have a low concentration of hardness impurities. It should be noted, however, that because the acid generally forms a small proportion of the combined acidified brine stream the specification need not be so tight as that for the brine itself. The hardness contamination acceptable in the acid can be calculated for each case based on the maximum acidification rate.

But the most insidious source of recontamination is from the downstream equipment and piping. Corrosion-resistant materials must be used because of the corrosive properties of pure brine. Examples include titanium, glass-reinforced polyester (GRP), polypropylene and rubber-lined carbon steel (RLCS). The last, in particular, can cause problems.

Rubber linings will normally be compounded using filler materials, one of the most commonly used being calcium carbonate. Pure brine is a powerful solvent for materials of this nature so that, particularly in the early years of the plant, sufficient calcium can recontaminate the brine to cause membrane damage. The remedy is to be more precise than is customarily the case when specifying rubber linings for new piping and equipment. In the case of a conversion project, however, there is normally an incentive to maximise the re-use of existing equipment. A specific example might be a rubber-lined brine head tank. Our experience suggests that it is unlikely that the plant records will include a detailed specification of the rubber lining material. Analysis of a sample could provide the necessary information but a lining that has been in service for some years will have lost its homogeneity so that a multiplicity of analyses would be required. Our recommendation is generally to reline the vessel or to provide a replacement one, using hardness-free materials throughout.

Another material sometimes used in brine service, though not by Stearns Catalytic, is PVC. Particularly in the USA this is sometimes compounded with a calcium filler. Again careful specification and quality control is essential.

10 CONCLUSION

The use of ion exchange is the industry standard for purification of membrane cell brine. Its effective and reliable operation is essential to the continuing economic performance of membrane cells. In this chapter we have endeavoured to provide an outline of the principles involved and an insight into the factors affecting design of brine ion-exchange units. The optimum design of such a unit depends on collaboration between an experienced contractor, a competent equipment fabricator and the resin supplier. Stearns Catalytic has demonstrated this with a series of successful applications.

REFERENCES

1. Electrochemical Society Symposium, Atlanta, 1977. Series of papers.
2. Watkins, J. M. & Maloney, D. E., *Proceedings of Oronzio de Nora Symposium on Chlorine Technology*, Venice, 1979.
3. Seko, M. *et al.*, *Modern Chlor-Alkali Technology*, Vol. 3, ed. K. Wall. Ellis Horwood, Chichester, 1986, pp. 178–95.

4. Dibble, G. J., ibid., pp. 350–64.
5. Langeland, O., Symposium on Ion Exchange and Solvent Extraction, Oslo, 1982.
6. Moser, C., this volume, pp. 325–32.
7. Carlyle, M. & White, I. F., private communication, 12 May 1988.
8. Peet, D. L. & Austin, J. H., *Proceedings of 29th Chlorine Plant Operations Seminar*, Tampa, 1986.

23

PRACTICAL EXPERIENCE IN MERCURY AND DIAPHRAGM CELL CONVERSIONS TO MEMBRANE TECHNOLOGY

Graham J. Dibble and Ian F. White

Badger Catalytic Ltd,* UK

1 INTRODUCTION

The reasons for consideration of conversion of an existing chlor-alkali plant to membrane technology are presented and an overview of the technical requirements provided. Finally, Stearns Catalytic's practical experience in the design and implementation of plant conversions is described.

2 WHY CONVERT?

This is not a simple question to answer because the justification varies considerably from country to country and even from plant to plant. The main aspects for consideration are:

  (i) Capital cost of conversion.
 (ii) Energy consumption.
(iii) Maintenance costs.
 (iv) Manpower requirement.
  (v) Environmental pollution.
 (vi) Product quality.
(vii) Loss of production during conversion.

Some countries, concerned about pollution risks, have legislated to enforce conversion whilst others are still prepared to consider further mercury cell installations.

Power consumption varies greatly from plant to plant, depending on the sophistication of the system. Similarly, power costs can vary almost by a factor of ten according to location.

Many other examples of wide differences can be cited between similar plants,

---

* Formerly Stearns Catalytic International Ltd.

demonstrating that each conversion project must be examined on an individual basis. All facets of operation of the existing plant and of the various conversion options must be carefully considered.

2.1 Basic Technical Parameters of Conversion

Several recent publications have discussed the technical requirements for converting both mercury and diaphragm cell processes to membrane technology [1–3]. It is therefore not the intention to go over these requirements. However, the following brief overview is included to allow a fuller understanding of Stearns Catalytic's practical experience.

The basic elements of a chlor-alkali process are shown in Fig. 1. The important changes to an existing plant when converting from mercury or diaphragm cells to membrane technology are:

  (i) Additional brine purification, including hardness removal ion exchange techniques.
 (ii) Additional brine dechlorination to protect the ion exchange resin.
(iii) Inclusion of a cell room caustic soda recirculation system.
 (iv) Change of electrolysers and details of the cell room system.

If mercury and diaphragm cell plant conversions to membrane are now considered separately, more specific modifications can be identified. (Note: The following

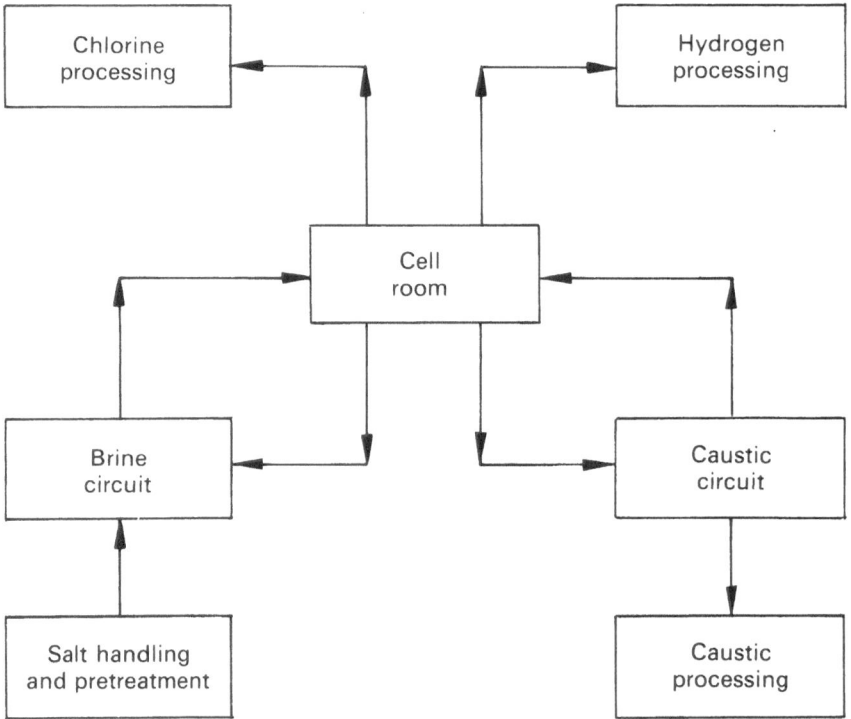

Fig. 1.   Chlor-alkali process units.

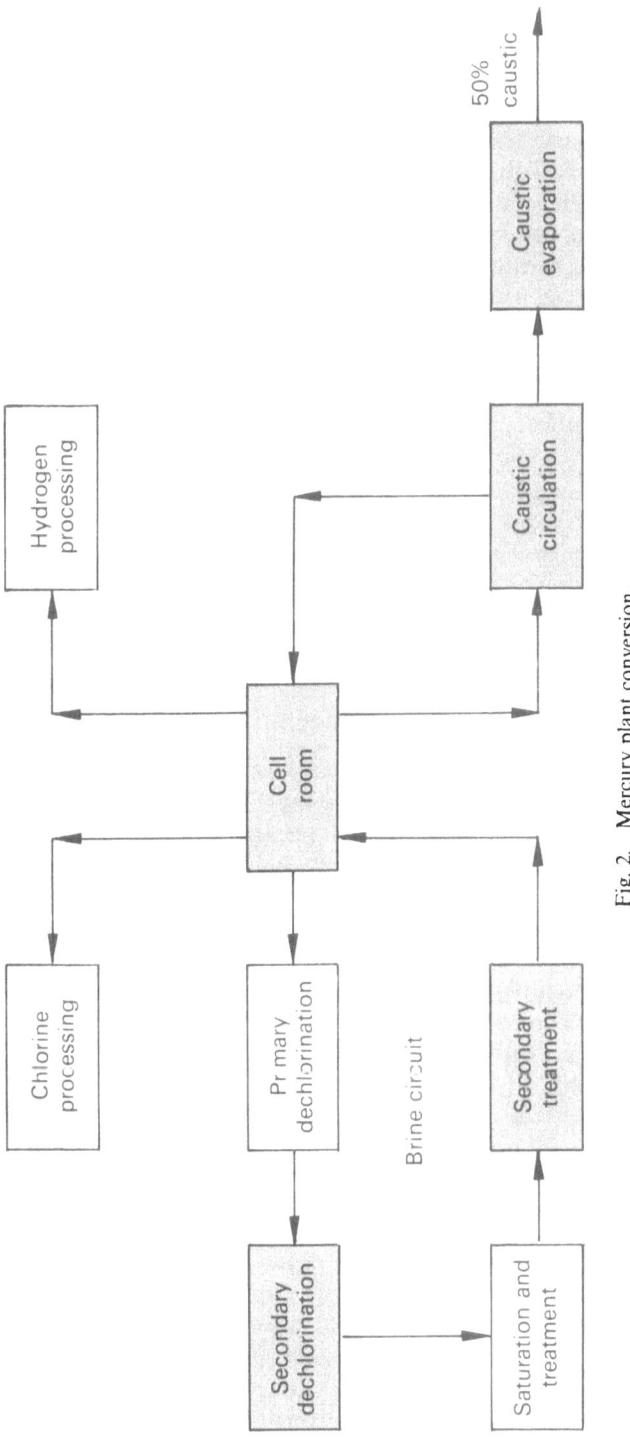

Fig. 2. Mercury plant conversion.

assumes conversion of all existing cells; additional changes would be required for partial conversion.)

2.1.1 Mercury Cells

From Fig. 2 it can be seen that brine processing requires the addition of a secondary treatment, ion-exchange unit. This, in turn, requires a secondary dechlorination step, subsequent to existing physical methods (vacuum or air stripping). Some other plant-specific revisions to the brine treatment system may be required to control the concentrations of other impurities such as sulphate.

The changes in flow rates in the brine system must be reviewed carefully, as for the same production capacity the brine flow for membrane cells will be approximately a quarter of that for mercury cells. This normally can be accommodated by minor changes to pumps and control valves.

The physical difference between the mercury cells and the membrane electrolysers will require that supports and piping be modified, or replaced. As the mercury cell is monopolar, use of a monopolar membrane electrolyser simplifies the changes, particularly in the electrical system.

Mercury cells typically produce 50% caustic soda, but membrane electrolyser product is usually between 32% and 35%. A recirculating caustic stream is therefore required, with associated heat exchange and dilution. If 50% product is required an evaporator must be added.

Gas processing systems usually require only minor modifications such as an accurate differential pressure control system.

2.1.2 Diaphragm Cells

Most of the modifications described above for mercury cell conversion also apply to a diaphragm cell plant conversion (see Fig. 3). However, further additions must usually be made to the brine system. Diaphragm cell plants operate with a once-through brine system, although many recirculate recovered salt from cell liquor evaporation. It is therefore necessary to complete the brine recirculation loop by adding primary dechlorination and possibly resaturation. Modifications to brine treatment may also be required.

The caustic product from a membrane cell room has a very low salt content and its caustic concentration is considerably greater than that of diaphragm cell liquor. In many instances no further treatment is required. However, if a 50% product is needed, it is possible to simplify cell liquor evaporators for membrane caustic evaporation.

The same modifications are required to the gas processing systems as for mercury cell conversions.

3 PRACTICAL ASPECTS OF CONVERSION

Stearns Catalytic has been involved in the design and implementation of conversion of both mercury and diaphragm cell plants to membrane technology. Thus our experience covers all aspects of a chlor-alkali plant. The following applies to both mercury and diaphragm cell conversions.

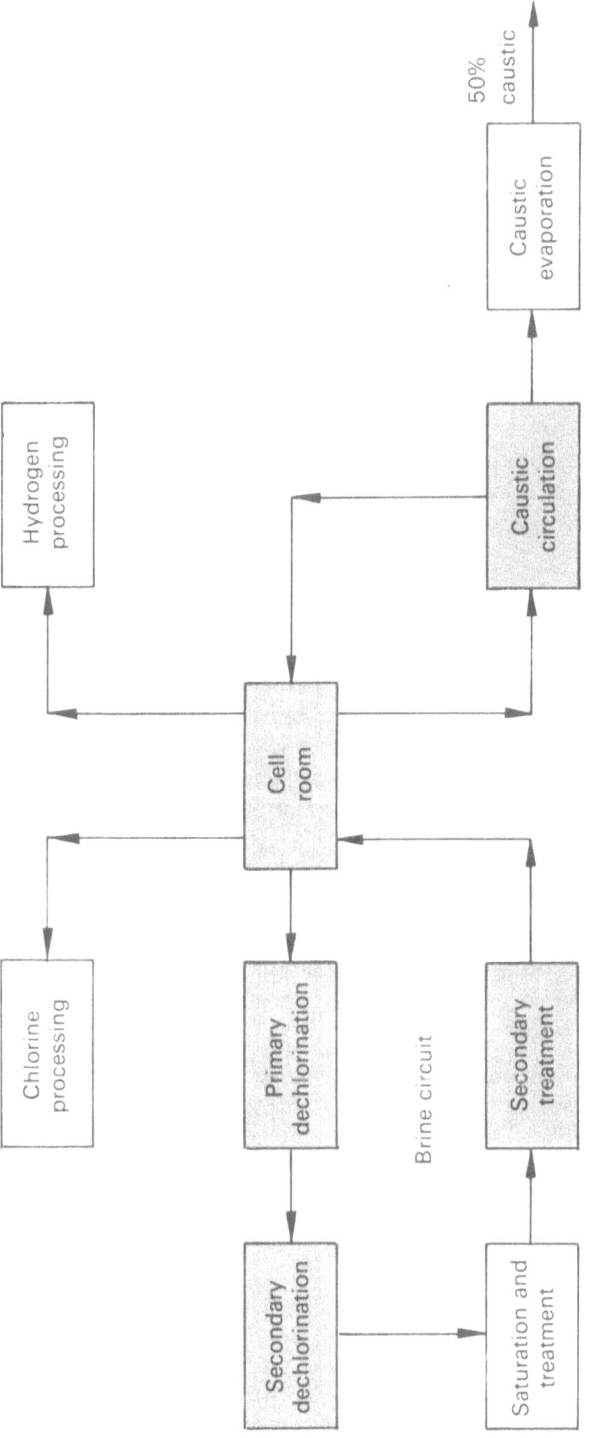

Fig. 3. Diaphragm plant conversion.

3.1 Brine System

3.1.1 Impurities

The main concern is the purity of the brine feed to the existing cell room system. In general membrane electrolysers need purer brine. A comparison of feed requirements is given in Table 1.

Hardness (i.e. calcium, magnesium and strontium) removal is described later. Few of the remaining impurity levels specified for membrane electrolysers are met in practice. The exceptions are sulphate and chlorate, which must normally be reduced in concentration for use with membrane technology. The usual methods of treatment are as follows.

(i) *Sulphate*. The simplest technique is to purge the excess sulphate from the depleted brine system on a continuous or periodic basis. This, however, results in a loss of salt and water, and assumes that effluent disposal is no problem.

Barium precipitation is normally achieved by treatment of the brine in a side stream. However, barium itself can affect the membrane and the anode coating so careful design and operation of such systems is essential. Barium treatment also has the disadvantages of high cost and toxicity problems in handling and final effluent disposal:

$$BaCl_2 + Na_2SO_4 \rightarrow BaSO_4 + 2NaCl$$

It is less common to employ calcium precipitation, but this has advantages: it is non-toxic and does not cause any additional problems with the membrane or anode coating performance. However, it is more expensive in capital cost owing to the higher flow and an additional brine settler (because of the relatively high solubility of calcium sulphate) may be necessary. A considerable increase in sodium carbonate consumption to remove the excess calcium content is another disadvantage:

$$CaCl_2 + Na_2SO_4 \rightarrow CaSO_4 + 2NaCl$$

Practical experience has shown that (except for caustic potash production) purging is often the best and most economic option. The purge has been directed either to associated plants (sodium chlorate production and a pulp mill) or to effluent discharge.

TABLE 1

Cell room brine feed quality (mg/kg except where stated)

Component	Mercury	Diaphragm	Membrane
Ca + Mg	10–20	3–10	<0·05
Sr	<1·0	<1·0	<0·5
Soluble SiO_2	—	—	<5·0
Al	—	—	<0·1
Heavy metals	<0·1	<0·1	<0·05 (each)
Iron	<0·5	<1·0	<0·5
Ba	<0·1	<0·1	<500/SO_4
SO_4	10–20 g/litre	2–5 g/litre	<4·7 g/litre
Solids	<3·0	<3·0	<0·1

(ii) *Chlorate*. As for sulphate control, the simplest method of chlorate control is purging. With most salts a brine purge for the control of sulphate concentration will also control the chlorate concentration to acceptable levels.

Another method used for chlorate control is destruction by acidification and heating, normally carried out on a side-stream of depleted brine. The outlet chlorate concentration is typically reduced to half that of the inlet:

$$NaClO_3 + 6HCl \rightarrow 3Cl_2 + 3H_2O + NaCl$$

For conversions both techniques have been used. Chlorate destruction was required in a plant using vacuum salt, with a low sulphate content and thus a low purge.

(iii) *Organics*. Problems have also been experienced with organic contamination of brine from lining materials of the existing brine circuit. This contamination adversely affected electrolyser performance, although this recovered when the lining material was replaced.

3.1.2 Secondary Brine Treatment—Ion Exchange

Salt and brine contain hardness elements such as calcium, magnesium and strontium. Mercury and diaphragm cell plants tolerate these impurities at parts-per-million levels. These levels are achieved, for solar and rock salts, by chemical precipitation using soda ash and caustic soda. However, for good performance of membrane electrolysers the hardness concentration must be reduced to less than 50 ppb. This requires the use of a chelating ion-exchange resin.

Stearns Catalytic has installed a number of these systems, each designed for the specific plant requirements. It has been found that they have high reliability and few problems have been encountered in maintaining the brine hardness concentration below the 50 ppb level.

The systems have had either two or three ion-exchange columns in series. Stearns Catalytic has predominantly used two column systems and would have no hesitation in recommending this type for most brines.

Brines containing high levels of strontium can lead to design problems for the secondary brine treatment unit. The selectivity of the chelating resin for strontium is considerably less than for calcium. Reducing high concentrations of strontium to levels acceptable to the membrane can lead to very large resin beds. To overcome this problem a three-column system was employed, with the third column operating as a guard. In practice it was found that high strontium levels rarely occurred, but the third column gave added security.

When converting mercury cell brine circuits care must be taken to remove mercury from the brine before it passes through the ion-exchange unit. The chelating resin will adsorb mercury, up to about one-third of its capacity, but it cannot be removed.

Prefabrication of the secondary brine treatment systems, together with pre-testing, can save considerable time for conversions (Fig. 4). Stearns Catalytic's practice has been to prefabricate the complete systems and then to test, as far as is practicable, the operating and regeneration sequences, dry and with water. This has undoubtedly minimised commissioning problems at site.

Fig. 4

Ion-exchange resin consumption was predicted by the resin manufacturers to be about 10% per year due to both attrition and chemical attack. Experience has shown that this consumption is approximately 5% per year.

3.1.3 Brine Hardness Analysis

Ensuring that the correct brine hardness level is being achieved requires monitoring. This monitoring is required in three forms:

(i)   on-line analysis   acceptable/not acceptable
(ii) operating laboratory analysis   acceptable/not acceptable
(iii) central laboratory analysis   accurate hardness concentration

Two types of on-line analysers have been tried, both based on colorimetric techniques. However, only the Hach analyser has given good reliability.

Laboratory analytical techniques have been substantially improved. A colorimetric technique can be used by plant operators, but accurate hardness concentrations can only be achieved by using the more sophisticated laboratory techniques such as atomic absorption spectrophotometry. Experience has shown, however, that the greatest source of inaccuracy has been in sampling where hardness has been introduced from the sample bottle, wash water or simply from the atmosphere. Sampling procedures must be carefully defined to overcome this problem.

3.1.4 Pure Brine Handling

Once the high purity brine has been produced it must be kept pure until it enters the electrolysers. Hardness contamination can be reintroduced if insufficient attention is given to design and maintenance. Particular care must be taken with material selection after the ion-exchange brine treatment. Many rubber linings, plastics and gasket materials contain calcium and magnesium, which can be leached out by the pure brine. This problem often precludes the re-use of existing cell room brine feed piping and head tank when a diaphragm or mercury cell conversion is considered. A number of brine feed problems caused by such contamination have been reported, but Stearns Catalytic's experience has been limited, by careful design, to a single instance of sample contamination. This was due to the fitting of PVC drain lines at site, which contained a calcium filler. It was believed that major problems had occurred with the ion-exchange unit until this simple problem was identified.

3.1.5 Secondary Brine Dechlorination

This subject has been covered in detail in Chapter 21 and therefore only brief consideration will be given here.

Primary dechlorination is normally a physical removal step by either use of vacuum or air stripping. This step would be existing for a mercury cell conversion, but needs to be added for a diaphragm cell conversion. Physical dechlorination methods normally achieve residual chlorine concentrations in brine of about 10–30 ppm. However, a residual chlorine level of less than 0·1 ppm is required for long ion-exchange resin life.

Thus one of a number of secondary dechlorination techniques is necessary:

   (i) Chemical reduction, e.g. using sodium sulphite or hydrogen peroxide.
  (ii) Catalytic.
 (iii) Activated carbon adsorption.

Stearns Catalytic has used each of these techniques, the selection being dependent on the type of salt and overall economics.

One beneficial factor often overlooked is the removal of free chlorine by reaction with naturally occurring organics in the salt.

3.1.6 Brine System Clean-up

A significant problem in the conversion of mercury cell plants may be the residual mercury in the existing equipment and facilities. This can create continuing process and environmental difficulties.

If equipment or piping is to be re-used a clean-out of the system is required, typically:

   (i) Draindown and flushing.
  (ii) Physical cleaning.
 (iii) Chemical precipitation.
 (iv) Ion exchange.

It is essential that this clean-out is done thoroughly if good performance of the membrane system is to be achieved. If mercury remains in the system the consequences will be:

   (i) Loss in capacity of ion-exchange resin.
  (ii) Loss of performance of activated cathodes.
 (iii) Mercury contamination of products.

The brine system, having the largest volume and the largest potential for re-use, is likely to present the most difficult clean-out problem. Carefully planned maintenance can reduce inventory levels and allow physical cleaning. The final mercury, in solution, can then be removed by chemical precipitation. This can be achieved by the addition of sodium sulphide and removal of the resultant precipitate on the existing brine filters.

Experience shows system clean-out takes several days.

3.2 Cell Room

The cell line must undergo substantial change when diaphragm or mercury cells are replaced with membrane electrolysers. The nature of the changes depend largely on the following design selections:

   (i) Monopolar or bipolar membrane electrolyser.
  (ii) New or existing transformer/rectifier(s).
 (iii) New or existing cell room.
 (iv) Electrolyser switching.
  (v) Re-use of existing process piping.
 (vi) Allowable loss of production during conversion.

3.2.1 *Monopolar or Bipolar*

This is a subject that has given rise to much discussion, with advantages and disadvantages claimed by suppliers of each technology. A detailed justification of why Stearns Catalytic believes monopolar technology is the best selection is outside the scope of this chapter.

If the fundamental design concepts of existing mercury and diaphragm cell chlorine plants are considered it is found, of course, that with few exceptions they are monopolar. The cell room system, including the transformer/rectifier, will have been optimised for the monopolar concept. Conversion using bipolar technology must lead to several compromises and thus seems illogical. A detailed examination of all the factors involved, including capital cost, operating cost, ease of operation, safety and production per unit area, demonstrates this conclusion.

The converse is not necessarily true. Indeed, in two instances, we have replaced bipolar cells with monopolar membrane electrolysers.

3.2.2 *New or Existing Transformer/Rectifiers*

Transformer/rectifiers and their associated a.c. and d.c. electrical systems are expensive, but also remain reliable over long periods. They are therefore usually available for re-use after conversion. Most of these systems are designed for monopolar applications, i.e. high current/low voltage, so monopolar membrane electrolysers can easily be made compatible with existing systems. This is particularly true of ICI's FM21 electrolyser which can be adapted to suit the current rating of every commercial plant that has been examined. The number of electrolysers can be selected to suit the existing circuit voltage.

Careful review of voltage range of the transformer/rectifier is required to ensure adequate control is achievable for the membrane technology application. This review must consider each stage of a stepwise conversion, when substantially reduced circuit voltages may occur due to operation of a reduced number of cells. All conversion projects implemented by Stearns Catalytic have re-used existing transformer/rectifiers except for one of the bipolar to monopolar conversions just referred to.

It is Stearns Catalytic's experience that if transformer/rectifiers are re-used, much of the existing a.c. and d.c. systems will also be re-useable, including most of the existing d.c. bus bars. Minor modifications will be required to adjust the bus bar connection to the membrane electrolyser cell line and to accommodate inter-cell connectors and sometimes to change the polarity.

Similarly, it has been possible to re-use almost all the existing cell room electrical systems, without compromising operability or safety. This has resulted in the most favourable economics due to the very high capital cost savings over system replacements.

3.2.3 New or Existing Cell Room Building

Most existing cell room buildings examined can easily be re-used to accommodate monopolar membrane electrolysers. In fact it is normally possible to install up to 400% more capacity with membrane technology in an existing building.

The decision to re-use an existing building is dependent on three main factors:

(a) the condition of the existing building;
(b) whether space for a new cell room is available;
(c) magnitude of savings from reduced loss of production.

Some cell room buildings examined have been in a poor condition and thus may justify total replacement. Environmental issues may also lead to consideration of building replacement. This particularly applies to mercury contamination of existing concrete structures, where some pollution will continue even after mercury cell replacement. Some chlor-alkali plants have space available close to existing cell rooms. This allows the erection of a new cell room building whilst re-using other plant systems *in situ*. The major advantage of a new building is the ability to minimise production losses during conversion, since it can be erected, and the electrolyser system installed (with only a very short shutdown required to make tie-ins) prior to being put into full production.

Where an existing building is re-used the FM21 electrolyser, with its small dimensions and low weight, can take full advantage of existing building features. Existing cell supports can be re-used, with a new lightweight flooring structure, if necessary. The cell room crane can often be re-used to move the electrolysers and/or jumper switch for maintenance. The FM21 can easily be removed from the cell room for maintenance purposes, thus leading to increased cleanliness and safety during maintenance.

Stearns Catalytic has implemented plant conversions using an existing cell room building and also using a new building immediately adjacent to the existing cell installation. In each case overall economics were closely examined. For one case where the building was re-used adequate production could be maintained during conversion from operation of other cell rooms elsewhere at the site. Hence the penalty of production loss was not significant. An example of a plant installing a new cell room building resulted from the potential value of loss of production on a site with a single large cell room.

3.2.4 Electrolyser Switching

Most diaphragm and mercury cell plants have the ability to take an individual cell out of operation by bypassing the electrical current using a jumper switch. This allows individual cell maintenance whilst the other cells remain in full production. A similar system can be used for FM21 monopolar membrane electrolysers.

Membrane electrolysers have now proven their reliability in commercial operation. In the event of minor failures a monopolar electrolyser can be allowed to continue to operate until it is convenient to carry out maintenance. A bipolar electrolyser must be shut down immediately, or major damage will occur, with the associated significant loss of production.

By studying plant operating requirements and the capital/operating costs of jumper switches we have been able to demonstrate that for many plants, particularly those with a capacity of less than 50 t/day Cl_2, a jumper switch is not justified. This assessment has been borne out in practice.

The inclusion of a new jumper switch system for membrane electrolysers is expensive. However, it can be justified on large capacity plants or where it significantly improves overall plant production capacity attainment. To reduce capital cost the jumper switch is usually moveable, as used for diaphragm cells, rather than the fixed switch used on mercury cell plants (Fig. 5).

Fixed mercury cell switches can be re-used in a mercury cell conversion. However, this has not been found to be an economic option, owing to the fact that the switches are located across the full width of the mercury cell and must be repositioned for the new duty or connected with expensive copper bars.

It is unlikely that the dimension of a diaphragm cell switch will allow its use without modification to its connectors and sometimes to its support structure.

Fig. 5

Stearns Catalytic has carried out conversions with and without jumper switches. Where a switch has been included it has been of the moveable type.

3.2.5 Re-use of Existing Process Piping

The physical location of the piping headers for mercury and diaphragm cells is often very different from that required by membrane electrolysers. Different services are at opposite ends of mercury and diaphragm cells. The spacing of mercury and diaphragm cells is also often different from that of membrane electrolysers so that the tie-ins are out of alignment. Thus to re-use existing cell room headers more distribution piping has to be added and work has to be carried out to locate new connection points to the headers. The final membrane electrolyser piping system is therefore not ideal. Many membrane cells require an extra caustic soda recirculation system, whilst demineralised water services are not required at each electrolyser. This results in further changes.

A final very significant consideration is the materials of construction of existing piping, which may be unsuitable for use with membrane electrolysers. The feed brine piping materials must not contain any hardness fillers, as mentioned previously. For electrical safety purposes non-metallic piping is preferred close to the electrolysers. Experience of several years of commercial operation of membrane electrolysers has shown that some materials, particularly GRP, which have given good service life with mercury and diaphragm cells, are not as applicable to this technology. Depleted brine piping has suffered some problems, probably due to a combination of higher temperature, pH cycling (between operation and shutdown) and higher hypochlorite concentrations towards the end of membrane life. An economic alternative material selection is available and has been proven in commercial operation.

In summary, experience has shown that the most economic and best solution normally is to replace the piping system close to the cells.

3.2.6 Allowable Loss of Production during Conversion

The cost of loss of production can play a significant part in conversion economics.

Stearns Catalytic has employed prefabrication techniques, developed from its small modular chlor-alkali plant design [5], to great effect in minimising plant shutdown time. The small dimensions and weight of the monopolar FM21 electrolyser lend themselves well to prefabrication techniques. Small lightweight support structures, with piping and instruments installed, can be prepared away from the existing cell room. They can then be brought into the cell room, erected in a very short time and the electrolysers finally installed.

In addition to prefabrication techniques a stepwise cell room conversion can also be undertaken. This is achieved by installing temporary bus bars during a short shutdown, to remove a number of cells from the circuit. These cells can then be removed and replaced by membrane electrolysers. This technique may be used more than once in a cell room provided the voltage of the remaining cells exceeds the lowest operating voltage of the existing transformer/rectifier.

3.3 Electrolyte Circulation

Re-use of equipment existing in mercury and diaphragm cell plants is very limited. On a mercury cell plant it is probable that the feed brine heater, the depleted brine tank and the depleted brine pumps can be re-used. The product caustic system is usually too small for use as the caustic recirculation equipment needed for the membrane electrolysers.

For a diaphragm cell plant conversion it is unlikely that much can be re-used except the brine heater. Cell liquor tanks and pumps are often found to be in poor condition or of unsuitable materials.

If coated cathodes are to be used in the membrane electrolysers it is essential that existing and new caustic soda circulation equipment is carefully checked to ensure iron cannot contaminate the caustic. In general this means that carbon steel and stainless steel must not be used in contact with the process fluid. Iron contaminates commercially used activated cathode coatings, such that most or all of their benefit is removed.

Good control of the temperature and concentration of streams around the membrane cell room is essential if optimum membrane performance is to be achieved. Poor control can lead to irreversible membrane damage if operating parameters fall outside the operating envelope. The use of an accurate on-line density meter is of significant assistance in monitoring cell room operation. Several types have been tried but the vibrating tube type has proved to be most successful.

In most cases the electrolytes system includes a heat exchanger for heat recovery. This can normally be justified from the value of energy savings. The method of heat recovery is by interchange of either feed brine with depleted brine or feed brine with recirculated caustic soda. In the latter case a careful selection must be made for the material of construction of the heat exchanger.

3.4 Product Gases

The existing product gas processing systems of both mercury and diaphragm cell plants can usually be re-used for the membrane facility with little change. The main concern for the new membrane cell room is the method of pressure control of the gases. Membrane technology requires a steady differential pressure of hydrogen over chlorine. This means that a differential pressure controller must usually be added to the existing control system.

If an existing pressure control system is to be re-used its operation must be carefully reviewed for cycling, which could ultimately lead to membrane failure from fatigue.

4 CONCLUSION

Stearns Catalytic's experience of studying and carrying out conversions of both mercury and diaphragm cell plants to membrane technology is extensive. It shows that in many cases conversion gives a good return on investment. However, a detailed investigation is necessary taking into account all the factors described above. Every conversion is different and a contractor specialising in chlor-alkali technology should always be selected.

ACKNOWLEDGEMENTS

The authors acknowledge with thanks the assistance of Messrs J. E. Harker, I. Barton and D. Hale in the preparation of this chapter.

REFERENCES

1. O'Brien, T. F., Considerations in conversion of existing chlor-alkali plants to membrane cell operation. *Modern Chlor-Alkali Technology*, Vol. 2, ed. C. Jackson. Ellis Horwood, Chichester, 1983.
2. Horvath, R. J., The Eltech membrane gap cell for the production of chlorine and caustic. *Modern Chlor-Alkali Technology*, Vol. 3, ed. K. Wall. Ellis Horwood, Chichester, 1986.
3. James, S. R., Retrofit chlorine technology. Institute of Chemical Engineering, NW Branch, Symposium, 1985.
4. O'Brien, T. F., Control of sulphates in membrane cell brine systems. *Modern Chlor-Alkali Technology*, Vol. 3, ed. K. Wall. Ellis Horwood, Chichester, 1986.
5. Dibble, G. J., Development of a modular chlor-alkali plant design. *Modern Chlor-Alkali Technology*, Vol. 3, ed. K. Wall. Ellis Horwood, Chichester, 1986.

24

MEMBRANE ELECTROLYZERS WITH GAS SUPPORTED INTERNAL CIRCULATION

Karl Lohrberg

Lurgi GmbH, FRG

For the operation of membrane cell plants many negative influences on membrane performance are known. Impurities in brine such as Ca, Mg and Si, and high depletion of salt, have to be avoided. The concentration of NaCl in the depleted brine has to be of the order of 200 g/liter.

Other negative factors on membrane performance are too high a concentration of caustic as well as gas accumulation in the electrolyzers, even if the latter is limited to small spots only.

One topic which has not been addressed widely in discussing membrane performance is the heat generation of the membrane itself. The amount of heat generated per square meter of membrane is as follows:

$$\text{Current density} = 4\,\text{kA/m}^2$$
$$K \text{ value of membrane} = 0.1\,\text{V}\left/\frac{\text{kA}}{\text{m}^2}\right.$$
$$\text{Generated heat } (4 \times 4 \times 0.1) = 1.6\,\text{kW/m}^2$$
$$= 1.6\,\text{kWh/m}^2\,\text{h}$$
$$= 1376\,\text{kcal/m}^2\,\text{h}$$

Under the given assumptions this clearly shows that the membrane can be looked at as a submerged heater.

All these points are addressed, neglecting differences in concentrations in the electrolyzer. However, in membrane cells concentration gradients do occur. The consequences of high concentration gradients are quite well known.

The negative effects of heat generation and concentration gradients could be solved by applying high velocities within the membrane electrolyzers in the vicinity of the electrodes and membranes respectively, e.g. in the catholyte between the membrane and the cathode in cells operating with an electrolyte gap.

Figure 1 shows the heat transfer coefficient in 50% caustic at 80°C in tubes. To achieve a reasonable heat transfer, velocities in the range of 0.1–0.2 m/s should be achieved.

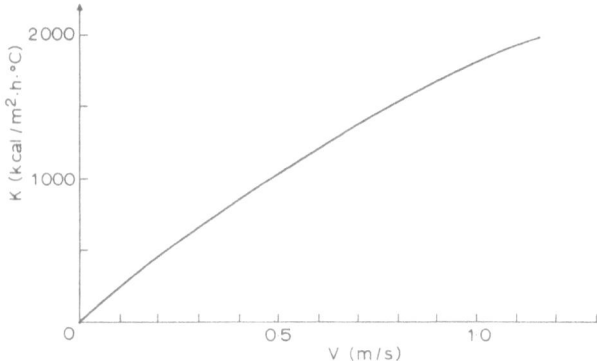

Fig. 1.   Heat transfer coefficient K of 50% NaOH at 80°C in tubes versus velocity V.

Figure 2 shows the mean velocity of the electrolyte plotted over the height of an electrolytic cell. The mean velocity in this figure clearly indicates that this velocity is not sufficient for a good heat transfer, the consequence of which is that the temperature in the membrane will increase so that the heat transfer will take place at a higher Δt. The fact that membrane cells work despite all these possible hazards can only be explained by assuming erratic stirring effects of the generated gas.

As mentioned before, the problems inherent with gas generation and concentration gradients could be solved by inducing high velocities.

In most electrolytic processes, however, the means available to produce a high relative velocity between electrolyte and electrode are more expensive than the value of the improvements which can be achieved. The only process of wide industrial use which for process reasons applies high velocities is chlorate electrolysis.

Figure 3 shows the principle of chlorate electrolysis. The hydrogen generated in the electrolytic cells is used as a gas lift to drive an external circulation.

The valve 'A' is only shown to clarify the expression 'circulation'. If valve 'A' were closed the efficiency of the circulation would be zero, although the stirring effect would be still there.

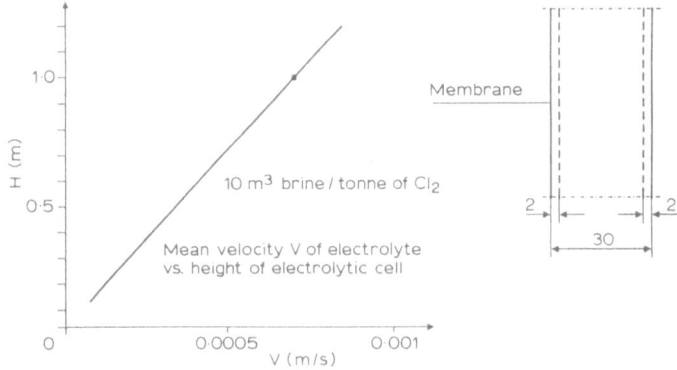

Fig. 2.   Mean velocity V of electrolyte versus height H of electrolytic cell.

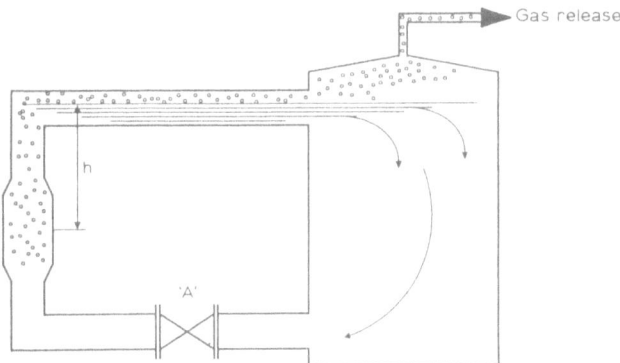

Fig. 3. Gas lift in chlorate electrolysis.

The energy available for such circulation can be calculated as follows:

$$E = V \times h \times y \left(\frac{m\,kg}{s} \right)$$

where V = volume of gas (m³/s)
  h = mean rising height of gas (m)
  y = specific weight of electrolyte (kg/m³)

In chlorate electrolysis the current concentration is 1 kAh/m³ electrolyte; 1 kAh generates approximately 0·4 Nm³ of H_2, which under the prevailing conditions is equivalent to about 0·8 m³ of gas, allowing for temperature and vapor pressure of the water. At a mean rising height of 2 m, 0·8 m³ of electrolyte can be lifted by 2 m or circulated at a pressure drop of 2 m gauge. Measurements have confirmed 1 kAh/m³ is equivalent to 1 m³/h. This means that the volume of circulated electrolyte is larger than the volume of the generated gas.

To apply such a system to a membrane electrolyzer two prerequisites have to be fulfilled:

(a) a 'closed loop' design is needed;
(b) a gas-rich and a gas-free space are needed to maintain a density difference between the gas-rich and the gas-free area.

Due to (b) a partition wall is an immediate requirement, i.e. a nonperforated plate type electrode is compulsory. One such design is shown in Fig. 4.

One strip of 0·1 m height and 1 m depth would yield, at 4 kA/m², 0·4 kA, equivalent to approximately 0·3 m³/h gas. A safe assumption is that the volume of circulated electrolyte is at minimum equivalent to the amount of generated gas. If the gap is 2 mm this would correspond to a velocity of 150 m/h or 0·04 m/s. This is to be compared to 0·001 m/s mean velocity for a 1-m high electrolyzer.

The principle of internal gas circulation shown in Fig. 4 has been used by Lurgi in several full size electrolyzers and in industrial plants. Measurements have shown that the height of the electrolyzer has virtually no influence on the current distribution. The upper half takes 49·7% of the current and the lower half takes

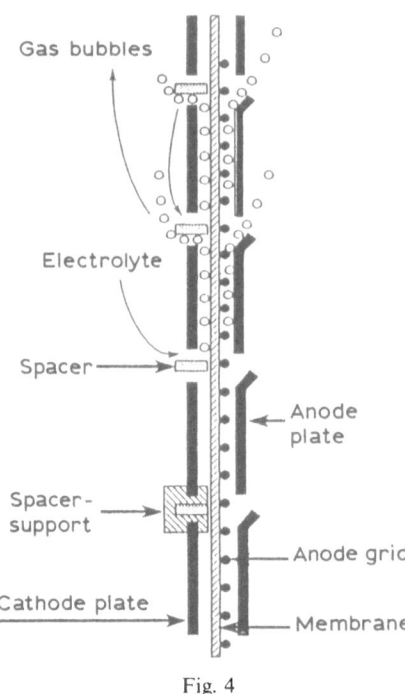

Fig. 4

50·3%. Figure 5 shows the voltage drops achieved with such membrane electrolyzers.

Of special interest in this context is line (b), where the results of the operation with NAFION 324 are shown. This NAFION 324 is a pure sulphonic membrane and consequently has the tendency to attract hydrogen which blinds part of the surface, resulting in a higher voltage drop.

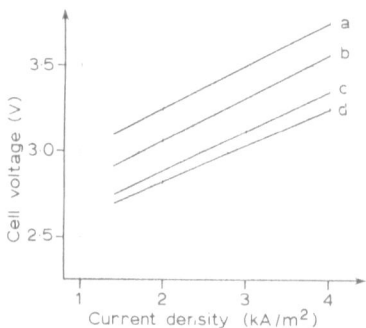

Fig. 5.   a—DM 14, Lurgi steel cathodes, NAFION 901; DSA, 32% NaOH. b—DM 14, Lurgi steel cathodes, NAFION 324; DSA 10% NaOH. c—Lurgi, Lurgi activated nickel cathodes, NAFION 902; DSA, 32% NaOH. d—Lurgi, Lurgi activated nickel cathodes, Flemion 865; DSA, 32% NaOH. Anolyte conditions: 200 g/liter NaCl, 90°C, 4·5 pH.

Impurity	Accumulation factor
Ba	4
Al	3
Si	1·3
Ca	0·1
Mg	0·3

Fig. 6.   Ratio of impurities in brine accumulated in membranes of electrolyzers with plate type cathodes versus membranes of electrolyzers with conventional perforated cathodes.

With standard mesh cathodes at 3·1 kA/m^2 one would have experienced a voltage drop of 3·6 V. In a 10 t/day plant we have measured 3·3 V, which proves the existence of internal circulation. The explanation is that the circulating liquid obviously sweeps the adhering bubbles away.

The consequence of this fact is that uncoated membranes can be used, which are less sensitive to the presence of silica and which are sold at a lower price than coated membranes. Further proof of the existence of the internal circulation is shown in Fig. 6.

Two electrolyzers with the same membrane, same brine and same current density were operated in parallel. One was equipped with electrodes allowing internal circulation, the other one was without such a possibility.

From long-term operation it was possible to develop what is called here an 'accumulation factor'. Obviously insoluble impurities like barium, aluminium, etc., are adsorbed or embedded into the membrane at a higher rate in an electrolyzer with circulation versus an electrolyzer without circulation. With soluble impurities the opposite is true.

The explanation again is that the supply of fresh undepleted electrolyte is unrestricted in the case of electrolyzers with internal circulation. The reduced accumulation of the soluble impurities calcium and magnesium is proof that the concentration gradients are reduced. The concentration of sodium ions in the vicinity of the membrane must be high enough to suppress competing reactions.

Barium is not normally present in membrane electrolyzer brine. Silica and aluminium have been purposely at rather high levels in this specific case, and are normally at much lower levels.

A positive influence of the circulation on other factors could be expected.

Blistering of membranes resulting from intermittent drying-out of membranes caused by overheating or uneven concentrations of caustic are not experienced in electrolyzers with internal circulation.

The current efficiency plotted over the days on line of an industrial plant equipped with plate type cathodes only is shown in Fig. 7. The current efficiency was quite stable, as can be seen on this figure.

Over a period of 500 days the oxygen content was virtually constant and then suddenly an unknown event appears to have caused the oxygen level to rise. This could have been caused by a high concentration of silica compounds in the brine.

Four years of operation with plate type uncoated cathodes gave sufficient data to prove the superiority of this configuration with respect to overall performance. The

Karl Lohrberg

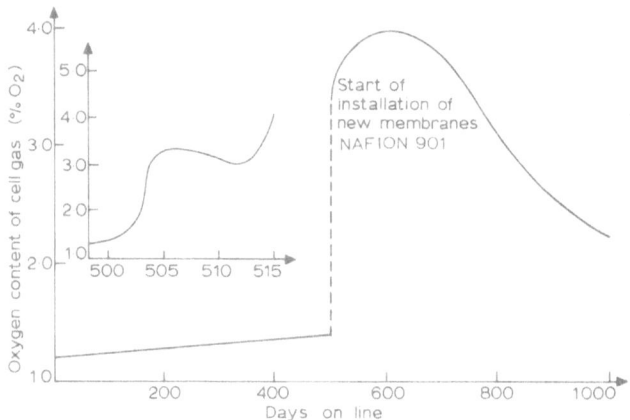

Fig. 7. Oxygen content in cell gas.

same holds true for electrolyzers with plate type coated cathodes and plate type anodes. Insufficient time has elapsed to define the lifetime of the cathode coating.

In modern Lurgi membrane electrolyzers coated cathodes are used. The coating is a Raney nickel coating which originally had been developed by Justi. The Raney nickel coating has to have a high actual surface. It is known that nickel is attacked by high caustic concentrations. Plate type cathodes should therefore also improve the lifetime of the coating by avoiding high local concentrations.

Calculations based on the nickel content in the caustic would lead to a lifetime of more than 20 years. This shows that with coated cathodes corrosion of nickel is not the decisive factor for the deterioration of the activity of the nickel coating. If there is a deterioration it is normally caused by impurities such as mercury in the brine, organic compounds or by recrystallization of the nickel coating.

What can be stated in any case is that the cathode coating will last for approximately 6 years or an equivalent of 250 t of caustic/m^2. This figure is based on long-term laboratory testing and on short-term industrial experience.

To summarize, it can be stated that to date experience with plate type electrodes has been very positive. More data will be available with growing penetration of the market.

25

OXIDATION OF CHLORIDE AND HYPOCHLORITE AT Pt AND RuO₂ ANODES

L. Czarnetzki and L. J. J. Janssen

Eindhoven University of Technology, The Netherlands

NOTATION

A_e	Surface area of electrode (m²)
c_i	Concentration of the species i (mol/m³)
F	Faraday constant (96487 C/mol)
h_4	Slope of the n_4/c_3 curve (m³/s)
i	Current density (kA/m²)
k_i	Mass transfer coefficient for the species i (m/s)
N	Number of electrons involved in a reaction
n_i	Rate of formation of the species i (mol/s)
t_e	Time of electrolysis (s)
T	Temperature (K)
v_s	Solution flow velocity (m/s)
η_i	Current efficiency of the species i

Subscripts

a	Average
e	Electrode, electrolysis
i	Species i
Ox	Oxygen
w	Water
0	Initial condition of electrolysis, $t_e = 0$

Product sequence: $Cl^- \rightarrow Cl_2 \rightarrow ClO^- \rightarrow ClO_3^-$
Subscript sequence: $1 \rightarrow 2 \rightarrow 3 \rightarrow 4$

1 INTRODUCTION

The on-site generation of hypochlorite becomes more and more attractive. Small-scale applications are of interest for the disinfection of private swimming pools.

Larger hypochlorite electrolysers are used to prevent fouling in cooling units of electricity power stations or to purify drinking water or sewage in municipal plants.

A comprehensive survey of both hypochlorite and chlorate formation was published some years ago [1]. The effect of some electrolysis parameters on both the chloride and the hypochlorite oxidation at anodes of different materials is not well understood. Current efficiency measurements have been carried out with undivided cells. Moreover, continuously changing conditions have been used in some experiments. The complexity of the hypochlorite electrolyser makes the interpretation of experimental data very difficult. Moreover, the dependence of the hypochlorite formation on various electrolysis parameters has not been examined extensively.

To avoid many of the named problems, a membrane cell is used for the hypochlorite experiments. The dependence of the formation and destruction of hypochlorite on various parameters is examined in the present research. Some results have already been published [2]. A more complete picture of the effect of many parameters is presented in this chapter.

2 EXPERIMENTAL

2.1 Cell Equipment

The cell is schematically given in Ref. 3. An acrylate cell, divided by a cation exchange membrane (NAFION 117) into an anodic and a cathodic compartment, was used for all current efficiency and potential measurements. The flat working electrode was placed against the back wall of the anodic compartment. A smooth platinum plate and a flat titanium plate coated with a ruthenium oxide layer were used as working electrodes. The preparation of the ruthenium oxide layer was carried out according to Ref. 4. The working electrode was 76·5 mm in length and 20·5 mm in width. Its geometric surface area was 1570 mm^2. The counter electrode was a perforated nickel plate with the same geometric dimensions as the working electrode. It was placed in the counter electrode compartment opposite to the working electrode and pressed against the membrane. A Luggin capillary was placed in the working electrode compartment.

2.2 Measurement of the Current Efficiency

All current efficiency experiments were carried out at a constant current supplied by a Delta Electronika power supply, type D050-10. A series of experiments started with 2000 cm^3 of anolyte containing 0·25–1·5M sodium chloride. The catholyte was 750 cm^3 of 1M sodium hydroxide. The pH, the temperature and the flow rates of both solutions were adjusted to fixed values before starting the electrolysis and were kept constant during the electrolysis.

Anolyte samples of 5 cm^3 were taken after a period of electrolysis time to determine the quantity of hypochlorite and chlorate. The length of this period depended on the electrolysis conditions and varied from 7 to 10 min. The samples were analysed potentiometrically using the AsO_3^{3-}/AsO_4^{3-} redox couple [5]. The average rate of hypochlorite formation, n_3, and of chlorate formation, n_4, were obtained in each period of electrolysis as described in Ref. 2.

3 RESULTS

3.1 Effect of Time of Electrolysis

Hypochlorite, chlorate, oxygen and hydrogen ions are formed in the anodic compartment by electrochemical and/or chemical reactions during the electrolysis of sodium chloride solutions of a pH varying between 7 and 12.

The gas evolved in the anodic compartment contained practically no chlorine. Consequently, loss of chlorine gas did not take place. For long-duration electrolysers the dependences of the hypochlorite and the chlorate concentration on the time of electrolysis are given by Fig. 1 in Ref. 2. The hypochlorite concentration reaches a maximum value and thereafter it decreases slowly with time of electrolysis, whereas the chlorate concentration increases with increasing rate with the time of electrolysis.

To minimise the effect of the changing composition of the anolyte upon the rate of hypochlorite and chlorate formation, respectively n_3 and n_4, the total time of electrolysis was usually short, viz. less than 1 h. Generally five samples of anolyte were taken during the electrolysis. A characteristic result is given in Fig. 1. This figure shows that the increase in hypochlorite concentration, c_3, with increasing time of electrolysis is practically linear with the time of electrolysis, t_e. The chlorate concentration, c_4, increases in increasing rate with time of electrolysis. From the hypochlorite and chlorate concentrations after various periods of electrolysis, and taking into account the effect of the anolyte volume at the start and at the end of a period of electrolysis, the addition of NaOH solution to the anolyte and the transportation of solution from the anodic to the cathodic compartments through the membrane, the average rate of hypochlorite and chlorate formation in a period were obtained, $n_{3,a}$ and $n_{4,a}$ respectively. For the electrolysis, for which results are given in Fig. 1, $n_{3,a}$ and $n_{4,a}$ are plotted as a function of $c_{3,a}$ in Fig. 2. It has been

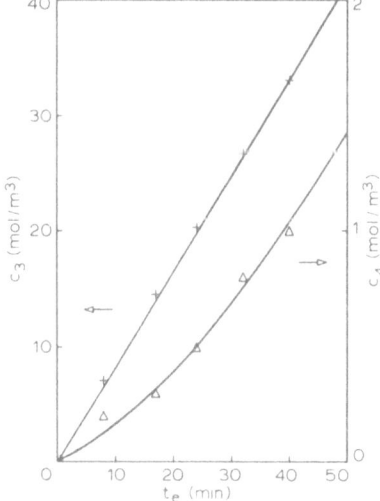

Fig. 1. Plot of the hypochlorite and chlorate concentration versus the time of electrolysis for a platinum anode at $i = 3.82\,\text{kA/m}^2$, $v_s = 0.075\,\text{m/s}$, $c_{1,0} = 0.75\text{M}$ and pH $= 10$. +, ClO⁻; △, ClO₃⁻.

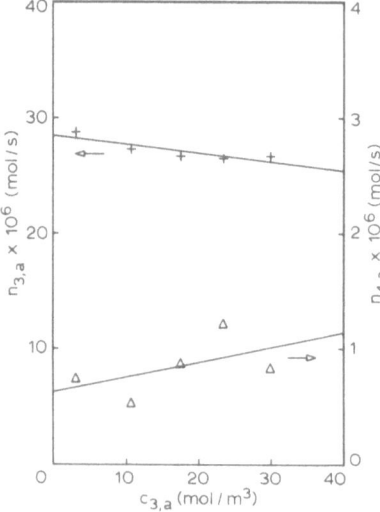

Fig. 2.   Plot of $n_{3,a}$ and $n_{4,a}$ versus $c_{3,a}$ for a platinum anode at $i = 3·82$ kA/m^2, $v_s = 0·075$ m/s, pH = 10 and $c_{1,0} = 0·75$M. +, $n_{3,a}$; △, $n_{4,a}$.

found that $n_{3,a}$ decreases linearly with increasing average concentration of hypochlorite. Linear extrapolation of the $n_{3,a}/c_{3,a}$ and of the $n_{4,a}/c_{3,a}$ curves to $c_3 = 0$ gives $n_{3,0}$ and $n_{4,0}$ respectively. The efficiency of hypochlorite formation at $t_e = 0$, $\eta_{3,0}$, is given by

$$\eta_{3,0} = \frac{N \times F \times n_{3,0}}{A_e \times i}$$

where i is the electric current density during the electrolysis, A_e is the surface area of the working electrode, N is the number of electrons and F is the Faraday constant.

To illustrate clearly the linear increase in rate of chlorate formation with increasing hypochlorite concentration, results for experiments at various initial concentrations of sodium chloride are given in Fig. 3. From this figure it follows that $n_{4,a}$ increases practically with increasing $c_{3,a}$. It should be noted that the chlorate concentration is extremely small so that the occurrence of relatively large deviations is unavoidable. Figure 3 shows clearly that the initial rate of chlorate formation, $n_{4,0}$, is well determinable and increases with decreasing concentration of NaCl. The efficiency of the chlorate formation at $t_e = 0$ is defined by

$$\eta_{4,0} = \frac{N \times F \times n_{4,0}}{A_e \times i}$$

The slope of the $n_{4,a}/c_{3,a}$ curve, h_4, is determined by the transport of hypochlorite from the bulk to the anode surface [2] and is used to calculate the mass-transfer coefficient for a chlorine–oxygen evolving electrode.

3.2 Effect of Current Density

Figure 4 shows the effect of the current density on the initial current efficiencies

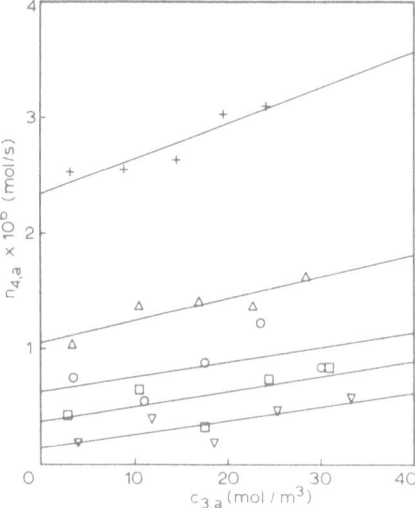

Fig. 3. Plot of $n_{4,a}$ versus $c_{3,a}$ for a platinum anode at $i = 3\cdot82$ kA/m^2, $v_s = 0\cdot075$ m/s, pH $= 10$ and at various concentrations of $c_{1,0}$: +, $0\cdot25$M; $\triangle$, $0\cdot5$M; $\bigcirc$, $0\cdot75$M; $\square$, 1M; $\triangledown$, $1\cdot5$M.

for the hypochlorite cell and the chlorate formation at a platinum anode in $0\cdot5$M NaCl at pH $= 7$, 298 K and a solution flow rate of $0\cdot075$ m/s. From Fig. 4 it follows that a small decrease in $\eta_{3,0}$ and a small increase in $\eta_{4,0}$ occurs with increasing current density. No significant effect of the current density on the slope h_4 of the $n_4/c_{3,a}$ curve has been found. Similar results were obtained for electrolyses at pH $= 10$ and 12.

It has been found that the initial current efficiency for the formation of

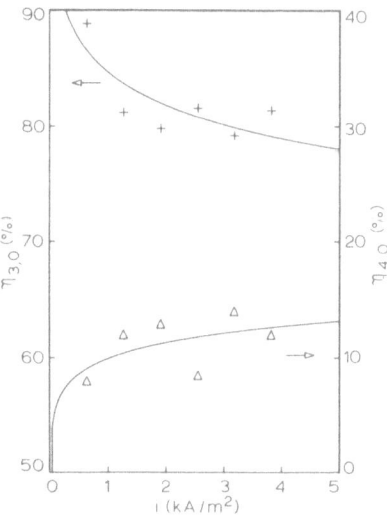

Fig. 4. The current efficiencies, $\eta_{3,0}$ and $\eta_{4,0}$, are plotted versus the current density for a platinum anode at $v_s = 0\cdot075$ m/s, $T = 298$ K, $c_{1,0} = 0\cdot5$M and pH $= 7$. +, $\eta_{3,0}$; $\triangle$, $\eta_{4,0}$.

hypochlorite on a ruthenium oxide electrode was almost 100% for electrolyses with 0.5M NaCl and at pH $= 8$, $T = 298$ K, a solution flow rate of 0.075 m/s and at current densities between 1.27 and 3.82 kA/m^2. The quantities of chlorate formed during these electrolyses were extremely low and undeterminable with reasonable reliability.

3.3 Effect of NaCl Concentration

In Figs 5 and 6 the initial current efficiency of hypochlorite and of chlorate formation is plotted versus the initial concentration of NaCl for a platinum anode at respectively $i = 0.64$ and 3.82 kA/m^2, $T = 298$ K, pH $= 10$ and $v_s = 0.075$. From these figures it follows that $\eta_{3,0}$ increases and $\eta_{4,0}$ decreases in decreasing rate with increasing initial NaCl concentration. For the experiments of Fig. 6, the slope h_4 of the $n_{4,a}/c_{3,a}$ curve is plotted in Fig. 7 versus the initial NaCl concentration and versus the initial current density used for oxygen evolution, $i_{Ox,0}$, where

$$i_{Ox,0} = (1 - \eta_{3,0})i$$

Oxygen is formed according to Foester's reaction, viz.

$$6ClO^- + 3H_2O \rightarrow 2ClO_3^- + 4Cl^- + 6H^+ + 1.5O_2 + 6e^-$$

and/or according to the reaction, viz.

$$2H_2O \rightarrow O_2 + 4H^+ + 4e^-$$

From Fig. 7 it follows that the slope h_4 increases with increasing $i_{Ox,0}$. Electrolyses with a current density of 0.64 kA/m^2 and an initial NaCl concentration varying between 0.1 and 1M showed similar results. For the ruthenium oxide anode at 298 K, however, chlorate is only formed in well determinable quantities in electrolyses with a solution with a small initial NaCl content and at high current densities. For an

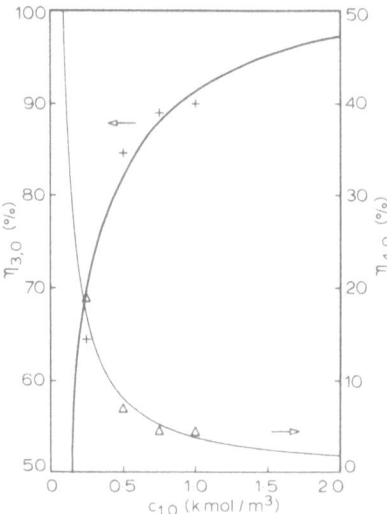

Fig. 5.   The current efficiencies, $\eta_{3,0}$ and $\eta_{4,0}$, are plotted as a function of the initial NaCl concentration, $c_{1,0}$, for a platinum anode at $i = 0.64$ kA/m^2, $v_s = 0.075$ m/s, $T = 298$ K and pH $= 10$. $+$, $\eta_{3,0}$; $\triangle$, $\eta_{4,0}$.

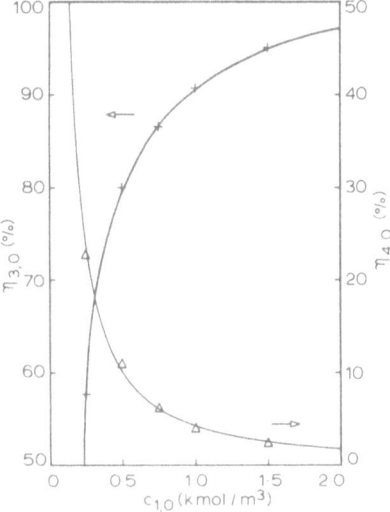

Fig. 6. The current efficiencies, $\eta_{3,0}$ and $\eta_{4,0}$, are plotted as a function of the initial NaCl concentration, $c_{1,0}$, for a platinum anode at $i = 3.82 \text{ kA/m}^2$, $v_s = 0.075 \text{ m/s}$, $T = 298 \text{ K}$ and pH $= 10$. $+$, $\eta_{3,0}$; $\triangle$, $\eta_{4,0}$.

electrolysis with $c_{1,0} = 0.1\text{M}$ NaCl and at $i = 3.2 \text{ kA/m}^2$, $T = 298 \text{ K}$ and $v_s = 0.075 \text{ m/s}$, the initial current efficiencies for hypochlorite and chlorate formation were 0.54 and 0.14 respectively. The slope h_4 was almost zero.

3.4 Effect of pH

The initial current efficiency for hypochlorite and chlorate formation on a platinum electrode in 0.5M NaCl at $T = 298 \text{ K}$, $i = 3.82 \text{ kA/m}^2$ and $v_s = 0.075 \text{ m/s}$ are plotted as a function of pH in Fig. 8. Moreover, it has been found that the slope h_4 of the $n_{4,a}/c_{3,a}$ curve is practically independent of pH.

Electrolyses with a ruthenium oxide anode are carried out at two different values of pH, viz. 8 and 12; the other conditions are equal to those for the platinum anode. It has been found for a ruthenium oxide electrode that the initial current efficiency for hypochlorite formation is 0.97 and the initial current efficiency for chlorate formation is less than 0.005.

3.5 Effect of Flow Rate of Solution

In Fig. 9 the initial current efficiencies for hypochlorite and chlorate formation on a platinum electrode are plotted versus the flow rate of solution. The further electrolytic conditions are $i = 3.82 \text{ kA/m}^2$, $T = 298 \text{ K}$, pH $= 8$ and $c_{1,0} = 0.5\text{M}$. This figure shows practically no effect of the flow rate of the solution on the initial current efficiencies at $v_s > 0.015 \text{ m/s}$, but a strong increase in the initial current efficiency for chlorate formation and a sharp decline in the current efficiency for hypochlorite formation at $v_s < 0.015 \text{ m/s}$.

A similar result was obtained for the ruthenium oxide anode. It has been found that the initial current efficiency for hypochlorite is practically independent of the solution flow rate at v_s between 0.004 and 0.075 m/s; the mean value of $\eta_{3,0}$ is about

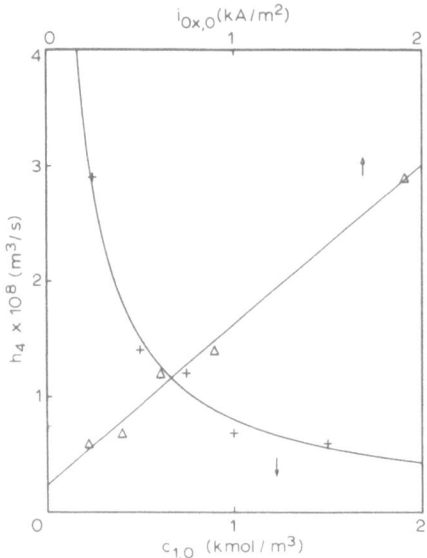

Fig. 7. The slope $h_{4,0}$ of the $n_{4,0}/c_{3,0}$ curve is plotted versus the initial NaCl concentration, $c_{1,0}$, and versus the initial current density used for oxygen evolution, $i_{Ox,0}$, for a platinum anode at $i = 3 \cdot 82 \, \text{kA/m}^2$, $T = 298 \, \text{K}$, $v_s = 0 \cdot 075 \, \text{m/s}$ and pH $= 10$. $+$, $h_{4,0}$ versus $c_{1,0}$; $\triangle$, $h_{4,0}$ versus $i_{Ox,0}$.

0·98. The current efficiency of chlorate formation is very small and no significant effect of solution flow rate has been found.

3.6 Effect of $K_2Cr_2O_7$

To investigate the effect of the presence of $Cr_2O_7^{2-}$ in the sodium chloride solution, one experiment was carried out with a $0 \cdot 5 \text{M}$ NaCl solution containing

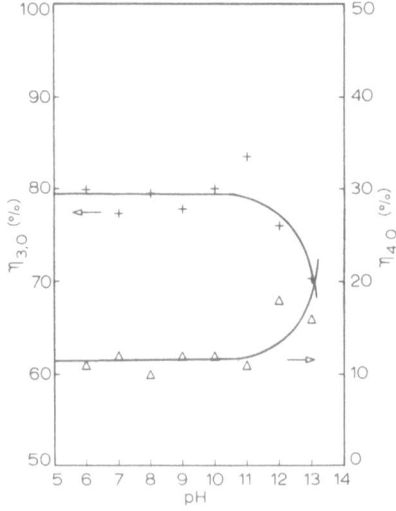

Fig. 8. The current efficiencies, $\eta_{3,0}$ and $\eta_{4,0}$, are plotted versus the pH for a platinum anode at $i = 3 \cdot 82 \, \text{kA/m}^2$, $v_s = 0 \cdot 075 \, \text{m/s}$, $T = 298 \, \text{K}$ and $c_{1,0} = 0 \cdot 5 \text{M}$. $+$, $\eta_{3,0}$; $\triangle$, $\eta_{4,0}$.

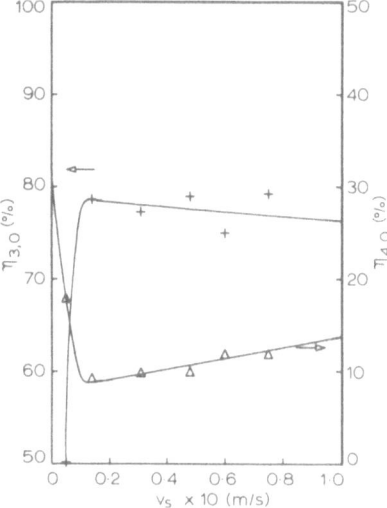

Fig. 9.   The current efficiencies, $\eta_{3,0}$ and $\eta_{4,0}$, are plotted versus the solution velocity, v_s, for a platinum anode at $i = 3.82\,\text{kA/m}^2$, $T = 298\,\text{K}$, $c_{1,0} = 0.5\text{M}$ and pH = 10. +, $\eta_{3,0}$; $\triangle$, $\eta_{4,0}$.

0.0076M $K_2Cr_2O_7$. For electrolysis at $i = 3.82\,\text{kA/m}^2$, $T = 298\,\text{K}$, $v_s = 0.075\,\text{m/s}$ and pH = 8, it has been found that the initial current efficiency of hypochlorite formation was 0.80. Comparison of this result with the one without $K_2Cr_2O_7$ addition shows that the presence of $Cr_2O_7^{2-}$ ions does not affect $\eta_{3,0}$ significantly. The current efficiency for chlorate formation was not determined, because of the high concentration of dichromate in the solution at the start of electrolysis.

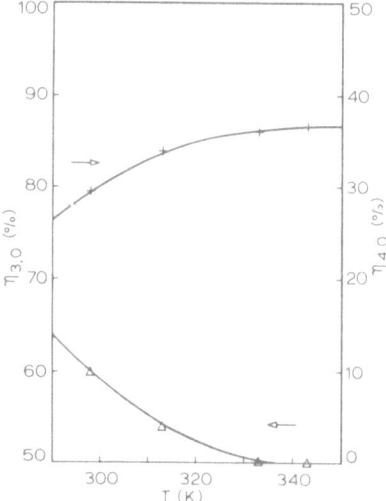

Fig. 10.   The current efficiencies, $\eta_{3,0}$ and $\eta_{4,0}$, are plotted versus the temperature, T, for a platinum electrode at $i = 3.82\,\text{kA/m}^2$, $v_s = 0.075\,\text{m/s}$, $c_{1,0} = 0.5\text{M}$ and pH = 8. +, $\eta_{3,0}$; $\triangle$, $\eta_{4,0}$.

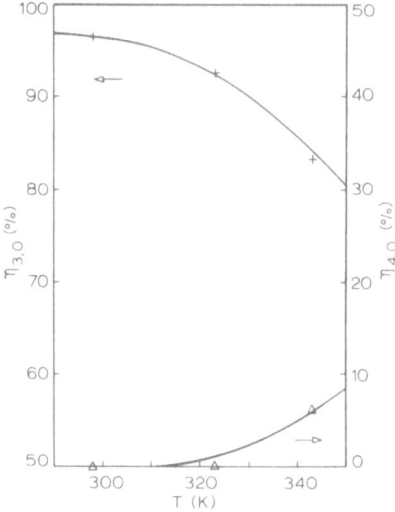

Fig. 11.   The current efficiencies, $\eta_{3,0}$ and $\eta_{4,0}$, are plotted versus the temperature, T, for a ruthenium
oxide electrode at $i = 3 \cdot 82 \, \text{kA/m}^2$, $v_s = 0 \cdot 075 \, \text{m/s}$, $c_{1,0} = 0 \cdot 5 \text{M}$ and pH $= 8$. $+$, $\eta_{3,0}$; $\triangle$, $\eta_{4,0}$.

3.7 Effect of Temperature

The initial current efficiencies for hypochlorite and chlorate formation are plotted
versus the temperature in Fig. 10 for experiments with an initial NaCl concentration
of $0 \cdot 5 \text{M}$ and pH $= 8$, $v_s = 0 \cdot 075 \, \text{m/s}$ and at various temperatures. From this figure it
follows that $\eta_{4,0}$ increases and $\eta_{3,0}$ decreases with increasing temperature. The
dependence of chlorate concentration on the average hypochlorite concentration
was strongly affected by the temperature. The rate of chlorate formation decreases
with increasing temperature during the first period of electrolysis where the average
hypochlorite concentration was lower than 20 mM and increases with increasing
temperature during the end period of electrolysis.

It is well known that the rate of the chemical conversion of hypochlorite into
chlorate increases strongly with increasing temperature. This conversion is
neglectable at temperatures lower than about 323 K and for hypochlorite
concentrations lower than about 40 mM [2].

The slope of the $\eta_{4,a}/c_{3,a}$ curve at temperatures higher than 313 K was about a
factor of 5 higher than at 298 K.

The effect of temperature on the initial current efficiencies for hypochlorite and
chlorate formation is given in Fig. 11 for a ruthenium oxide electrode. From Figs 10
and 11 it follows that for the platinum and the ruthenium oxide anode the
temperature has the opposite effect on the initial current efficiencies for hypochlorite
and chlorate formation.

4 DISCUSSION

Figures 4–10 show that as the initial rate of formation of hypochlorite decreases,
the initial rate of formation of chlorate simultaneously increases, and vice versa.

Moreover, it follows from Fig. 3 that the rate of formation of chlorate increases with increasing hypochlorite concentration in the bulk solution. But, on the other hand, the formation of chlorate occurs also when no hypochlorite is present in the solution. Consequently, chlorate is formed in two ways, viz. hypochlorite diffuses from the bulk solution to the anode where it is oxidised and chlorate is formed either by direct oxidation of chloride ions at the electrode or by oxidation of hypochlorite which is formed by hydrolysis of chlorine within the diffusion layer.

In Ref. 2 it is shown that the chlorate formation at a platinum anode in a 0.5M NaCl solution at current densities greater than $2\,\text{kA/m}^2$ and at 323 K is determined by the diffusion of hypochlorite from the bulk of solution to the anode surface. Assuming that for the experiments with varying NaCl concentration at 303 K and at $i = 3.8\,\text{kA/m}^2$ the hypochlorite diffusion determines the increase in the current efficiency for chlorate formation with increasing concentration of hypochlorite in the bulk of solution, the mass-transfer coefficient for hypochlorite is obtained from the slope h_4 (Fig. 7).

Chloride ions are transported to the anode by migration and diffusion. Assuming that the chloride concentration at the electrode surface is practically zero for the experiments with a low concentration of NaCl, viz. 0.5M, the mass-transfer coefficient for chloride ions is calculated from the initial current efficiency for hypochlorite formation (Fig. 4). In Fig. 12 the mass-transfer coefficients for chloride ions, k_1, and for hypochlorite, k_3, in a $0.5\,\text{M}$ NaCl solution are plotted versus the current density. This figure shows that the mass-transfer coefficients for hypochlorite and chloride ions are almost the same. Consequently, it is likely that hypochlorite and chloride transported from the bulk of the solution to the anode surface are oxidised at their limiting current.

The rate of chlorate formation according to the second method depends on the

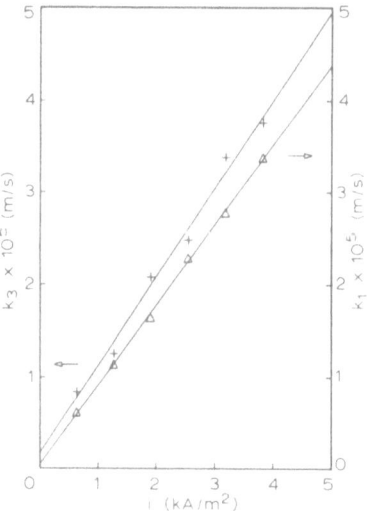

Fig. 12. The mass-transfer coefficient of hypochlorite, k_3, and of chloride ions, k_1, are plotted versus the current density at $v_s = 0.075\,\text{m/s}$, $T = 298\,\text{K}$, $c_{1,0} = 0.5\text{M}$ and pH $= 7$. +, k_3; △, k_1.

chloride concentration, as shown in Fig. 5, and it follows that the initial rate of chlorate formation at $t_e = 0$ decreases exponentially with increasing chloride concentration. Moreover, it has been found that the anode potential also decreases exponentially with increasing chloride concentration. It is unlikely that the rate of chlorine hydrolysis within the diffusion layer also decreases exponentially with increasing NaCl concentration in the bulk. Furthermore, it can be shown that the chloride concentration at the electrode surface is practically zero for the experiments of varying NaCl concentration and a very high current density, viz. $i = 3.8 \, \text{kA/m}^2$. Consequently, it is probable that the chlorate formation by the second method may be explained by direct oxidation of chloride ions at the anode surface.

As shown in Figs 10 and 11, the temperature has an opposite effect on the initial formation rates of hypochlorite and chlorate at platinum and ruthenium oxide anodes respectively. The current used for the oxygen evolution by the oxidation of water is given by

$$i_{Ox,w,0} = i - i_{3,0} - i_{4,0}$$

At the platinum anode, $i_{Ox,w,0}$ decreases with increasing temperature, whereas $i_{Ox,w,0}$ at the ruthenium oxide anode increases with increasing temperature. The dependence of $i_{Ox,w,0}$ and $i_{4,0}$ on temperature are similar. Probably the mechanisms for water oxidation and for chlorate formation by direct oxidation of chloride ions may be related to each other.

REFERENCES

1. Ibl, N. & Vogt, H., in *Comprehensive Treatise of Electrochemistry*, Vol. 2, ed. J. O'M. Bockris *et al.* Plenum Press, New York, 1981, p. 167.
2. Janssen, L. J. J. & Barendrecht, E., in *Modern Chlor-Alkali Technology*, Vol. 3, ed. K. Wall. Ellis Horwood, Chichester, 1986, p. 430.
3. Janssen, L. J. J. & Barendrecht, E., *Electrochim. Acta*, **30** (1985) 683.
4. Trasatti, S. & Lodi, G., in *Electrodes of Conductive Metallic Oxides, Part A*, ed. S. Trasatti. Elsevier, London, 1980, p. 315.
5. Ibl, N. & Landolt, D., *J. Electrochem. Soc.*, **115** (1968) 713.

26

CONTROL OF SULPHATE IN THE PRODUCTION OF CRYSTAL SODIUM CHLORATE

C. Moser

Chemetics International Co. Ltd, Canada

Chemetics International, as a preferred contractor for ICI's FM21 chlor-alkali membrane electrolyzer, have in the last 5 years obtained ten contracts for chlor-alkali plants in seven different countries. The future trend for chlorine plants is in the development of small onsite facilities. Today Chemetics is the leader in the supply of such facilities.

Chemetics is also involved in several other chemical industries, and has pioneered the development of high temperature sodium chlorate electrolyzers. We now offer to merchant crystal chlorate producers a new process to allow total sodium chlorate production in crystal form without a liquor purge or chemical treatment to control sulphate levels.

Commercial salt, when used as a feedstock for the electrolytic production of chlorate, brings impurities such as calcium, magnesium and sulphate ions into the chlorate plant. These impurities must be removed from the system to minimize precipitation on the electrodes used for sodium chlorate production as the formation of deposits increase the chlorate cell voltage and thus increase the cost of production due to the additional electrical energy consumed.

Techniques for the chemical treatment of salt solutions with sodium carbonate and sodium hydroxide to precipitate calcium and magnesium are well known. In addition, in recent years, with the introduction of membrane electrolyzers for the production of chlorine and caustic soda, ion-exchange resins have been developed for the removal of residual calcium and magnesium ions from chemically treated salt solutions. The same ion-exchange technology is now more commonly used in chlorate electrolysis plants.

Sulphate ions in the feed salt solution usually exist at such low levels that it is impractical to use chemical treatment to precipitate sulphate compounds. Thus in a crystal chlorate plant the sulphate accumulates and progressively increases in concentration in the system unless removed. As sulphate has been found to precipitate between the chlorate cell electrodes before saturation concentrations are reached in a standard crystallizer, steps must be taken to control the sulphate concentration. Existing technologies control the sulphate concentration by either

(i)   a chlorate liquor purge or
(ii)   chemical treatment of chlorate liquor using barium or calcium salts.

Neither of these two methods is entirely satisfactory for reasons to be explained shortly. Chemetics has patented a third method using co-crystallization of sodium sulphate and sodium chlorate, and this is what will be presented in this chapter.

Referring to Fig. 1, we will quickly review crystal chlorate production technology.

In the process a saturated purified salt brine is introduced into the cell room, where it is electrolyzed with water to produce sodium chlorate and hydrogen. The product electrolyte contains sodium chloride and sodium dichromate, a compound which improves the efficiency of the electrolysis. It is not possible to electrolyze all of the sodium chloride to sodium chlorate because of the increased wear on the precious metal coatings applied to the titanium anodes.

The hot product liquor is treated to destroy residual sodium hypochlorite and is then transferred to an evaporative crystallizer in which water is evaporated, causing subsequent cooling of the liquor and crystallization of sodium chlorate. By selection of the process operating conditions the sodium chloride may be kept in solution so that after subsequent separation of the essentially pure crystal sodium chlorate from the mother liquor the sodium chloride may be recycled back to the electrolytic cells in the mother liquor.

Thus sulphate entering the chlorate plant with the commercial salt and sulphate produced during hypo destruction, if sulphites are used, has no route to exit as it is recycled from the evaporator back to the electrolytic cells in the mother liquor. However, sulphate levels can be controlled by:

(1)   a chlorate liquor purge;
(2)   chemical treatment using barium or calcium;
(3)   co-crystallization of sodium sulphate and sodium chlorate.

The sulphate concentration in liquor leaving the electrolysis section can be controlled by purging a small portion of this liquor. The rate at which liquor is purged will be that required to maintain the sulphate at a maximum acceptable concentration in the electrolyte. Obviously rates higher than the maximum can be employed.

A liquor purge is the most commonly used method of sulphate control. Reference to Tables 1a and 1b shows the effect of varying sulphate concentrations in the raw

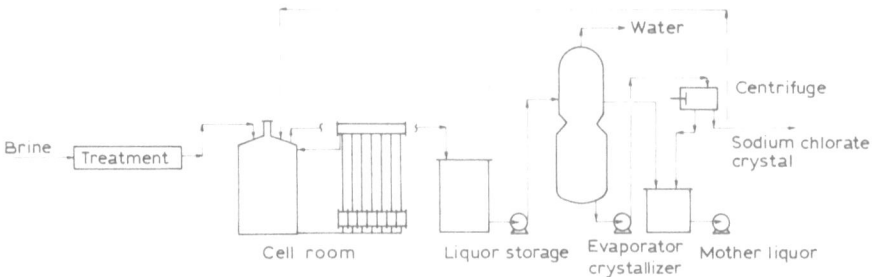

Fig. 1.   Simplified flow diagram—crystal sodium chlorate production.

TABLE 1a
100 t/day crystal $NaClO_3$ plant: sulphate control by purge

SO_4 as Na_2SO_4 in salt (%)	Salt required (t/day)	Crystal $NaClO_3$ (t/day)
0·0	55	100
0·2	55·6	96·5
0·5	56·6	91·2
1·0	58·2	82·0
2·0	61·8	61·7
3·0	65·9	38·7

TABLE 1b
100 t/day crystal $NaClO_3$ plant: sulphate control by purge

Chlorate and sulphate in liquor purge (t/day)	Crystal Na_2SO_4 (t/day)	Salt loss (t/day)	Dichromate loss (t/day)
0 + 0		0·0	0·0
3·5 + 0·11		0·6	0·02
8·8 + 0·28		1·6	0·06
1·8 + 0·58		3·2	0·12
38·3 + 1·24		6·8	0·25
61·3 + 1·98		10·9	0·40

salt on the consumption of salt and the maximum quantity of crystal sodium chlorate that can be produced in a nominal 100 t/day operation. It can be seen that as the sulphate concentration in the salt increases the proportion of production available as crystal drops significantly. For example, with 2% sulphate as sodium sulphate in the raw salt, that is approximately 1·3% as sulphate ion, only 61·7 t/day of the sodium chlorate produced is available for sale as crystal. The balance of 38·3 t/day must be sold in the form of liquor which will contain 0·25 t/day of sodium dichromate and 6·8 t/day of salt.

This method of sulphate control has the major advantage of low capital costs (only a load out facility for liquor is required) and is simple to operate (as sulphate concentrations change relatively slowly due to the large volume of liquor in the system). However, there are several disadvantages with this approach:

(1) The sulphate content of the salt, not market demand, determines the crystal production of the plant.
(2) As the sulphate content of the salt increases so does the salt consumption, as more salt is lost in the liquor purge.
(3) In the same way the dichromate consumption also increases.
(4) A long-term user of chlorate liquor must be found.
(5) The cost of shipping the chlorate liquor is higher than that of shipping chlorate crystal because of the associated weight of salt and water.

Referring back to the three methods of sulphate control, the second method is by chemical treatment using barium or calcium.

It is well known that the addition of a barium compound, such as barium chloride or barium carbonate, to a solution containing sulphate ions will precipitate barium sulphate. However, the barium compounds will also precipitate barium chromate by reaction with sodium dichromate which is present as sodium chromate under alkaline conditions.

An alternative to treatment with barium compounds is the addition of calcium compounds, typically calcium chloride. The reaction precipitates calcium sulphate. The reaction requires the addition of greater than stoichiometric quantities of calcium to the liquor to control the sulphate solubility. This additional calcium, which if not removed would deposit on the cathodes and increase the cell voltage, may be reacted with sodium carbonate to produce insoluble calcium carbonate. However, not all the calcium can be removed by this method and the use of ion-exchange resins in chlorate solutions to remove the remaining calcium has not yet been commercially established.

Commercially chemical treatment is not generally practised for several reasons:

(1)   barium compounds are generally considered toxic and are also expensive;
(2)   sodium dichromate is lost from the process;
(3)   the sludge, consisting of barium sulphate and barium chromate, must be disposed of;
(4)   excess chemical addition may result in precipitation on the anodes.

A third method using co-crystallization of sodium sulphate and sodium chlorate has been invented and patented by Dr Ian Warren and Mr Jack Burkell of Chemetics.

CO-CRYSTALLIZATION OF SODIUM SULPHATE/SODIUM CHLORATE

The solubility phase diagram for a solution containing sodium chlorate, sodium chloride, sodium sulphate and water is complex. By performing multiple experiments to determine the characteristics of the phase diagram it has been discovered that a rapid decrease in sodium sulphate solubility occurs at temperatures below 0°C in a liquor of composition similar to that of crystallizer mother liquor, resulting in the crystallization of a mixture of sodium sulphate decahydrate and sodium chlorate. As can be seen in Table 2, the solubility of sodium sulphate is essentially constant at all normal operating temperatures but drops significantly below 0°C. As noted previously, it is not possible to increase sulphate concentration to the saturation value in normal operation as precipitation between the electrodes occurs before saturation is reached in the crystallizer.

However, this low temperature co-crystallization of sodium sulphate decahydrate and sodium chlorate allows sodium sulphate to be removed from a crystal chlorate plant, as shown in Fig. 2. A small sidestream of mother liquor is cooled in a secondary crystallizer to crystallize sodium sulphate decahydrate and sodium chlorate. The crystal is removed and the mother liquor from the secondary

TABLE 2
Solubility of sodium sulphate in mother liquor of varying compositions

Temperature (°C)	NaClO₃ (%)	NaCl (%)	Na₂SO₄ (%)
60	34·7	11·1	2·1
	36·2	9·1	2·1
	37·0	7·5	2·1
20	31·2	12·0	2·2
	31·9	11·7	2·2
	32·8	10·0	2·2
	33·9	9·5	2·1
	37·3	6·2	2·1
10	32·2	10·0	2·1
	33·8	8·0	2·1
	38·2	5·7	2·1
4			1·25
0			1·05
−3			0·83
−5			0·70
−7	32·3	9·0	0·55
	34·3	6·6	0·45

crystallizer returned to the normal mother liquor system. The sidestream flow rate and exit temperature from the secondary crystallizer are adjusted so that the sulphate removed balances the sulphate entering the process.

Depending on the quantity of sulphate entering the system, the maximum efficiency of sulphate removal may occur if the secondary crystallizer is run in shifts with the sulphate concentration in the mother liquor at the highest tolerable levels in the electrolysis plant.

The crystals produced in the secondary crystallizer are 90% NaClO₃ and 10% Na₂SO₄ (on a dry basis). In most cases the chlorate and sulphate crystals produced in the secondary crystallizer will only be a small fraction of the total crystal product. Therefore the crystal size of the material produced in the secondary crystallizer is only important in that it can reasonably be mechanically handled.

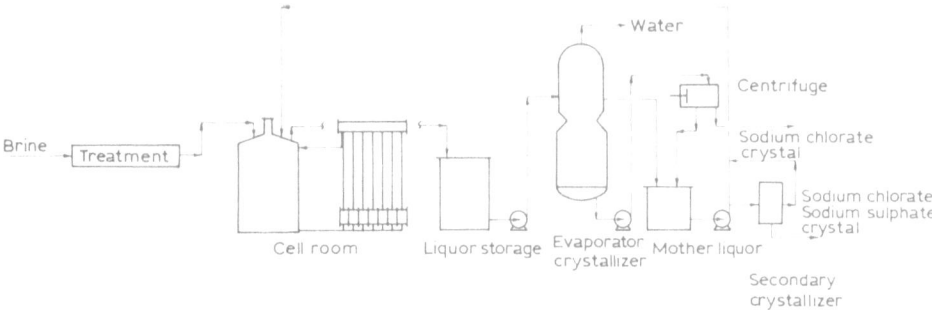

Fig. 2.   Simplified flow diagram—crystal sodium chlorate production with co-crystallization of sodium chlorate and sodium sulphate.

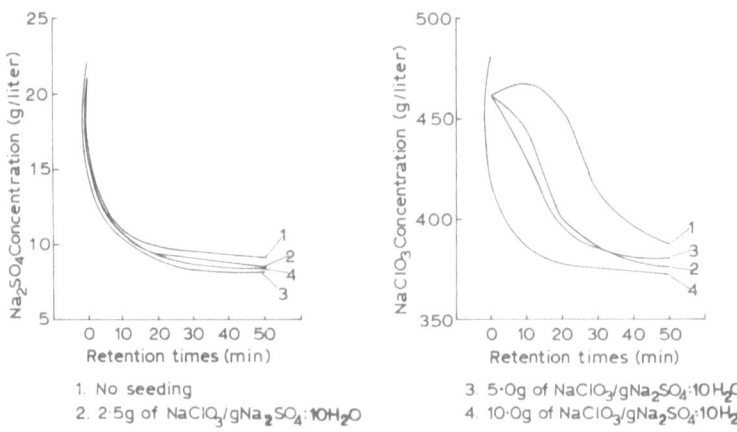

1. No seeding
2. 2·5g of $NaClO_3/gNa_2SO_4:10H_2O$
3. 5·0g of $NaClO_3/gNa_2SO_4:10H_2O$
4. 10·0g of $NaClO_3/gNa_2SO_4:10H_2O$

Fig. 3

From the initial lab work performed to determine the characteristics of low temperature sodium sulphate solubility further work was performed to determine the operating parameters for the secondary crystallizer. Fundamental crystal growth equations state that among other relevant factors crystal growth is positively dependent on:

(1) the degree of supersaturation of the liquor in contact with the crystals;
(2) the length of time that the crystals are in contact with the supersaturated liquor.

To maintain supersaturation of the liquor in a continuous operation the liquor residence time must be less than that required for saturation conditions to be reached. To evaluate this multiple tests were performed to see how rapidly sodium sulphate and sodium chlorate attain saturation when liquor is chilled to $-7°C$ and allowed to supersaturate. Crystallization was initiated by agitation and varying amounts of seed material were added to enhance crystallization. As can be seen from Fig. 3, both Na_2SO_4 and $NaClO_3$ approached equilibrium concentration within 30 min of the initiation of crystallization for all cases except where no seeding occurred.

Analysis of the crystals produced during the batch tests showed an average chlorate crystal diameter of 214–428 μm. While this reflects the adverse effects of low temperature crystal growth and co-crystallization of two species combined with operating conditions in violation of the two factors previously noted (i.e. minimum residence time at maximum supersaturation), this crystal can easily be handled by a pusher-type centrifuge. The sodium sulphate decahydrate formed fine needle-like crystals that by itself would be hard to handle, but in the presence of a greater mass of sodium chlorate crystals will be carried along by the bulk of the crystals.

Once separation of the sodium sulphate/sodium chlorate crystal mixture from the liquor is performed several alternatives are available:

(a) The crystal may be blended with chlorate from the main crystallizer, in which case the sodium sulphate will be present in the crystal product. This is not normally a problem for the customer as most sodium chlorate is converted to sodium sulphate in chlorine dioxide generators and used as a byproduct in the pulp mill.

(b) The crystal may be dissolved and, since the solution is now relatively free of dichromate, treated with barium or calcium compounds to precipitate sulphate. In the case of barium compounds there will be a lower chemical consumption as the precipitation of chromate is avoided.

In relation to the effectiveness of the removal of sodium sulphate via co-crystallization versus removal via a chlorate liquor purge a comparison of the figures listed previously in Table 1 is useful. In Tables 3a and 3b it can be seen that there is no effect in the crystal chlorate production as sulphate levels increase in the salt. Salt and dichromate requirements remain constant as neither is lost in a chlorate liquor purge.

Referring to the same example discussed previously, for 2% sulphate as sodium sulphate in the raw salt, the crystal chlorate production is split, 89% occurring in the primary crystallizer and 11% occurring in the secondary crystallizer.

The total crystal sodium sulphate production is 1·24 t/day and there are no losses of chlorate, salt or dichromate.

TABLE 3a
100 t/day crystal $NaClO_3$ plant: sulphate control by secondary crystallization

SO_4 as Na_2SO_4 in salt (%)	Salt required (t/day)	Crystal $NaClO_3$ (t/day)
0·0	55 (55)	100 (100 + 0)
0·2	55·6 (55)	96·5 (98·9 + 1·1)
0·5	56·6 (55)	91·2 (97·2 + 2)
1·0	58·2 (55)	82·0 (94·5 + 5·5)
2·0	61·8 (55)	61·7 (89 + 11)
3·0	65·9 (55)	38·7 (83·5 + 16·5)

Primary + secondary in parentheses.

TABLE 3b
100 t/day crystal $NaClO_3$ plant: sulphate control by secondary crystallization

Chlorate and sulphate in liquor purge (t/day)	Crystal Na_2SO_4 (t/day)	Salt loss (t/day)	Dichromate loss (t/day)
0 + 0 (0)	(0)	0·0 (0)	0·0 (0)
3·5 + 0·11 (0)	(0·11)	0·6 (0)	0·02 (0)
8·8 + 0·28 (0)	(0·28)	1·6 (0)	0·06 (0)
1·8 + 0·58 (0)	(0·58)	3·2 (0)	0·12 (0)
38·3 + 1·24 (0)	(1·24)	6·8 (0)	0·25 (0)
61·3 + 1·98 (0)	(1·98)	10·9 (0)	0·40 (0)

Primary + secondary in parentheses.

In conclusion, the control of sodium sulphate by low temperature co-crystallization offers several advantages:

(1) The process provides a continuous method of producing sodium chlorate totally as crystal.
(2) The salt consumption is reduced by elimination of losses in the liquor purge.
(3) In the same way the dichromate consumption is reduced.
(4) The costs of chemical treatment and sludge are removed or reduced to those associated with the removal of calcium and magnesium.
(5) Lower cost salts with higher sulphate content may be used.

27

POLYRAMIX®: A DEPOSITABLE REPLACEMENT FOR ASBESTOS DIAPHRAGMS

L. C. Curlin, T. F. Florkiewicz

OxyTech Systems Inc., USA

and

R. C. Matousek

ELTECH Systems Corporation, USA

INTRODUCTION

OxyTech Systems is a joint venture company formed between Occidental Chemical Company (formerly Hooker Chemical and Diamond Shamrock Chemical Companies) and ELTECH Systems to license, market and continue to improve upon the technology and rich heritage of the parent companies in the field of chlor-alkali electrolysis. Hooker Chemical introduced the first deposited asbestos diaphragms in 1929, and Diamond Shamrock introduced the Modified Diaphragm® and Expandable Anode® in the early 1970s. ELTECH Systems is a leader in the field of DSA® anodes. These technologies are used alone or together for over 70% of the world's chlorine production.

In recent years papers presented at symposia have focused on developments in the field of membrane cell technology. It is recognized that most new grass roots chlorine plants will utilize membrane cells. We at OxyTech Systems are very active in this field having licensed 27 membrane cell plants. However, OxyTech Systems also has 151 diaphragm cell licensees producing more than 8m t of chlorine per year. This chapter, for a change, is about a new diaphragm cell technology.

In North America alone diaphragm cells account for 76% of the chlorine capacity. Sixty percent of the diaphragm cell chlorine capacity is in plants having cogeneration of electricity and steam.

The use of asbestos has become a concern for all producers of chlorine with diaphragm cells, as the use of asbestos may be banned. However, even if it is not banned, asbestos may become too difficult to obtain, or too expensive, or it may become economically impossible to dispose of the asbestos waste. The record of the chlor-alkali industry in compliance with asbestos handling regulations is very good. Therefore it is doubtful that the use of asbestos specifically in the chlorine industry will be banned. The other economic situations are at this time more of a factor. Thus any alternative technology to replace asbestos diaphragm cells must stand on its own economically to be attractive to today's chlor-alkali industry.

The present alternatives to the asbestos diaphragm cell are membrane cells or nonasbestos diaphragms. One would first consider conversion to membrane cells. However, PEI Associates, in a report to the US Environmental Protection Agency, has estimated that it would require a capital investment of 2 billion US dollars to convert the US diaphragm cells to membrane cells! Next, one might consider retrofit of membranes into existing diaphragm cells. This has already been accomplished in most of Japan's diaphragm cells. However, the conversion in Japan was accomplished with large government subsidies. All known studies in the remainder of the world have been concluded that the retrofit technology is uneconomical compared to continued operation of the existing diaphragm cells.

For these reasons OxyTech set out to produce an economical replacement for asbestos in diaphragm cells. The result of this effort is called POLYRAMIX.

WHAT IS POLYRAMIX?

The name POLYRAMIX is derived from the words polymer and ceramic. POLYRAMIX, or as it is also called 'PMX'℠, is a unique (patent pending) fibrous material produced from a physical combination of metal oxide particles and a fluorocarbon resin. The preferred metal oxide is zirconium oxide, although other metal oxides such as titanium dioxide can be used, and the preferred polymer is PTFE; however, other fluorocarbons could be substituted. The PMX fibers are

Fig. 1.   Branch chain structure of PMX.

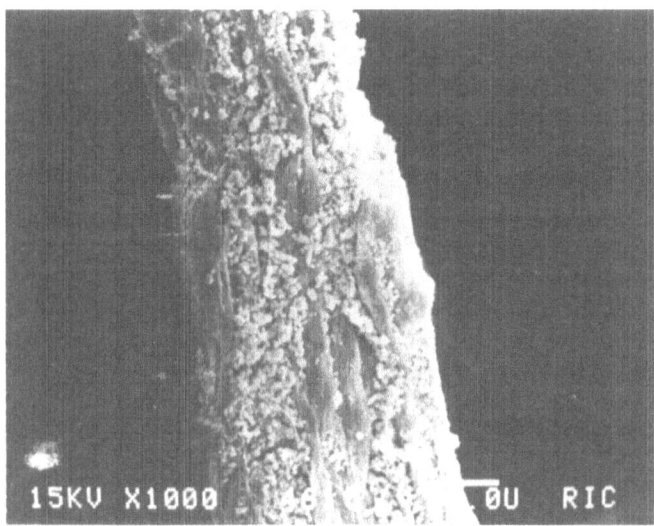

Fig. 2. Micrograph of PMX fiber showing attached metal oxide particles.

branched, irregular and flexible, with lengths of 1–7 mm and diameters of 10–100 μm (Fig. 1). The metal oxide particles are not only dispersed throughout the fiber but also attached to the fiber (Fig. 2). The density is twice that of asbestos.

Other nonasbestos diaphragms have been developed in the past. Fluorocarbons have been the most successful. OxyChem developed the 'Microporous Diaphragm'®, a sheet of PTFE containing sodium carbonate as a pore former. The difficulties of wetting the fluorocarbon were overcome, and the operating results were satisfactory; however, the product was manufactured in sheet form and the labor required to apply the sheets to the cathode tubes, the relatively high cost of the PTFE and the lack of commercial production technology prevented its commercialization. The knowledge gained about wetting of the Microporous Diaphragm by OxyChem and the experience with conductive ceramics developed for aluminium cells of ELTECH were significant factors in the successful development of POLYRAMIX.

APPLICATION OF POLYRAMIX

The fibrous and flexible nature of POLYRAMIX accounts for its advantage over previously developed nonasbestos diaphragms. POLYRAMIX is vacuum deposited on a diaphragm cell cathode from a water-based slurry in the same manner, using the same equipment, as an asbestos or modified asbestos diaphragm. In most cases only minor equipment changes would be required to convert an existing plant to use of POLYRAMIX. After deposition and vacuum drying, where drying vacuum in excess of 500 mm Hg is applied to produce a smooth tight diaphragm, the POLYRAMIX diaphragm is heat cured. The same oven and the same temperatures as used to cure an OxyTech SM-2 Modified Diaphragm are utilized.

Since a hydrophobic fluorocarbon constitutes a portion of POLYRAMIX, the diaphragm is treated with a surfactant prior to assembly into the cell. This is accomplished by simply immersing the cathode in a tank filled with water and the wetting agent. The depositing system vacuum pump is used to thoroughly and evenly wet the diaphragm.

Due to the prewetted condition of the POLYRAMIX diaphragm, which accelerates the even percolation of brine through the entire diaphragm, brine flow rates at startup are somewhat higher than for Modified Diaphragms. Within an hour after startup, however, the surfactant is washed from the diaphragm and brine flow rates are set to achieve the desired caustic strength.

The POLYRAMIX diaphragm does not require rewetting after each circuit shutdown as long as the diaphragm remains covered by brine.

ENVIRONMENTAL CONCERNS

Since POLYRAMIX has been developed due to the problems with asbestos, one of the more commonly asked questions is: 'What about the health and environmental status of POLYRAMIX?'.

The comparison of asbestos and POLYRAMIX fibers is:

	PMX	*Chrysotile asbestos*
Fiber diameter (μm)	10–100	0·1–3
Fiber length (μm)	1000–7000	0·1–60

POLYRAMIX is a much longer, coarser fiber than asbestos. Asbestos is carcinogenic due to its physical properties. The length and diameter fall within the range considered peak for carcinogenic potency. Figure 3 is an illustration of this, and the comparison to POLYRAMIX. Thus POLYRAMIX falls well outside the carcinogenic range. It has not been possible to conduct inhalation studies with POLYRAMIX as its size and density prevent its suspension in air.

The raw materials comprising POLYRAMIX (PTFE, ZrO_2 and NaCl) are also noncarcinogenic.

Disposal of waste asbestos is now required to be conducted in approved toxic landfills; such treatment should not be required of POLYRAMIX.

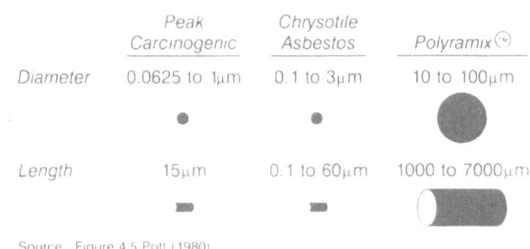

Source Figure 4 5 Pott (1980)

Fig. 3.   Carcinogenic potency of fibers as a function of diameter and length.

STATUS OF PMX WORK

POLYRAMIX is now being deposited and operated on commercial sized electrolytic diaphragm cells in actual commercial cell rooms. To this date POLYRAMIX performance has been demonstrated in four different sized cells, at four different current densities, in four different OxyChem plants.

Cell type	Cathode area (m^2)	Current density (kA/m^2)
MDC-21	23	1·50
MDC-29	32	2·59
MDC-55	60	2·20
H-4	74	2·32

A total of more than 25 POLYRAMIX cells are currently in operation. Development efforts have succeeded in steadily improving the performance levels of the diaphragms. Recent performance of POLYRAMIX diaphragm is presented in Fig. 4.

The diaphragm age of the oldest commercial sized PMX cell is currently 570 days (19 months). Based upon the data from the oldest cell (Fig. 5) and laboratory experience OxyTech is confident that the life of an average POLYRAMIX diaphragm will exceed 3 years.

An area of continued development work is in optimization of the diaphragm weight. Because POLYRAMIX is more dense than asbestos, the resulting POLYRAMIX diaphragms are heavier if the total number of fibers is kept the same. The thickness is required to obtain satisfactory current efficiency and anolyte level. The thickness, however, contributes to the voltage. As usual with diaphragm cells, we have been balancing the diaphragm weight to achieve optimum performance. Figure 6 illustrates, however, that we have been able to reduce the diaphragm loading and thus the voltage while maintaining anolyte level and good current efficiencies. At present the slope of the voltage versus current density graph is somewhat higher than a standard SM-2® polymer modified asbestos diaphragm for the commercial cells, although laboratory results indicate an equal slope. Thus at low (1·5 kA/m²) current densities the commercial POLYRAMIX cells demonstrate similar voltages to the modified asbestos cells, but at the higher current densities the

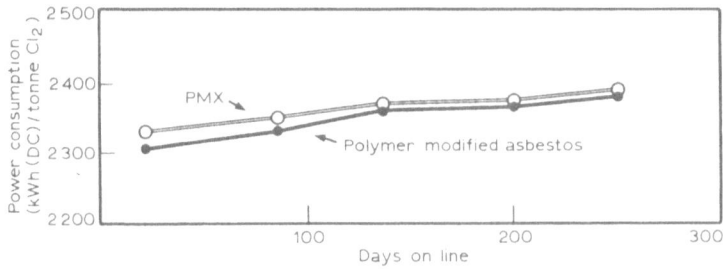

Fig. 4. Power consumption—PMX versus polymer modified asbestos diaphragm. MDC-20 cell, 1·55 kA/m², 5·3 kg/m² diaphragm weight, 130 g/liter NaOH cell liquor.

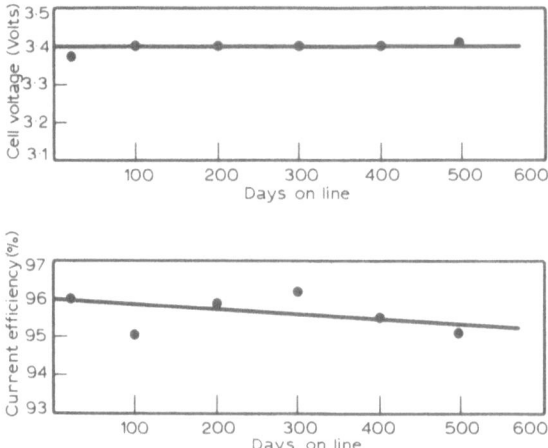

Fig. 5. PMX life versus cell voltage (top); PMX life versus current efficiency (bottom). MDC-20 cell, $1.55 \, kA/m^2$, $9.6 \, kg/m^2$ diaphragm weight.

POLYRAMIX cells have had slightly higher voltages. At all current densities, however, the current efficiencies of POLYRAMIX cells are equivalent to the efficiency of the standard cells.

Laboratory work and now plant experience have demonstrated that POLY-RAMIX diaphragms are significantly more resilient to current load fluctuations, power outages and onpeak–offpeak operation than modified asbestos diaphragms. This can be explained by the resistance of POLYRAMIX to chemical attack. The magnesium hydroxide in the asbestos is chemically attacked by acids and alkalis. During power outages or even current density changes the pH gradient across the diaphragm changes. In time the asbestos disintegrates and the current efficiency is reduced. POLYRAMIX is made of PTFE and zirconium oxide. These compounds are not affected by pH changes. This also accounts for the long life projections for POLYRAMIX diaphragms. POLYRAMIX should be economical today for a plant operating with onpeak–offpeak power or frequent outages.

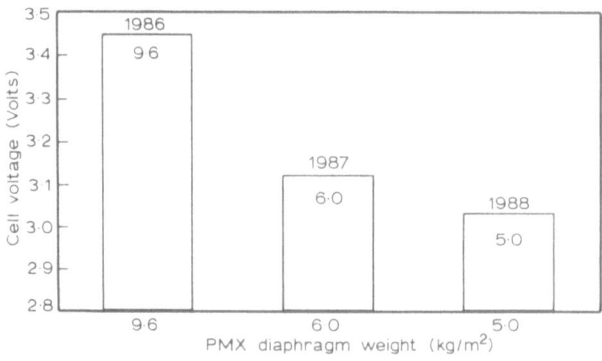

Fig. 6. PMX cell voltage versus PMX diaphragm weight. MDC-20 cell, $1.55 \, kA/m^2$, $130 \, g/liter$ NaOH cell liquor.

POLYRAMIX diaphragms can be plugged with brine impurities just like modified asbestos diaphragms. However, POLYRAMIX's ability to withstand pH changes without degradation of the diaphragm results in an added benefit. Shutdown of POLYRAMIX cells often results in an improved current efficiency and a lower voltage as the impurities are washed from the diaphragm by the acidic brine. This does not mean that poor quality brine is satisfactory for PMX operation. The ppb hardness level required for membrane cells is not necessary. Brine containing less than 4 ppm calcium and less than 0·5 ppm magnesium is satisfactory.

ECONOMICS

It is OxyTech's opinion that a commercially successful nonasbestos diaphragm must economically compete with the modified asbestos diaphragm. Possible future governmental regulations and increasing public concern about the environment and toxic waste disposal will increase the cost of producing, transporting and disposal of asbestos. As these costs increase, conversion to the use of a nonasbestos diaphragm such as POLYRAMIX will be economically justified. This will occur some time in the future. The economics and timing of conversion to POLYRAMIX will be determined by the analysis by each individual chlorine plant of the following factors:

—cost of a POLYRAMIX diaphragm;
—life of a POLYRAMIX diaphragm and operational stability;
—power consumption;
—asbestos and asbestos related costs.

As previously discussed, a POLYRAMIX diaphragm is approximately twice as heavy as a polymer modified asbestos diaphragm. In addition, PTFE polymer and zirconium oxide are more expensive on a per unit of weight basis than the present cost of asbestos. The combination of the greater diaphragm weight and the higher cost per unit of weight makes the present cost of a POLYRAMIX diaphragm approximately ten times higher than the present cost of a polymer modified asbestos diaphragm. The present cost of POLYRAMIX is $20/kg; however, because NaCl is used to keep the fibers from agglomerating, the effective cost is $40/kg. Present optimization has resulted in a diaphragm weight of approximately 5 kg/m² of cathode area. This results in a present cost of $200/m² of cathode area.

The second factor to be considered is the life of the POLYRAMIX diaphragm. As mentioned earlier, the life of a POLYRAMIX diaphragm is projected to be at least 1000 days as compared to a polymer modified asbestos diaphragm life of approximately 300 days. Since the life of a POLYRAMIX diaphragm is greater than three times longer and the other cell renewal costs such as labor are proportionally less, the effective cost of a POLYRAMIX diaphragm is reduced to approximately three times that of the polymer modified asbestos diaphragm.

The cost of using asbestos today and in the future is very dependent upon the costs associated with handling and disposal of asbestos. The present conservative estimate of this cost is $0·50/t of chlorine. This cost includes depreciation on the

TABLE 1
PMX economics: analysis of operating costs
(operating cost, $/t Cl_2)

Increases		*Decreases*	
PMX	$3·18	Modifier	$(0·91)
Chemicals	0·06	Asbestos	(0·22)
Fixed charges	0·02	Asbestos handling	
		costs	(0·50)
Operating cost = $3·26		Labor	(0·30)
			$(1·93)
			+ 3·26
		Incremental operating cost = $ 1·33	

Basis: MDC-55 at 2·2 kA/m².
       5·0 kg/m² PMX weight.
       PMX life = 1000 days.
       Polymer modified asbestos.
       Life = 300 days.

fixed capital investment, insurance, special work clothes, asbestos recovery, waste asbestos packaging and disposal at an approved landfill.

Taking the above four factors into account, the economics of POLYRAMIX versus present Modified Diaphragms is presented as Table 1. Using the assumptions of a 5 kg/m² PMX diaphragm, a 1000-day PMX diaphragm life and equivalent power consumption, the incremental operating cost of POLYRAMIX technology on an MDC-55 diaphragm cell is $1·33/t of chlorine more than present Modified Diaphragm Technology costs. Analysis of various different plant economics has revealed that the range of increased operating costs under the present asbestos cost situation is between $1·00 and $2·50/t of chlorine depending on cell type and current density.

Future advances in POLYRAMIX technology or changes in the cost of mining, transporting, storing and disposing of asbestos would shift the economics in POLYRAMIX's favor. For example, a three- to four-fold increase in asbestos related costs would make the economics equivalent. Such a cost increase scenario is reasonable considering the present concern with proposed regulations concerning the use and disposal of asbestos.

FUTURE WORK

OxyTech has been pleased with the progress in development of the POLY-RAMIX technology in the economic replacement of asbestos. The POLYRAMIX development program continues to receive high priority and support of OxyTech's directors and parent companies.

The present POLYRAMIX diaphragms operate with similar power consumption as polymer modified asbestos diaphragms. We believe that with further developmental work the power consumption of POLYRAMIX cells will be lower than conventional Modified Diaphragm cells today.

The next program objectives are:

—reduce power consumption to less than Modified Diaphragms;
—reduce POLYRAMIX manufacturing costs;
—demonstrate commercial POLYRAMIX performance in different cells and plants.

With modifications to depositing procedures, reductions in diaphragm weight appear achievable. The lower weight will reduce cell voltage and reduce the usage of PMX required and thus also the cost.

Presently, POLYRAMIX is manufactured in a pilot plant facility. Projections are now being made on the manufacturing costs if a commercial sized production facility is constructed utilizing economies of scale including bulk purchasing and handling of the raw materials.

OxyTech operates an independent POLYRAMIX depositing plant within the grounds of Occidental Chemical's Battleground Texas plant. This facility can deposit cathodes of all known sizes. OxyTech is presently scheduling deposits of cathodes for several companies to demonstrate the performance of POLYRAMIX, and to give these companies experience with POLYRAMIX long before full plant conversion is contemplated or required.

In summary, OxyTech Systems has developed a depositable replacement for asbestos in the chlor-alkali industry. POLYRAMIX and the materials it is manufactured from are noncarcinogenic. Existing diaphragm cell plants could convert to POLYRAMIX with minimum additional capital investment.

POLYRAMIX is available today for in-plant demonstrations. POLYRAMIX will be an attractive economic alternative to asbestos when the chlor-alkali industry requires one.

28

WORLD CHLOR-ALKALI MARKETS AND TECHNOLOGY: AN ICI PERSPECTIVE

B. Appleton and S. M. Andrew

ICI Chemicals and Polymers Ltd, UK

1 INTRODUCTION

In this chapter it is proposed to range fairly widely around the very broad subject of the world chlor-alkali industry, often taking a historical perspective in the belief that this does something to illuminate the future. Also, it will focus in on three main themes: this company's position in the industry; the evolution of chlorine and caustic soda markets and the crucial 'balance' between them; and chlor-alkali production technology and its past and future influence on the industry.

2 ICI IN CHLOR-ALKALI

First, ICI. Most of you will be familiar with this company in general terms, but you may not be so clear as to its position in the chlor-alkali business. In fact, through its predecessor companies in the UK, ICI is one of the true pioneers of the chlor-alkali industry, with production dating back to that major breakthrough in technology, the introduction of the Castner–Kellner mercury cells at Runcorn, England, in 1897, the same year as the first commercial plant in the USA at Niagara Falls. Interestingly, this same site at Runcorn, now known as Castner–Kellner Works, is still the location of ICI's largest chlorine operation, though on a scale (over 700 000 t/year) that would have seemed unimaginably large to our nineteenth-century ancestors (500 t were made in 1897, increasing to 3500 t in 1900). This electrolytic process, made possible by the increasing availability of electric generation capacity, was essentially the beginning of the modern chlor-alkali industry, replacing a variety of inefficient and expensive chlorine manufacturing processes based on oxidation of by-product HCl which had prevailed before, and providing a new source of high quality caustic soda, another chemical in increasing demand at the time. In the long period from the beginning of the century to the 1970s ICI continued to remain in the forefront of mercury cell technology (the existing cell rooms at Runcorn are among the most modern in the world), but it has also established a strong position in the other chlor-alkali processes, including its own modern diaphragm cell technology developed in the 1970s and applied at Teesside in the UK and at two sites in North America. This diaphragm cell was the first designed uniquely for metal anodes, and incorporated a number of important innovations. There is now an installed capacity in ICI diaphragm cells equivalent to 500 000 t per annum.

With this background in mercury and diaphragm cells it was natural for ICI to turn its mind to membrane cells as well, and early work on these was done in the 1960s. However, the real boost to membrane cells came in the 1970s with the development by Du Pont of perfluorinated membranes and the rapid technological developments in Japan in response to the need to replace mercury cells, and ICI considerably accelerated its membrane cell work at that time—a process given a further fillip by the energy crisis. Both monopolar and bipolar types were brought into operation in the late 1970s, and a choice was then made in favour of monopolar designs, leading to the emergence of the FM21 electrolyser. The design of this successful cell, which has evolved in detail but is still essentially the same in principle as the original version, was a major step forward and the result of a deliberate target of simplicity, energy efficiency and operability, together with the ability fully to exploit new developments in the membranes themselves. As with ICI's earlier technologies, development of the FM21 was carried out with the intention of using it in-house within the ICI Group, but it was recognised in the market conditions of the early 1980s that there were opportunities to sell it outside ICI. Accordingly the ICI Board approved the marketing of the technology outside the Group and the FM21 was officially launched onto the market in July 1981. Since then it has established itself as one of the most widely used membrane cell technologies outside Japan, having so far been committed for 20 installations, representing over 400 000 t/year of capacity.

But technology is only one part, albeit an important one, of the ICI story. Equally important has been the development of markets and of manufacturing capacity— abetted all the time of course by the availability of competitive in-house technology. Over the years here the story has been one of expansion and diversification both geographically and in terms of chlorine and caustic soda markets. Thus while in the early part of the century chlorine went almost entirely into bleaching powder, over the years as chlorine costs progressed down the learning curve and scale increased a wide range of 'chlorine derivatives' began to be introduced, several of them pioneered by ICI. The biggest breakthrough was the advent of various chlorinated organic compounds in the 1930s and 1940s, with the hydrocarbon content coming first from acetylene (from coal) and later in the 1960s from oil-based petrochemicals such as ethylene. The dramatic growth of petrochemicals in the 1960s in particular caused specially rapid expansion in chlorinated hydrocarbons, with PVC leading the way. As a consequence of its very broad chemical base, going back ultimately to its origins as a merger of four widely different companies in the 1920s and its strong record of innovation since, ICI possesses the widest range of different chlorine derivatives of all the chemical majors, a great contributor to its stability in this business. At the same time it has developed a similarly diverse position in caustic soda markets, important in keeping the whole business in balance. As the portfolio of chlor-alkali and derivative products has spread so has the business's geographic base. From being basically a UK business in the 1950s and 1960s, ICI now has a major stake in continental European manufacture through its Wilhelmshaven complex in Germany and a global position through its overseas subsidiaries, of which those in North America and Australia are the most important in this context. As a consequence ICI now ranks No. 2 in Europe and No. 3 worldwide in terms of chlorine capacity and sales, with an equally strong position in many derivatives. For example, ICI's PVC business is incorporated in EVC (the European Vinyls Corporation), a joint venture with EniChem, which is now the clear leader in the European PVC industry. This combination of worldwide manufacturing and market strength with a strong position in the technology business gives ICI the opportunity to take a truly global view of the industry!

3 GENERAL CHARACTERISTICS OF THE WORLD CHLOR-ALKALI INDUSTRY

It is hoped that what has been said so far has helped to put ICI's chlor-alkali position in perspective, and at the same time to introduce a number of general themes of universal importance to this industry. Among these are the questions of chlorine and caustic soda growth rates and the balance between them, which will now be enlarged on. Later more will be said about the new technology and its effect on industry investment and development.

In talking about chlorine and caustic soda growth rates and balances, it is perhaps best to start by reminding ourselves about the dimensions and geographical spread of the world industry. Thus in 1987 total world chlorine production was about 36 million t, of which just under 10 million t (27%) was in West Europe and nearly 12 million t (33%) in North America. West Europe and North America therefore have 60% of the total between them (it was 66% in 1975), so it is still the performance of these two regions that has the dominant effect, though other parts of the world are growing faster. Outside West Europe and North America, Japan with just under 3 million t is by far the biggest producer in the non-communist world, while the rest of the Far East is still relatively low in tonnage but is showing much faster growth. In terms of chlorine outlets the largest is PVC, with more than 30% in Europe though only about 20% in the USA. About 30% of chlorine goes into other chlorinated organics including solvents and fluorocarbons, while a further 15–20% is used in processes to make organic compounds such as propylene oxide and MDI, which do not contain chlorine but where chlorine is an intermediate. Inorganic derivatives account for about 10%, and the balance is sold as 'elemental' chlorine to pulp and paper and water treatment outlets; this category is much more important in North America (20%) than in West Europe (7·5%)—the inverse of PVC—mainly because of the importance of the American pulp and paper industry. Caustic soda production is of necessity about 10% greater than that of chlorine, i.e. almost 40 million t worldwide, and goes into a very wide range of outlets, many of which are chemicals. The only individual area commanding more than 10% of the total is pulp and paper, which constitutes some 12% of consumption in Europe and 25% in the

USA. Alumina, despite its importance in world caustic trade, represents less than 5% of the West European and USA markets.

It is hardly necessary to remind you of the two special characteristics of the chlor-alkali industry which differentiate it from other bulk chemicals businesses and have had (and will continue to have) a profound effect on its operation and evolution. First, there is the co-product aspect. With the overwhelming bulk of chlorine and caustic soda made by electrolysis, in which they are produced in fixed proportions to one another, sales outlets for both 'halves' have ultimately to be found in the right ratio: this of course is the famous 'chlor-alkali balance' problem which has preoccupied the industry, and to a large extent sorted out the winners from the losers, over much of its history. The fact that chlorine as such is relatively expensive and hazardous to transport, while caustic moves much more easily, has tended to establish a broad pattern, especially in Europe, where chlorine is converted into (higher value and more transportable) derivatives at source and movement of caustic soda establishes the balance. And, second point, chlorine–caustic is unique among large-scale chemicals in its dependence on electric power as its principal 'raw material'. Thus growth has been strongest when power is relatively cheap, and the industry has tended to concentrate in low energy cost areas—such as regions with hydroelectric power, or the USA Gulf Coast area when cheap gas became available for power generation. A more recent example, which has had quite an effect on the world industry, is the SADAF project in Saudi Arabia. The escalation of power costs in the 1970s and 1980s was a powerful influence in moving the technological emphasis towards energy saving—leading for example to the intensification of effort and interest in membrane cell development which has already been referred to.

4 CHLORINE AND CAUSTIC SODA GROWTH RATES AND 'THE BALANCE'

Now something will be said about chlorine and caustic soda growth rates and the balance between them, also referring to a number of graphs. These look at West Europe and the USA, which account between them for 55% of world chlorine production, and they all cover the period 1955 to the present day. They are all plotted on the same scale on 'semi-logarithmic' coordinates (so that constant growth rates show up as straight lines), and are therefore easily comparable with each other. They show some very interesting historical trends, an understanding of which gives us some pointers to the future.

The first two graphs (Figs 1 and 2) cover West Europe and show total chlorine consumption (for practical purposes the same as chlorine production) and caustic soda consumption respectively. A number of interesting features are apparent. First, up to 1973 there was very steady growth in both chlorine and caustic soda, but significantly faster in the case of chlorine: the 'best fit' over the period 1955–73 shows average growth of just over 10% for chlorine and 7% for caustic, a differential of 3%. It is no surprise therefore that a great worry in the industry in the 1950s and 1960s was the coming 'caustic surplus'. For although historically the demand for caustic soda was much more than that for its chlorine equivalent (it was about twice

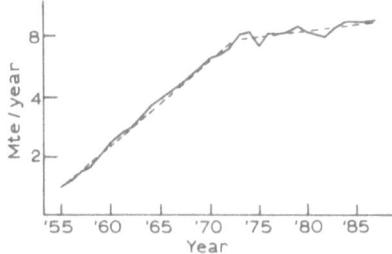

Fig. 1. W. Europe chlorine consumption (Mte/year).

as much in the 1950s), and the balance was made by converting soda ash to caustic soda by the lime–soda process (which does not involve co-production of alkali and chlorine), it looked as if the differential in growth rate, mainly due to the very rapid development of PVC, would soon lead to the point where *all* the caustic needed was co-produced with chlorine. The lime–soda process would therefore disappear as a balancing mechanism—as indeed it did during the 1960s—and after that the balance would be reversed. Ever-increasing quantities of caustic in excess of the 'natural' market would have to be 'disposed of'.

Three mechanisms appeared to be available for this 'disposal': development of new markets (perhaps exports); conversion of caustic soda to soda ash by carbonation (the reverse of the lime–soda process); and the substitution of caustic soda for soda ash in end-use markets. These latter two mechanisms reflect the relationship between caustic soda and soda ash which is fundamental to the chlor-alkali industry. The function of both in most outlets is to supply either alkalinity (for neutralisation) or sodium atoms (to make sodium compounds), so in theory they are often interchangeable, and the choice between them is a matter of economics and technology. In this context the concept of 'total alkali', i.e. the sum of caustic and soda ash sales expressed on an equivalent sodium basis, is a useful one in understanding the chlorine–caustic balance. In practice, however, particular industries are often firmly based on one or the other, and considerable investment may be required to change over, so rapid switching is not possible. For example, the glass industry is based on soda ash, the alkali historically more readily available, and a sustained price differential in favour of caustic much greater than has ever happened in practice would be necessary to tempt it to change. At the opposite

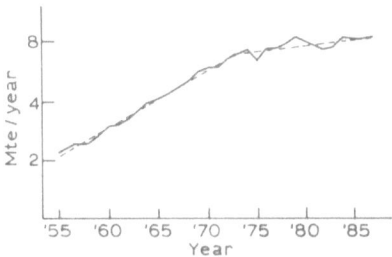

Fig. 2. W. Europe caustic soda consumption (Mte/year).

extreme, the export alumina industry, which developed as a caustic importer in the 1960s and 1970s in response to the expected caustic surplus, is firmly based on caustic liquor. In between comes, for example, the polyphosphate detergent industry, which can and does switch its allegiance from time to time in response to relative availability and price of ash and caustic. It is reasonable to think therefore of a 'hierarchy of interchangeability' between caustic and soda ash which includes direct conversion of one to the other as well as switching of outlets.

In the later 1960s, with the lime–soda process already gone and chlorine still growing much faster than caustic, it looked as if the balance would be strained to the limit, and that co-produced caustic would soon start catching up with the total alkali market. This would lead to catastrophically weak alkali demand and large-scale carbonation of caustic to soda ash, with consequent loss of profitability. But the industry was 'rescued' from this fate by the events of 1973 onwards.

Returning to Figs 1 and 2, it is evident that something dramatic happened to chlorine and caustic markets after the first 'great oil crunch' in 1973/74. The growth rates of 7% caustic and 10% chlorine ceased abruptly, and both entered a period of low growth and (compared with earlier years) short-term instability. Most interestingly, the 3% differential in growth between caustic and chlorine disappeared at the same time. The trend slopes over the period 1973–87 are barely distinguishable from each other at about 1% per annum each, and although these 'straight lines' are not nearly such close fits as they were before 1973, the general flatness (with perhaps a slight upward trend) is clear—this despite the erratic large amplitude variations which are hardly regular enough to deserve the name 'cycles'. Indeed the dips are clearly identifiable with the two energy price increase 'shocks' in 1973/74 and 1979/80. The chlor-alkali balance, therefore, has not been such a serious problem in the last dozen years or so as it was expected to be. Why has the caustic–chlorine growth differential disappeared, and what does it all mean for the future?

The different nature of chlorine and alkali outlets gives a clue to the answer to the first question. Thus the strong growth of chlorine in the 1950s and 1960s was due to the rapid development of the petrochemical industry and associated chloro-organics, particularly PVC. This was linked to the increasing availability and falling real price of oil, and a general feeling of optimism that good times would continue— a 'golden age' of growth indeed for most energy or hydrocarbon intensive industries. But it could not last, and most of this 'energy push' high growth came to an abrupt halt when oil prices shot up by a factor of five in 1973–74. Effects similar to those in Fig. 1 show up if production levels for virtually any energy-intensive industry are plotted over the same period. But chlorine and its derivatives also had some problems peculiar to themselves in the environmental and toxicological fields which began to have an influence about the same time and have held down growth relative to many other commodities since then and will continue to do so in the future. The VCM carcinogenicity scare in the later 1970s and the fluorocarbon–ozone issue leading to the recent Montreal protocol to reduce production are two of the best known examples, and there are many more. Chlorine in its nature therefore is sensitive to energy and environmental effects, and both have been holding it back since the earlier 1970s.

In contrast, caustic soda has most of its outlets in more traditional industries, which did indeed benefit from the high economic growth up to 1973, but not so directly from the oil or energy factor. Thus when the reckoning came in 1973–74, caustic did not have so far to fall as chlorine did, and its range of outlets is also less sensitive to environmental issues. It is not surprising therefore that the growth differential between chlorine and caustic is less than it used to be.

As for the second question—'the future'—the flat trend, which contrasts so strongly with the earlier high growth rates, has now been followed for some 14 years, and 1987 and 1988 in this context look quite 'typical' of the post-1973 pattern. So the question that has to be asked now is: will these trends continue or will there be a move to something different, perhaps higher growth rates, in response to a generally brighter economic outlook (with lower oil prices) than seemed likely in the earlier 1980s; or a fall-off in chlorine production because of intensified environmental demands? We cannot know for certain, but our experience is that established trends tend to continue until something definite happens to change them. There is certainly no sign of a move away from the post-1973 pattern, despite the dip in oil prices a couple of years ago, and probably the best bet is that the pattern will continue in much the same way for the next decade. This means caustic and chlorine growth in Europe of the order of 1% per annum each. If anything the environmental and toxicity effects could push chlorine down a little, but it should be remembered that these, though they have a high profile at present, are not new. The history of the chlorine industry in the last decade-and-a-half has been one where such problems have been repeatedly faced up to and dealt with, with a resulting downward pressure on chlorine volumes—so growth has been less, as it were, than the 'natural' growth of the outlets would suggest. This prospect of slow and roughly equal growth of caustic and chlorine in West Europe means that an essentially 'balanced' situation will be maintained. This implies that the amount of caustic available for export from West Europe will remain similar to what it is now, a subject to be returned to later.

Caustic soda and chlorine markets may also be looked at in the context of penetration of the economy, and the industrial production index (IIP) is a useful parameter for this purpose given the wide range of manufacturing outlets that both products go into. Figure 3 shows West European IIP over the period 1955–87, indicating that this has a growth pattern, with a 'change of slope' in 1973–74,

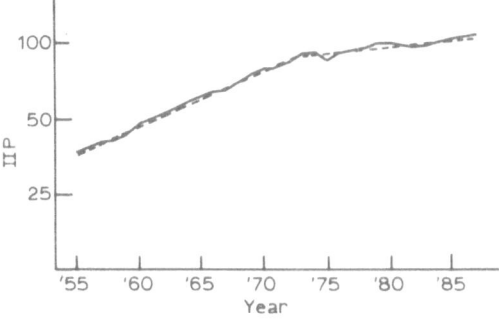

Fig. 3.   W. Europe IIP (1980 = 100).

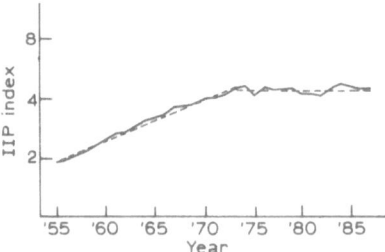

Fig. 4.   W. Europe chlorine/IIP index.

generally similar in shape to that of chlorine. However, in the period 1955–73 IIP grew at just over 5% per annum compared with chlorine's 10% and caustic's 7%, indicating rapid penetration by both, but with chlorine outperforming caustic for the 'energy-linked' reasons discussed earlier. After 1973, however, the IIP itself goes flat, and shows an average growth over the 1973–87 period of just over 1%, similar to that of chlorine and caustic. So evidently chlorine and caustic both ceased to penetrate the economy ('matured'?) after the 1973 energy crisis and have been moving more or less in line with industrial production—perhaps a little below it— since. This perhaps shows up more clearly in Figs 4 and 5, where a more direct measure of 'penetration' is shown as the ratio of chlorine and caustic consumptions to IIP. It seems reasonable to expect this relationship between chlorine–caustic and IIP to continue, so the expectation of only 1% p.a. growth in chlorine and caustic suggested earlier implies a similarly dull performance from industrial production. Is this excessively pessimistic or will West European industrial production do significantly better in the next decade than it has since 1973? This is a question for the economic forecasters, who have been having a difficult time recently with erratic oil price behaviour, stock market crashes, etc. However, even if IIP growth is 2% + rather than 1%, and chlorine and caustic growth rates in Europe are proportionately higher as a consequence, this makes little difference to the essentially flat, balanced nature of the industry.

It is worth adding that the apparently sudden maturation of the industry c. 1973–74 is by no means unique to chlor-alkali: the trends for virtually all bulk petrochemicals (ethylene, polythene, PVC, etc.) show a similar effect at the same time. It is illustrative of the robustness and long-term strength of chlor-alkali that

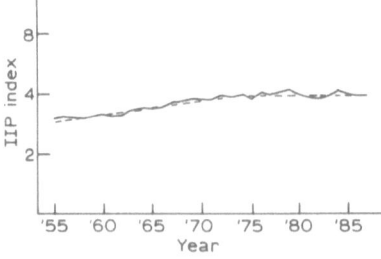

Fig. 5.   W. Europe caustic soda/IIP index.

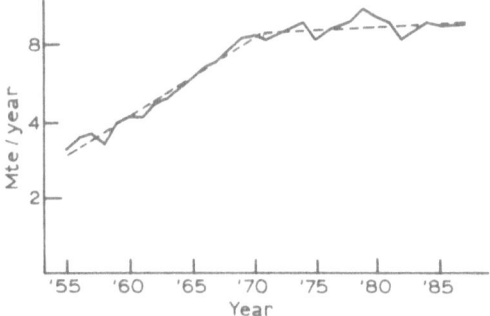

Fig. 6.   US chlorine consumption (Mte/year).

this sudden maturity came so late in its life—after a century or so, compared with a mere 20 years for petrochemicals!

We have talked mainly about Europe so far, taking the point of view of a European-based (if globally minded) producer. However, a similar look at the US industry (which is about the same size as the West European one) is of at least passing interest! In particular, it is interesting and instructive to note the differences and similarities between these two very large industries. Accordingly, Figs 6–10 show the same data for the USA as Figs 1–5 do for West Europe. Looking first at Figs 6 and 7, which show chlorine and caustic consumption, it will be seen that the picture resembles West Europe in that both chlorine and caustic show a strong growth trend in the 1950s and 1960s which is followed by much flatter 'growth', with massive short-term fluctuations (again hardly regular enough to be called 'cycles'), in more recent years. However, a difference from Europe is that the change in trend happened about 3 years earlier, which seems to have been due to the recession in 1970–71 which anticipated the oil crisis and was not nearly so severe in Europe (compare Figs 3 and 8), and to the earlier impact of some of the environmental effects.

As for the actual growth rates, the average trend for chlorine consumption in the USA over the period 1955–70 was about 7%, 3% less than the European figure at the same time. This reflects the fact that the USA started this period with a much higher level of chlorine demand than Europe did, so Europe had a lot of catching up

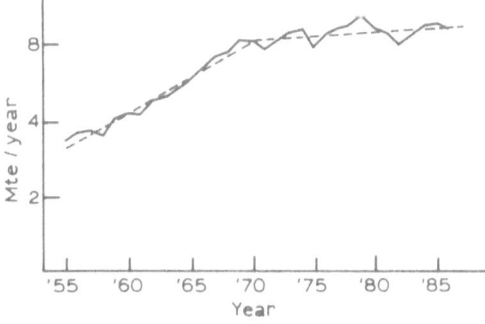

Fig. 7.   US caustic soda consumption (Mte/year).

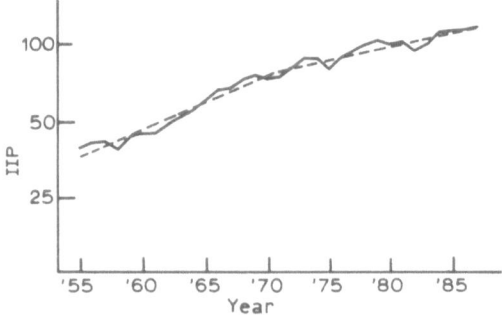

Fig. 8.   US IIP (1980 = 100).

to do; thus West European chlorine demand was only 45% of American in 1955, but had reached 70% by 1970. That the US industry was considerably more mature than the European then is also indicated by the fact that the chlorine growth rate exceeded that of industrial production by only 2% (see Fig. 8), compared with a 5% 'premium' in Europe. Caustic soda consumption in the USA over the same period was growing at a rate hardly slower than chlorine—a quite different relationship from that in Europe at the same time where, as we have seen, chlorine was growing 3% p.a. faster than caustic. As total alkali consumption, which reflects a broad spectrum of industries, was probably growing at a similar rate to IIP in this period (about 5% in both the USA and W. Europe), it would appear that caustic soda was already displacing soda ash in some markets, a process which will have slowed down since as the USA soda ash industry has improved its economics dramatically by moving from 'synthetic' to 'natural' ash production.

However, whatever the reasons for the differences in earlier years, since the early 1970s the US chlorine–caustic industry has followed a pattern similar to that in West Europe. Thus both chlorine and caustic soda appear to have shown growth rates of rather less than 1% p.a. over the period 1970–87, indicating an essentially balanced situation as far as chlorine–caustic co-production is concerned. As in Europe, there seems no obvious reason why this should change. It should be noted, however, that in the USA this 1% represents a significant 'depenetration' of the economy, which expressed as IIP has averaged something like 2·5% growth over the period 1970–87, considerably higher than Europe. This seems to be something of a

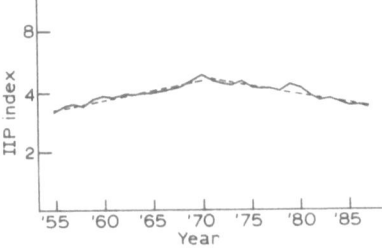

Fig. 9.   US chlorine/IIP index.

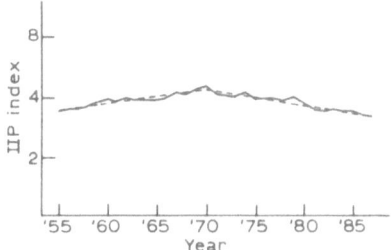

Fig. 10.   US caustic soda/IIP index.

reaction from the high chlorine usages of earlier years, when less severe environmental attitudes and low energy and power costs led to relatively extravagant usage of chlorine which is now being corrected for—a process that is likely to continue.

While talking about chlor-alkali in West Europe and the USA reference has been made to the industry in both regions as 'balanced' in the sense that caustic soda and chlorine consumption are growing at roughly the same (albeit low) rates. As in both regions the amount of co-produced caustic exceeds local market needs, there is a 'surplus' for export into 'deep-sea' markets that will remain at something like its present level. However, experience suggests that this 'balanced' situation normally represents an apparently weak caustic market; for example, deep-sea caustic soda export prices as deduced from trade statistics have been trending downwards in real terms for some years despite the sharp peaks experienced after the two oil price shocks of 1973–74 and 1979–80—and a new and less easily explained peak that appears to be being climbed now. To understand the longer term trend it is therefore necessary to return to the concept of 'total alkali' (caustic soda and soda ash together), and to consider total alkali growth rates since the first oil crisis in the 1970s. This is difficult to do with precision owing to a lack of published data, but using our own best guesses, and noting particularly the changes that have been going on in the glass industry, it appears that the growth rate of soda ash consumption is now significantly less than that of caustic, and so therefore is that of 'total alkali'. Therefore if caustic consumption is apparently growing at the same rate as chlorine, then total alkali is growing more slowly, and the alkali market as a totality is normally 'long'. With relatively low growth rates throughout, the amount of material that has to cross the boundary between 'natural' caustic and ash markets is small, and well within the bounds of outlets that can be regarded as 'easily interchangeable'. Thus it is entirely natural that caustic growth rates should appear to be the same as chlorine. They have to be, as all the co-produced caustic has to be moved: the slack is taken up in soda ash production, and alkali as a whole remains weak.

This statement about 'weak' alkali may seem surprising when everybody knows that caustic is booming at present, but remember we are talking about the long-term drift of a fluctuating industry, not the situation at any particular time. At present (April 1988) caustic is on an 'up', while if this chapter had been written 18 months

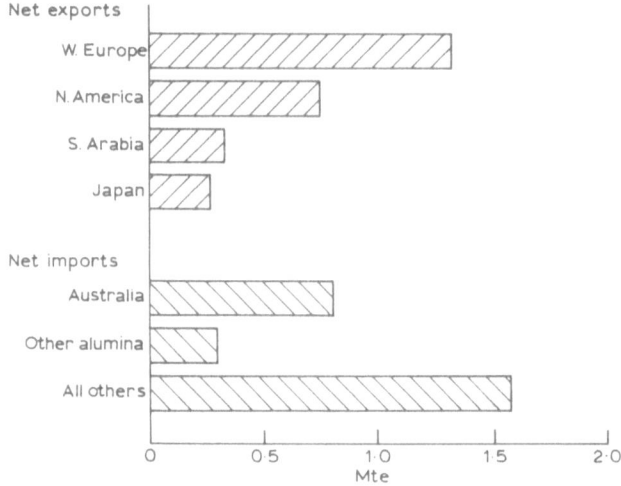

Fig. 11.   World trade in caustic soda 1986.

ago everybody would have agreed that is was 'down' and that a long-term down-ward drift was in line with experience. It is worth saying something therefore about these shorter term fluctuations, and this brings us back to the subject of world trade.

Broadly speaking, the pattern of the 'balancing' world caustic trade, in which regions with a 'surplus' export generally to developing countries and to the global alumina industry, has remained fairly steady for some years at a level of about 2·5 million t/year, of which rather over 1 million goes into alumina. This 2·5 million is a small fraction of the 40 million-odd manufactured worldwide and mostly consumed in producers' domestic markets, but it has a disproportionate influence on the world business as the producing areas rely on it for balancing their production and it is a vital raw material for the very large alumina refining industry, much of which (most notably in Australia) is situated remote from caustic soda production on a commensurate scale. Figure 11 shows our estimate of the pattern of this trade in 1986, with West Europe supplying 1·3 million t. Twice in the last 15 years this balance has been disrupted by the energy price shocks, which caused an immediate dip in petrochemicals (hence chlorine and caustic) production, while caustic demand, including that for alumina, reacted more slowly, leading to a sudden and drastic shortage of caustic for export markets. The 'peaking' effect this has on caustic prices is well enough known, though the effect in real terms was more dramatic in 1974 than in 1981, and a downward drift is discernible over the whole period. However, in the 1985 period prices began falling below sensible 'trend' levels, and this was probably ascribable to the beginning of the introduction into the market of 300 kt/year of caustic from the new Saudi Arabian operation. This was not a very big addition to total world chlor-alkali production, but was a much larger proportion of trade, and because the chlorine went to different markets (to Japan as EDC, previously mainly supplied from North America) from the caustic (mainly Australia, previously supplied from Europe) the effect on the delicate balance was considerable. An impression of surplus was created, and export prices fell to extremely low levels—which inevitably led to shortage as the geographical balance, as it had to, reasserted itself. This has led to the present over-reaction in the opposite direction towards very high prices, intensified by the now high occupacities in the industry. It will be interesting to see how long caustic stays tight in this present rather unusual 'cycle'. But eventually it will move back to something closer to the longer term trend. Nothing fundamental has changed.

5 TECHNOLOGY AND THE FUTURE WORLD CHLOR-ALKALI SCENE

Having dwelt at some length on the subjects of chlorine–caustic growth rates, balances and trade, we shall conclude by returning to the subject of technology and its position and influence in the future world chlor-alkali scene. As we have seen, reliable membrane cell technology is now commercially available from ICI and several other sources, and there can be no doubt that—at least until the next major technological breakthrough—such cells are going to dominate in all new installations and upgradings. Their freedom from the environmental problems associated with mercury and asbestos, low power consumption, pure caustic, and relatively low capital cost will see to this. Indeed, over the past 5 years almost all new capacity worldwide has been membrane cells, the few exceptions being projects which have been many years in planning or which are direct extensions to existing cell rooms using older technologies. The need for large-scale conversions in Japan following the decision to eliminate mercury cells of course played a major part in helping membrane cell technology to become established.

But how rapid will the penetration of the technology be, and where will the principal action take place? Here we need to consider the supply and demand situation, and the state of existing cell rooms. First, we should remember that the chlor-alkali heartland of North America and West Europe was until recently suffering from severe overcapacity which was leading to generally poor financial performance, and only in the past couple of years have supply and demand come into reasonable balance—in America mainly through substantial capacity closures and in Europe through a creeping up of chlorine demand (led by PVC) combined with a more modest reduction of capacity. With forecast growth rates of no more than 1% p.a. few producers are going to risk spoiling this by adding significant new capacity, and in Europe in particular the likely applications of the new technology will be in conversions or upgradings for environmental or cost saving reasons rather than for expansion. However, even here the amount of activity seems likely to be relatively limited. For the existing capacity in Europe, divided between diaphragm and mercury installations, mostly has a number of years of useful life ahead of it and is capable of meeting current and immediate future environmental standards. Also, with the easing of energy cost escalation rates in the last few years it has become increasingly difficult to justify conversion on cost saving grounds if the existing plant is relatively modern and in good condition. No doubt the time will come when such conversions become commonplace, but not yet.

It is different, however, elsewhere in the world, and here we need to look at the changing shape of the world industry. As said earlier, North America and West

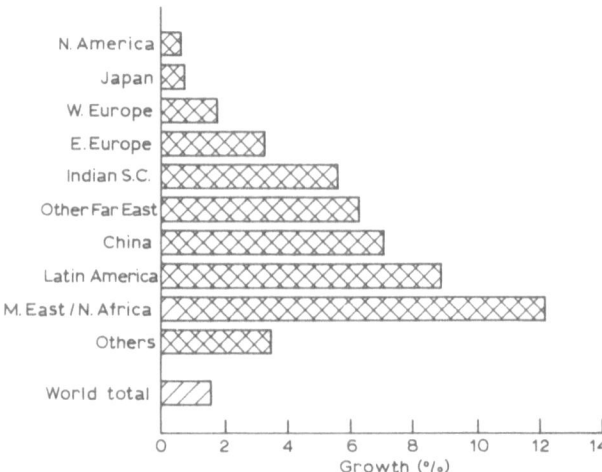

Fig. 12.   Chlorine growth trend 1975–87 (% annual rate).

Europe still dominate with 60% of the total, but they can muster barely 1% p.a. growth in a world total that is moving at nearly twice that—see Fig. 12 for a comparison of growth rates in different regions. Thus the Far East (outside Japan) has managed 6%, the Indian sub-continent 5% and the Middle East (led by Saudi Arabia) 12% p.a. in 1975–87. All this is from a low base of course, but these regions are surely catching up, and with their high economic growth and programmes for industrial development this growth differential will continue. The statistics of new membrane cell installations themselves tell the same story: over 80% of membrane cell contracts in the last 5 years have been outside West Europe and North America. Further, the very adaptability of the technology to relatively small-scale plants is a boost to its application in smaller, growing markets. So there will be opportunities aplenty for further application of membrane cell technology—and no shortage of suppliers ready and willing to provide it—but they will be concentrated in the next few years where the industry is expanding fast: the developing world, and the newly industrialised countries of the West Pacific area.

6 CONCLUSION AND SUMMARY

We have ranged through a series of aspects of the industry which are considered particularly important—looked at from the point of view of a company that has been in chlor-alkali since the last century and expects to remain there as long again!

We see this as a basically healthy industry, able to cope with, as before, all the problems that come its way. If growth in the traditional regions is flat at best, the old 'balance' problem has gone, and there is opportunity for a continuing profitable existence—provided things are not spoiled, as they were for the petrochemicals industry in the 1970s, by overexpansion based on excessively optimistic market forecasts. Such expansion as there is, which will be on a smaller scale than hitherto, should be left to the 'newer' industrial regions where the growth is.

The new membrane cell technology will have its part to play in these expansions, and in due course in renewing the major plants in the traditional industrial regions and in improving environmental and safety standards yet further. Membrane cells are typical of the type of innovation that has come along from time to time to renew the vitality of this dynamic industry—continuing a tradition that includes the Leblanc and Solvay processes for alkali in the nineteenth century; the Castner mercury cell process at the turn of the century; and numerous technical improvements, such as metal anodes, in the course of the twentieth.

29

OPERATIONAL SURVEILLANCE FOR CELL-TO-CELL UNIFORMITY

T. C. Jeffery

PPG Industries Inc., USA

1 INTRODUCTION

The Lake Charles plant of PPG Industries produced in excess of 3000 t/day chlorine as a daily production average for the year 1987. Stable steady-state operation at reliable rates are of utmost importance. Couple that with reliability at high current efficiencies and you have the possibility of lowest cost operation.

To narrow the scope of this chapter to fit the space available, the topic to be covered is controlling cell-to-cell uniformity by using analyses of individual cell liquor samples.

Figures 4–9 show the number of occurrences of weekly cell liquor samples in 5 g/liter categories for each of 11 cells of an electrolyzer. This bivariate histogram of Figs 4 and 8 for a high current efficiency electrolyzer shows results very close to a normal probability curve, whereas low current efficiency electrolyzers (Figs 7 and 9) bear no resemblance to a normal probability curve.

Using a plant the size of PPG's Lake Charles facility, the difference between all electrolyzers operating as in Fig. 4 (good) and all operating as in Fig. 7 (bad) would result in a cost saving of $2 000 000/year.

Given this history of operation, yesterday's results can be used to greater advantage for decisions related to today's operational adjustments causing tomorrow's analyses to fall within a more narrow range.

2 APPLICATIONS OF PERSONAL COMPUTER TECHNOLOGY

Closed-loop computer control in the chemical process industries is a 25-year-old subject. In an operation so fragmented as chlorine cells where you have so many individual units, computer control per se, has not gained wide acceptance, i.e. it has not been economic. The idea here is to allow computer generated data to be extracted, sorted and plotted to show in graphic form a cell's history enabling you to take remedial action to correct the individual cell's abnormalities.

An IBM AT clone with color monitor, modem, 70-MB hard disk, data base, spreadsheet and three-dimensional color graphics software is used for this technique. This combination of software technology has only become available within the last year.

For meaningful progress with this type of project you must have two types of people, one who knows computer technology and one who can apply the results. Either without the other is helpless.

3 DESCRIPTION OF CELL TYPE

The technique being discussed herein has been used with PPG V-1161 Bipolar Electrolyzers. It has also been used with the V-1144 type and is applicable to membrane cells.

Figure 1 shows one-half of a circuit of electrolyzers. Figure 2 shows an electrolyzer in transit during the renewal process but more importantly indicates the 11-cell liquor sample points referred to later in the graphics description. Figure 3 is taken from the anode end of the electrolyzer and shows the memorial to previously used monopolar technology.

This section is included mainly for the benefit of R&D personnel and academia, so they may see how recent developments are put into practice in commercial operation.

4 SAMPLE HANDLING AND ANALYSIS

Each cell is sampled once per week. With two circuits the 2300–0700 shift operator catches either the north or south side of either circuit in a predetermined sequence and leaves the samples at a pickup point.

Fig. 1

Fig. 2

Fig. 3

During the latter part of the 2300–0700 shift the samples are picked up and delivered to the plant laboratory, where concentrations are determined by an auto-analyzer. Results are reported on a floppy disk.

5 DATA COMMUNICATION

Once data are in an electronic form any further telecommunication or manipulation is fast. Since the laboratory is physically located some distance away the results are transmitted by modem to the operations office. For that matter it would be no more difficult to transmit the results to a different country.

One of the greatest values of computer use is that repetitious activity helps pay for the initial difficulty of set-up and accuracy of data handling is assured.

6 DATA BASES AND GRAPHICS

Same-day results are downloaded to the data base, which includes operating information on all electrolyzers.

A command is written to extract, for the electrolyzer to be studied, from the date of its cut-in to present, and this is copied to a file named, say 16-12.

A macro is then used to activate data distribution and count occurrences. This is copied and imported into the graphics package.

Identification in Fig. 4, whose sub-title is 16-12 Electrolyzer, means Circuit 16, Electrolyzer Number 12. Each electrolyzer, of the ones sited in this study, have 11 cells which are the numbers along the x-axis. Fourteen categories of cell liquor strength of 5 g/liter NaOH each are found on the y-axis. The z-axis counts the number of occurrences of samples in each 5 g/liter category, thus a bivariate histogram. All three-dimensional graphics are in color.

Albert Einstein once talked of the fourth dimension being time. Anyone reading this chapter who would like to tackle software programming to allow the date of the sample to be reflected in a color sequence, or a shade of grey with the last sample being black and the first one being white and all others sorted accordingly, is welcome to inquire for details. Then the most recent, being the most important, would be the boldest. Simplistically, this may not be what Einstein had in mind, but it would be a logical application of the fourth dimension or time variable.

In Figs 4 and 8 all 11 cells form approximate normal probability curves. The goal here is to make these curves as narrow and steep as possible.

Figures 4 and 6 are 'old' electrolyzers, i.e. over 300 days life, and Figs 5 and 7 are 'new' electrolyzers, i.e. under 200 days life, with the first named judged as 'good' and the second 'bad', respectively.

Electrolyzer	Age class	Rated performance
16-12	Old	Good
16-18	New	Good
15-03	Old	Bad
16-02	New	Bad

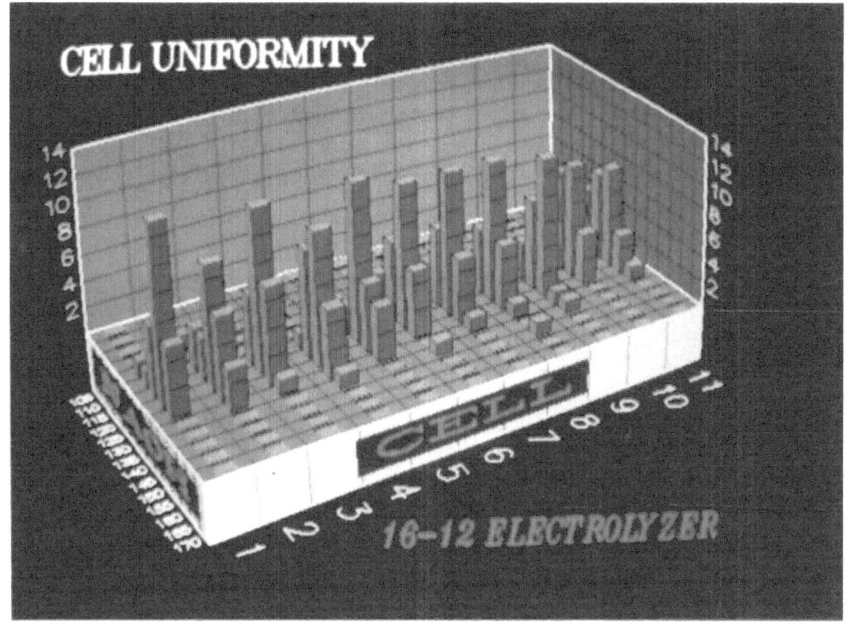

Fig. 4

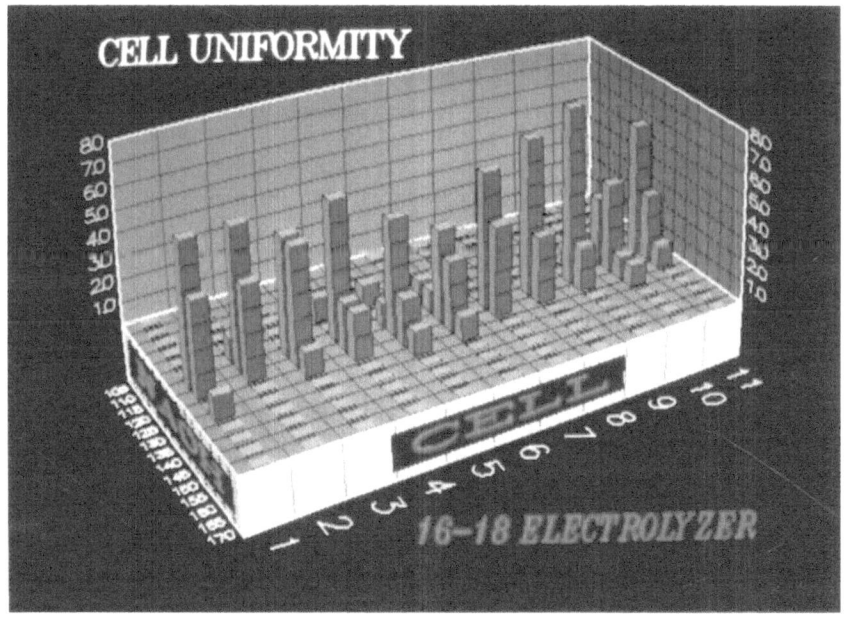

Fig. 5

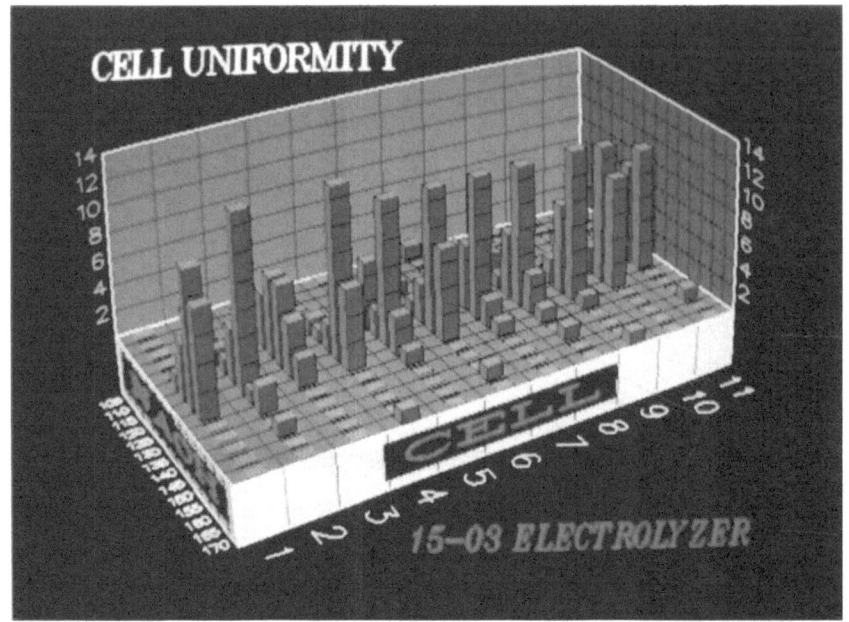

Fig. 6

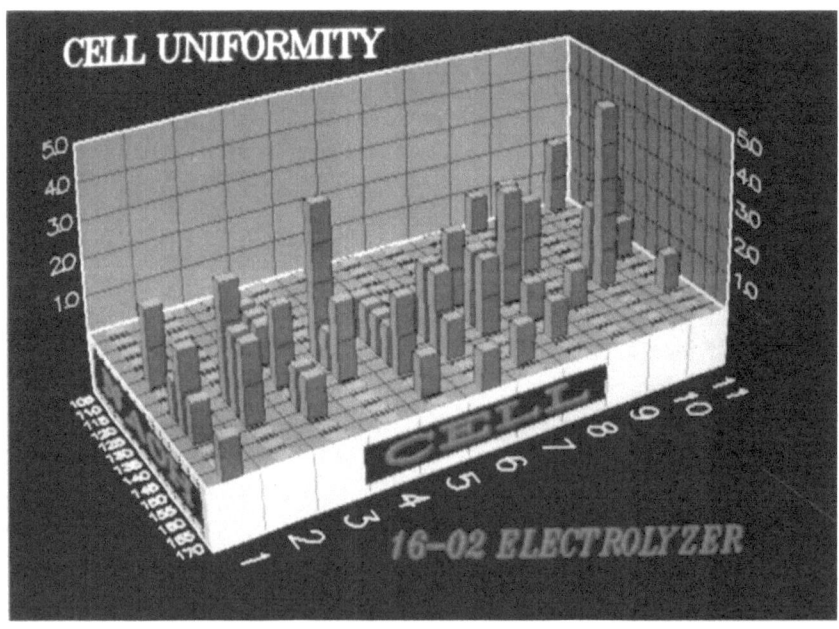

Fig. 7

Fig. 8

Fig. 9

The difference in profit between all electrolyzers running 'good' or 'bad' for a plant our size is $2 000 000/year. Good and bad is a relative term; it depends on one's standards. We know of plants where all electrolyzers are running below what we consider bad. Cost savings for implementing this type of program are therefore on individual cases.

When looking at the extremes presented here, it does not take one schooled in the art to detect a gross difference in the bivariate histograms of Fig. 4 (16-12) and Fig. 7 (16-02) as regards cell-to-cell uniformity. Figures 8 and 9 are the same data as Figs 4 and 7, respectively, only the plotting format has been changed for emphasis.

7 HARDCOPY FOR OPERATOR

After downloading results of samples taken earlier in the day by modem from the laboratory, the data are reduced to hardcopy by printing. This sheet is then marked with the required remedial action for individual cells based on all knowledge available.

US Patent 4 174 266 [1], entitled Method of Operating an Electrolytic Cell Having an Asbestos Diaphragm by T. C. Jeffery, outlines many of the procedures used at this time.

Table 1 is an example of this printout. There are nine electrolyzers on the north side of Circuit 15 and the results of the 99 cells are shown. Laboratory results for four variables are checked for each sample. With four sheets per week you have an accumulated total of 80 000 values per year and within recent memory there have been no renewals of electrolyzers under 1 year of service.

The beauty of this system is that it allows an engineer and an operator to look at, in many cases, well over a year's operating results in a single picture. (Confucius says 'a picture is worth a thousand words'.)

There is only so much time in a day and you can afford only so much expense for operating surveillance. The trick is to make changes by exception and allow all 'in-spec' operation to continue with minimum attention.

We feel additional work could be done with serial calculations using recent data to predict future values and if any variation between predicted and actual values occur then an investigation could be started and corrections made.

8 ELECTROLYZER RENEWAL

Due to the very extended operating life of this cell technology there are rather few cell renewal decisions. The downside to that is that with fewer decisions to make it behoves you to make those as accurately as possible. In these instances electrolyzer renewal candidates, having been selected by longstanding criteria, are reviewed by the three-dimensional format.

As much as we hate to admit it there are times that, try as we may, there is no operational response to remedial action. If this is the case and the uniformity is less than standard, the electrolyzer is removed from service [2].

TABLE 1

Plant A—Circuit: 15 NORTH 05/01/88

1 Cell	NaOH (g/liter)	NaCl (g/liter)	$NaClO_3$ (g/liter)	NaOCl (mg/liter)	2 Cell	NaOH (g/liter)	NaCl (g/liter)	$NaClO_3$ (g/liter)	NaOCl (mg/liter)	3 Cell	NaOH (g/liter)	NaCl (g/liter)	$NaClO_3$ (g/liter)	NaOCl (mg/liter)
1	135	199	0·15	2	1	139	198	0·08	4	1	142	194	0·11	2
2	142	199	0·05	1	2	135	198	0·04	4	2	137	198	0·12	4
3	138	199	0·10	3	3	136	198	0·04	3	3	126	205	0·16	6
4	140	199	0·05	1	4	135	199	0·04	3	4	139	196	0·08	1
5	135	201	0·11	1	5	134	201	0·02	3	5	136	201	0·13	3
6	134	200	0·10	2	6	137	199	0·04	5	6	140	197	0·09	5
7	134	201	0·07	1	7	139	199	0·02	4	7	141	198	0·12	3
8	124	205	0·13	3	8	139	198	0·02	3	8	139	198	0·11	2
9	136	199	0·11	3	9	138	197	0·04	4	9	144	196	0·12	2
10	140	197	0·13	4	10	138	199	0·07	10	10	145	197	0·10	2
11	135	198	0·10	2	11	139	194	0·05	4	11	142	195	0·13	2
Avg	136	200	0·10	2	Avg	137	198	0·04	4	Avg	139	198	0·12	3

4 Cell	NaOH (g/liter)	NaCl (g/liter)	$NaClO_3$ (g/liter)	NaOCl (mg/liter)	5 Cell	NaOH (g/liter)	NaCl (g/liter)	$NaClO_3$ (g/liter)	NaOCl (mg/liter)	6 Cell	NaOH (g/liter)	NaCl (g/liter)	$NaClO_3$ (g/liter)	NaOCl (mg/liter)
1	129	207	0·25	18	1	145	197	0·11	1	1	133	199	0·11	6
2	136	201	0·12	2	2	142	200	0·07	0	2	128	202	0·13	6
3	140	201	0·05	0	3	141	200	0·08	1	3	123	202	0·07	9
4	141	199	0·07	1	4	139	202	0·13	1	4	132	197	0·06	3
5	140	201	0·07	1	5	141	201	0·09	0	5	141	177	0·10	5
6	137	202	0·10	0	6	140	199	0·10	0	6+	133	196	0·14	23
7	149	198	0·14	1	7	141	195	0·10	1	7	133	171	0·16	11
8	143	201	0·09	1	8	139	201	0·07	0	8	126	192	0·09	11
9	138	202	0·09	1	9	135	201	0·07	0	9	135	195	0·05	10
10	124	206	0·13	5	10	135	202	0·11	1	10	139	197	0·08	3
11	137	201	0·10	2	11	141	196	0·12	1	11	137	195	0·08	4
Avg	138	202	0·11	3	Avg	140	199	0·10	0	Avg	133	193	0·10	8

Cell	7 NaOH (g/liter)	NaCl (g/liter)	NaClO$_3$ (g/liter)	NaOCl (mg/liter)	Cell	8 NaOH (g/liter)	NaCl (g/liter)	NaClO$_3$ (g/liter)	NaOCl (mg/liter)	Cell	9 NaOH (g/liter)	NaCl (g/liter)	NaClO$_3$ (g/liter)	NaOCl (mg/liter)
1	135	199	0·07	3	1	137	197	0·12	2	1	138	195	0·10	6
2	137	199	0·09	6	2	135	200	0·15	16	2	135	197	0·09	6
3	140	197	0·07	5	3	139	200	0·13	4	3	134	198	0·06	6
4	133	200	0·05	5	4	143	196	0·08	1	4	134	199	0·08	8
5	136	199	0·07	6	5	137	199	0·12	8	5	139	196	0·09	8
6	136	201	0·10	7	6	136	198	0·10	3	6	131	200	0·09	15
7	137	199	0·06	5	7	138	195	0·08	2	7	136	197	0·09	7
8	141	198	0·04	1	8	137	198	0·07	2	8	146	193	0·07	7
9	140	197	0·07	6	9	136	198	0·07	1	9	143	194	0·07	5
10	131	201	0·08	6	10	137	196	0·12	2	10	132	198	0·10	5
11	136	199	0·04	3	11	142	194	0·18	4	11	134	195	0·14	11
Avg	136	199	0·07	5	Avg	138	197	0·11	4	Avg	136	196	0·09	8

Limits: half-circuit cell liquor composites

	First level boundary condition	Second level boundary condition	Technicon NaOH (g/liter)	NaCl (g/liter)	NaClO$_3$ (g/liter)	NaOCl (mg/liter)	Decom. eff. (%)	Fe (% AB)	Titration NaOH (g/liter)	NaCl (g/liter)
NaOH	120–150	110–160	135	193	0·08	6	50·5	0·00000	0	0
NaCl	170–210	160–220	135	196	0·08	9	50·1	0·00000	0	0
NaClO$_3$	0–0·25	0–0·35	136	198	0·11	8	50·0	0·00000	0	0
NaOCl	0–20	0–30	135	198	0·08	11	50·0	0·00000	0	0

This, of course, opens up another complete realm of inspection opportunities upon disassembly. Armed with the previous run operational results many obvious maintenance steps present themselves.

9 CONCLUSIONS

1. Cell-to-cell uniformity is very important in achieving quality performance.
2. Computer technology has progressed to the point that data bases of this size may be handled by personal computers.
3. Three-dimensional color graphics may be used to see in excess of a year's data at once forming the basis for valid decisions.
4. Significant monetary gain can result from its use.

ACKNOWLEDGEMENTS

Acknowledgement is made to Mark A. Wilson, Kirk M. Mellard, H. Kevin Owens and PPG Industries, Chemical Division, for permission to publish this chapter.

REFERENCES

1. Jeffery, T. C., US Patent 4 174 266, November 1979.
2. Jeffery, T. C., Commercial application of PPG bipolar electrolyzer technology. *Modern Chlor-Alkali Technology*, Ellis Horwood, Chichester, 1980.

INDEX